Springer Series in
OPTICAL SCIENCES 77

founded by H.K.V. Lotsch

Springer
Berlin
Heidelberg
New York
Barcelona
Hong Kong
London
Milan
Paris
Singapore
Tokyo

Springer Series in
OPTICAL SCIENCES

The Springer Series in Optical Sciences, under the leadership of Editor-in-Chief *William T. Rhodes*, Georgia Institute of Technology, USA, and Georgia Tech Lorraine, France, provides an expanding selection of research monographs in all major areas of optics: lasers and quantum optics, ultrafast phenomena, optical spectroscopy techniques, optoelectronics, information optics, applied laser technology, industrial applications, and other topics of contemporary interest.
With this broad coverage of topics, the series is of use to all research scientists and engineers who need up-to-date reference books.

The editors encourage prospective authors to correspond with them in advance of submitting a manuscript. Submission of manuscripts should be made to the Editor-in-Chief or one of the Editors. See also http://www.springer.de/phys/books/optical_science/os.htm

Reinhard Iffländer

Solid-State Lasers for Materials Processing

Fundamental Relations and Technical Realizations

Translated by Stefan Weber

With 230 Figures, 73 Tables and Numerous Examples

Springer

Reinhard Iffländer, Dipl.-Phys.
Akazienweg 4
78713 Schramberg
Germany

Translator
Dr. Stefan Weber
New York University
Courant Institute
of Mathematical Sciences
New York, NY 10012-1185
USA

Title of the original German edition:
Festkörperlaser zur Materialbearbeitung.

ISSN 0342-4111

ISBN 3-540-66980-9 Springer-Verlag Berlin Heidelberg New York

Library of Congress Cataloging-in-Publication Data
Iffländer, Reinhard. [Festkörperlaser zur Materialbearbeitung. English] Solid-state lasers for materials processing : fundamental relations and technical realizations / Reinhard Iffländer ; translated by Stefan Weber. p.cm. – (Springer series in optical sciences ; v. 77) Includes bibliographical references and index.
ISBN 3540669809 (alk. paper) 1. Solid-state lasers. I. Title. II. Series.
TA1705.I3313 2000 621.36'61–dc21 00-049702

Springer-Verlag Berlin Heidelberg New York
a member of BertelsmannSpringer Science+Business Media GmbH

Typesetting: Camera-ready by the author
Cover concept by eStudio Calamar Steinen using a background picture from The Optics Project. Courtesy of John T. Foley, Professor, Department of Physics and Astronomy, Mississippi State University, USA.
Cover production: *design & production* GmbH, Heidelberg

Printed on acid-free paper SPIN 10712942 57/3141/di 5 4 3 2 1 0

Preface

Basically, the first edition was expanded and errors corrected. The aim remains to provide a reference book for technical information.

I would like to thank all my colleagues for constructive comments. Also, I acknowledge support by projects 13 EU 104 and EU 226 of the German Federal Ministry for Research and Technology. In addition, results from the project BE 7997, supported by the European Commission, have been included.

Specifically, I thank Dr. Petermann, Institut für Laserphysik, University of Hamburg and Dr. Ackermann of the Research Institute in Idar-Oberstein for the proof-reading of Chap. 11 and 12, Dipl. Phys. Luft of Siemens AG, Regensburg for the constructive reading of Chap. 7, my colleagues S. Benz and Dr. Krönert of Heraeus Quarz-Schmelze, Hanau for checking Chap. 6, Dr. H.J. Hoffmann of Schott Glas, Mainz for looking at Chap. 4 and 13, and above all Prof. Dr. H. Weber of the Optical Institute, Technical University Berlin.

Schramberg, January 2001 *Reinhard Iffländer*

Preface to the German Edition

This book was written during my work in the field of solid-state laser development for material processing. The main emphasis therefore lies in the compilation of physical and technical fundamentals of these lasers.

The purpose of this book is to provide a specialized introduction to the field for engineers and technicians. It is not intended as a substitute for more detailed textbooks and specialized literature The bibliography gives details of many textbooks in the field of study.

For the permission to publish results, I would like to thank HAAS-Laser, which has supported me generously during the preparation of this manuscript.

I would like to thank all those who have contributed to this book, in particular my wife who showed a lot of patience during its writing.

Special thanks go to Professor Weber. Without his support and constructive criticism this book would not have been written. Essential parts are based on his lectures and publications as well as on numerous discussions with him.

Parts of this book are based on work that was financially supported by the German Federal Ministry for research and technology under the contract numbers 13 N 5387/8, 13 EU 28 and 13 EU 0069.

The material has been compiled to my best knowledge. Mistakes, however, cannot be excluded.

Schramberg, January 1990 *Reinhard Iffländer*

Contents

Part III Data and Specifications

List of Symbols

Roman Letters

A	amplitude of the electrical field, matrix element, zoom factor
a	matrix element, aperture radius
B	matrix element
b	matrix element
c	matrix element, specific heat, vacuum light velocity
C	matrix element, tension coefficient, capacity
C_{Q}	system constant for Q-switch
d	matrix element, diameter
D	refractive power, diameter, matrix element, fiber diameter
D_i	critical refractive power, $i = a, b, c, d$
$\mathrm{d}n/\mathrm{d}T$	thermal refractive variation
E	energy, modulus of elasticity, laser energy
ΔE	energy difference
E_{p}	pumping energy
F	factor, area
f	focal length, area, frequency, cross section of the radiation field
F_+	factor
f_i	focal length of the i-th lens (starting at the resonator)
G	gain factor
g	g-parameter of the stability diagram, constant
g_0	small signal gain coefficient
g_{s}	saturated gain coefficient
G	saturated gain factor
G_0	small signal gain
g_{e}	degeneration factor of the lower laser energy level
h	main level distance, Planck constant
H_{m}	Hermite polynomials
H_{s}	saturation energy density
I	current
J	intensity, normalized current
J_+	intensity of the travelling wave inside the resonator (positive z direction)
J_-	intensity of the travelling wave (negative z direction)
J_0	intensity at $z = 0$
J_{cw}	stationary intensity
J_{in}	incident intensity
J_{mn}	intensity of the mode TEM_{mn}
J_n	Bessel function of order n

J_{out}	output intensity
J_p	intensity of the pumping light
J_s	saturation intensity
J_0	initial intensity
K	constant
k	heat conductivity, wave number
K_0	lamp parameter
k_B	Boltzmann constant
l	azimuthal mode order, length, geometric length of the active medium, rod length, arc length
L	geometric resonator length, inductance
L_{pl}	Laguerre polynomials of order p, l
M	material parameter, optical matrix, molecular weight
M	generalized mode order/beam propagation factor
m	mode order for Cartesian coordinates in x direction, zoom factor of a telescope
N	number, inversion density
n	refractive index, mode order for Cartesian coordinates in y direction, number
N_0	ion density of the ground state, total density of the active ions
N_1	ion density of the lower laser level
N_2	ion density of the upper laser level
NA	numerical aperture
N_a	initial inversion
ΔN	inversion density difference
N_e	final inversion
N_{Gr}	Grashof number
N_i	ion density of the level i
N_{max}	maximum inversion density
N_{Pr}	Prandtl number
N_{Re}	Reynolds number
N_T	thermal density of the lower laser level (4-level system)
N_{th}	threshold inversion density
P	power, laser power
p	radial mode order, resonator parameter, lamp pressure, polarization
P_0	initial power
P_{av}	average laser power
P_{el}	electrical pumping power
P_H	heating power
p_{ij}	photoelastic coefficient
P_{mn}	power of mode TEM_{mn}
P_p	pumping power
P_s	system constant
P_{th}	threshold power
q	complex beam parameter
Q	heat
Q_m	melting heat

Q_v evaporation heat
r differential resistance, pulse repetition rate, radius, beam radius
R radius, bending radius, reflection factor, mirror radius, flow resistance, gas constant, resistance
r_0 rod radius
r_f flow tube radius
R_i reflection factor of surface i
R_{opt} optimal reflection factor
R_T thermal shock parameter
s depth, length, material strength, cut width
T pulse duration, temperature
t time
T_K temperature of the coolant
T_m temperature in the center, melting temperature
T_0 temperature at the surface
T_p duration of pump pulse
T_r recovering time
T_{th} pulse delay till threshold
T_v evaporation temperature
u velocity
U voltage
U_B booster voltage
U_{th} reverse voltage
U_Z trigger voltage
V loss factor
V volume
V_m mode volume
v velocity
w beam radius
W pump rate
w_0 beam radius of TEM_{00} mode
w_B processing radius
w_F beam radius in the focus
w_M beam radius of mode order M
w_T waist radius of the laser beam
w_{T0} waist radius of TEM_{00} mode
W_{th} threshold pump rate
w_{TM} waist radius of TEM_{mn} mode order M
x coefficient
z z coordinate, processing depth
Z_0 complex resistance
z_F focal depth
z_i Rayleigh length at position i
z_R Rayleigh length, cutting width
z_s focal depth

Greek Letters

α ray slope in geometrical optics, loss coefficient, expansion coefficient, absorption coefficient, exponent
α_{abs} absorption loss coefficient
α_{sc} scattering loss coefficient
β characteristic length of the thermal lens, normalized resistance, exponent
δ azimuthal angle
ΔM displacement
ΔV deflection
θ divergence
Γ cubic expansion coefficient
κ specific heat coefficient
λ heat-transfer coefficient, wavelength, thermal conductivity
ρ specific weight, gas density
σ surface tension, effective cross section of the stimulated emission, absorption cross section
σ_{max} maximal stress
τ time constant, lifetime of the upper laser level, pulse-train duration
τ_R resonator decay time
τ_{ij} transition time from level i to level j
τ_d characteristic delay time
τ_r recovering time
η efficiency
η_{excit} excitation efficiency
η_{extr} extraction efficiency
η_{plug} plug efficiency
η_{slope} slope efficiency
η_{stor} storage efficiency
η_{tot} total efficiency
η_{ESA} efficiency for excited state absorption
η_{ASE} efficiency for amplified spontaneous emission
η_{Cav} cavity efficiency
η_{Lamp} lamp efficiency
η_{Med} efficiency of the laser medium
η_{Power} efficiency of the power supply
η_{Niv} efficiency of the upper laser level
η_{Opt} efficiency of the optical system
η_{Proc} processing efficiency
η_Q quantum efficiency
η_{QD} quantum defect
η_{Res} resonator efficiency
η_H heat efficiency
ν Poisson number, viscosity, light frequency

Introduction

The last few years have seen a steady increase of solid-state lasers for material processing in various applications. This trend is continuing. With respect to other processing methods solid-state lasers have several advantages, and also partially over other laser types:

- The process is contact-free and therefore almost no force is exerted on the work piece.
- Due to the specific wavelength it is possible to transmit the light by fiber to the application station. Therefore solid-state lasers are excellent for use in flexible production lines.
- Some processing methods are only possible with a laser, e.g. welding of electro-optical parts in vacuum through the glass of the electron tube.
- No vacuum or gas is necessary, except for shielding or blow out of the melt during processing.
- The wavelength allows for use of standard optics for transmission.

Nevertheless, solid-state lasers do have some disadvantages.

- A rather low efficiency.
- The lifetime is limited, particularly due to lamp failure at higher power. Recent developments using diode laser promise much longer lifetime.
- The beam quality is limited due to thermal properties of the laser material or due to multimode oscillation. Meanwhile, concepts were developed which allow considerably improved beam quality.
- The average power, and thus also the processing velocity, is in some cases less than for a CO_2 laser.

Therefore the construction has to be optimized with respect to the particular application, in order to make use of all the advantages of this laser type and to be economically competitive compared to conventional processing methods.

In the following, the fundamentals and techniques are provided for the construction of solid-state lasers. Necessary accessories are discussed in simplified form in some detail. Care was taken that the equations and results are of use in practice, despite some simplifications.

The first part of the book mainly considers the necessary fundamentals and their technical consequences. The technical explanations which are necessary for solid-state lasers are discussed in Sect. II of the book. Section III is essentially a data compendium concerning laser mediums, materials and excitation sources.

Part I

Theoretical Background

1 Fundamentals of Solid-State Lasers

The most important solid-state laser material used for material processing is Nd:YAG; followed by Cr:ruby and Nd:glass. Possible new materials are Cr:alexandrite, Nd:GSGG, Nd:GGG, Nd:Cr:GGG and Yb:YAG. For medical applications, holmium and erbium are used at different wavelengths in various host crystals or glasses.

Except for materials doped with Yb or Cr (YAG, alexandrite, ruby), they can be described as 4-level systems. Therefore the derivation of the equations is based on a 4-level system and the corresponding equations for the 3-level system will be given without derivation.

The material properties necessary for amplifiers and oscillators will be described. The emphasis is on the efficiency of the systems, which depends on the threshold and slope efficiency.

Given the results, estimations for new materials and systems can easily be carried out. Simple experiments make a comparison and an assessment of the materials possible.

1.1 Energy-Level Diagram and Rate Equations

Solid-state lasers are normally designed according to Fig. 1.1. The laser material interacts with the radiation field of its own or another resonator and is pumped by one or more excitation sources (lamps or diode lasers). The pump light may be incident on the laser material from any direction relative to the laser radiation.

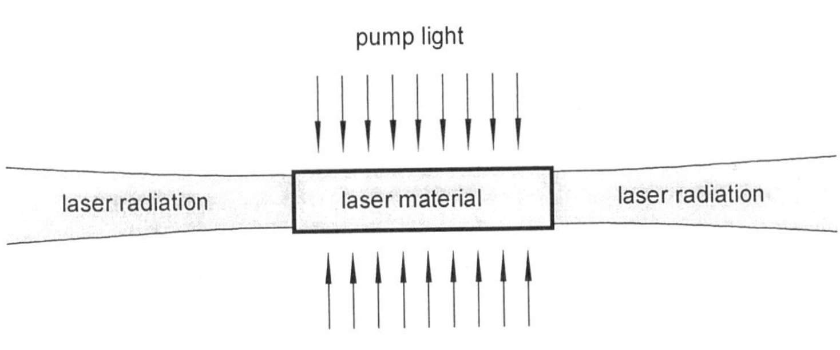

Fig. 1.1. Schematic setup of a solid-state laser

Rate Equations for a 4-Level System

The stimulated and spontaneous emission processes are described by the simplified energy-level diagram of Fig. 1.2 and the corresponding rate equations.

By absorbing the pump light, atoms are excited from the ground state 0 to the absorption band 3. From there, a fast transition (10 ns) takes place to the upper laser state 2. At this level the atoms stay for a long time (1 to 1000 μs) and are available for stimulated emission.

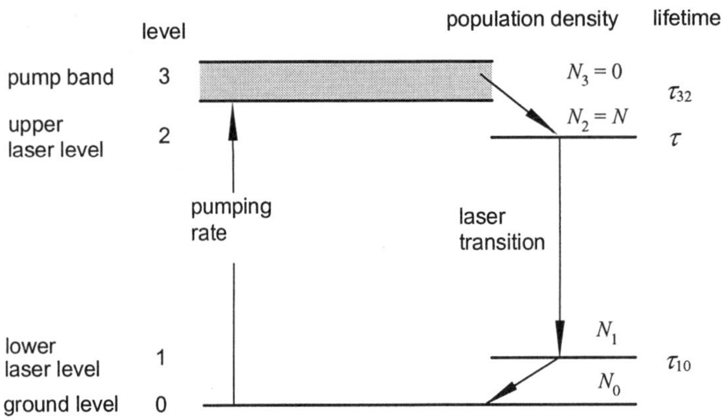

Fig. 1.2. Simplified energy-level diagram of a 4-level system

The following simplifications are invoked to derive the rate equations for a 4-level system:

- the transition from the upper absorption bands is rapid, and their populations can be neglected
- the laser transition is homogeneously broadened
- the intensity of the radiation field is assumed to be spatially constant
- the density N_0 of the ground state population is large compared to the other states, so that the depletion by the pump can be neglected.

The rate equation for the upper level is

$$\frac{dN_2}{dt} = WN_0 - \frac{N_2}{\tau} - \frac{\sigma J}{h\nu}(N_2 - N_1), \tag{1.1}$$

$\dot{N}_i$ = density of the active ions in level i,
W = pump rate,
τ = lifetime of the upper laser level, consists of the fluorescence lifetime τ_{21} and of non-radiative transitions ($1/\tau \approx 1/\tau_{21} + 1/\tau_{\text{phonon}}$),
J = intensity of the radiation field,
h = Planck constant = 6.6×10^{-34} J s,
ν = resonance frequency of the transition,
σ = cross section of the ion for induced emission.

Three mechanisms induce a change of the ion density N_2:

- an increase caused by pumping, proportional to the ground-state density,
- a decrease due to the finite lifetime of the upper laser level,
- a decrease due to the stimulated emission, proportional to the interacting radiation field, the cross section and the inversion ($N_2 - N_1$). The energy difference is supplied to the radiation field.

For the lower laser level, one has

$$\frac{dN_1}{dt} = \frac{\sigma J}{h\nu}\left(N_2 - N_1\right) + \frac{N_2}{\tau_{21}} + \frac{N_1}{\tau_{10}}, \tag{1.2}$$

τ_{10} = lifetime of the lower laser level

The lower laser level population changes by spontaneous and stimulated emission from the upper state as well as by phonon transitions to the ground state.

The intensity of the radiation field in the medium is described by

$$\frac{dJ}{dt} = \frac{\sigma cJ\left(N_2 - N_1\right)}{n} - \frac{\sigma cJN_{th}}{n} \tag{1.3}$$

$N_{th} = \alpha/\sigma$ = term containing all losses,
α = total loss coefficient,
c = light velocity in vacuum,
n = refractive index of the laser medium.

The intensity increases due to the inversion and decreases due to losses like absorption, scattering and outcoupled power, which are taken into account by the absorption coefficient.

If one assumes that the depletion of the lower laser level is very rapid, the rate equation (1.1) simplifies to

$$\frac{dN}{dt} = WN_0 - \frac{N}{\tau} - \frac{\sigma NJ}{h\nu} \qquad \text{for } \tau_{10} << \tau \text{ and } N := N_2 - N_1 \approx N_2. \tag{1.4}$$

If one introduces the saturation flux J_s, at which the population by spontaneous emission and phonon transitions equals stimulated emission,

$$J_s = \frac{h\nu}{\sigma\tau}, \tag{1.5}$$

J_s = saturation flux,

then (1.4) becomes

$$\frac{dN}{dt} = WN_0 - \frac{N}{\tau}\left(1 + \frac{J}{J_s}\right), \qquad \textit{change of the inversion}, \tag{1.6}$$

and with $N_2 = N$, (1.3) is given as

$$\frac{\mathrm{d}J}{\mathrm{d}t} = \frac{\sigma c J(N - N_{\mathrm{th}})}{n}, \qquad \textit{change of the intensity.} \tag{1.7}$$

Equations (1.6) and (1.7) are the rate equations for a 4-level system.

1.2 Amplifiers

Up to now the spatial dependence for inversion and intensity has been neglected. Taking into account the dependence in one coordinate (z direction), (1.6) and (1.7) change into

$$\frac{\partial}{\partial t} N(z,t) = WN_0 - \frac{N(z,t)}{\tau}\left(1 + \frac{J(z,t)}{J_\mathrm{s}}\right), \tag{1.8}$$

$$\left(\frac{c}{n}\frac{\partial}{\partial z} + \frac{\partial}{\partial t}\right) J(z,t) = \frac{\sigma c J(z,t)}{n}\left(N(z,t) - N_{\mathrm{th}}\right). \tag{1.9}$$

Stationary Case for Low Intensities ($J << J_\mathrm{s}$)

The stationary case is even fulfilled for pulsed operation, provided the pulse duration T is large compared to the upper laser level lifetime. Lasers used for spot welding operate with pulse durations T in the ms range. This is long compared with the upper laser level lifetime of Nd-YAG, which is approximately 200 µs.

At low intensities of the input wave ($J << J_\mathrm{s}$) its influence on inversion is small. The inversion $N(z, t) = N$ can therefore be considered as constant and (1.8) simplifies to

$$0 = WN_0 - N/\tau, \qquad N = \tau W N_0 = \mathrm{const}, \tag{1.10}$$

and (1.9) to

$$\frac{\partial}{\partial z} J(z) = \sigma J(z)(N - N_{\mathrm{th}}), \tag{1.11}$$

having the solution

$$J(z) = J_0 \mathrm{e}^{\sigma(N - N_{\mathrm{th}})z}. \tag{1.12}$$

An incident field with the intensity J_{in} in Fig. 1.3 will be amplified by an active medium with the length l to the intensity J_{out},

$$J_{out} = J_{in} e^{\sigma(N - N_{th})l}, \qquad \textit{intensity amplification for } J << J_s. \tag{1.13}$$

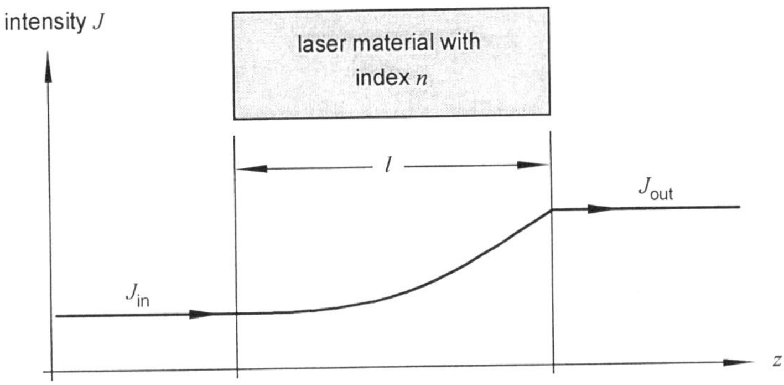

Fig. 1.3. Amplifier principle

One therefore defines

$$GV = J_{out}/J_{in} \qquad \textit{gain factor including the loss factor,} \tag{1.14}$$

$$G_0 = e^{\sigma N l} \qquad \textit{small signal gain factor,} \tag{1.15}$$

$$g_0 = \sigma N \qquad \textit{small signal gain coefficient,} \tag{1.16}$$

and, analogously,

$$V = e^{-\sigma N_{th} l} \qquad \textit{loss factor (in amplifiers),} \tag{1.17}$$

$$\alpha = \sigma N_{th} \qquad \textit{loss coefficient (in amplifiers).} \tag{1.18}$$

Stationary Case (or $T >> \tau$)

At high input intensity the inversion N is reduced by the radiation field and becomes z dependent. Equation (1.8) reads:

$$0 = WN_0 - \frac{N(z)}{t}\left(1 + \frac{J(z)}{J_s}\right) \tag{1.19}$$

and delivers for the inversion:

$$N_s = \frac{WN_0\tau}{1 + J(z)/J_s}. \tag{1.20}$$

Introducing the small signal gain coefficient g_0 and the saturated gain coefficient g_s, the above equation becomes:

$$g_s = \frac{g_0}{1 + J/J_s}, \qquad \textit{saturated gain coefficient.} \tag{1.21}$$

The z dependence of the intensity is given by (1.11) and reads in the stationary case

$$\frac{\partial J(z)}{\partial z} = J(z)(g_s - \alpha). \tag{1.22}$$

Neglecting the losses ($\alpha = 0$), (1.22) can be integrated, considering the intensity dependent gain coefficient g_s and the boundary condition $J_{out} = J_{in}$ for $G_0 = 1$

$$\ln \frac{J_{out}}{J_{in}} + \frac{J_{out} - J_{in}}{J_s} = \ln G_0 . \tag{1.23}$$

For $G_0 - 1 << 1$ an approximate solution holds

$$J_{out} = J_{in} G_0^{1/(1+J/J_s)} \qquad \textit{intensity amplification.} \tag{1.24}$$

For $J_{in} >> J_0$ (1.24) can be approximated by [1.1]

$$J_{out} \approx J_{in} + J_s g_0 l . \tag{1.25}$$

The high input intensity saturates the gain completely. The stored inversion energy is transformed into radiation energy and allows a high extraction efficiency. The maximum output intensity which can be extracted depends on the internal losses α of the amplifier and the gain coefficient g_0 [1.1]

$$J_{out,max} = \left(\frac{g_0}{\alpha} - 1\right) J_s . \tag{1.26}$$

The results in both cases apply analogously for pulsed mode operation, provided the pulse duration T is large compared to the lifetime τ of the upper laser level. For the energy E of a rectangular pulse one has

$$E = f \times J \times T , \tag{1.27}$$

f = cross section of the laser beam
T = pulse duration.

Pulse Operation for Short Pulses ($T << \tau$)

The energy amplification of short rectangular pulses is given in [1.1] as

$$E_{out} = fH_s \ln\left(1 + \left[e^{E_{in}/fH_s} - 1\right] G_0\right) \tag{1.28}$$

E_{in} = input energy
E_{out} = output energy
$H_s = h\nu/\sigma$ = saturation energy density.

Losses have not been taken into account. Two limiting cases can be distinguished:

$$E_{in} << fH_s \quad \text{results in} \quad E_{out} = G_0 \, E_{in}. \tag{1.29}$$

The input energy is amplified linearly.

$$E_{in} >> fH_s \quad \text{results in} \quad E_{out} = E_{in} + fH_s \, gl = E_{in} + h\nu NV \tag{1.30}$$

V = volume of the radiation field in the active medium.

In the latter case the output energy increases linearly with the pump length. The efficiency is more favorable than in the low energy case, since the input energy is increased by the full stored energy $h\nu NV$ of the volume V.

The small-signal gain factor and the amplification coefficient can be determined experimentally from (1.28)

$$G_0 = \frac{e^{E_{out}/fH_s} - 1}{e^{E_{in}/fH_s} - 1}. \tag{1.31}$$

No losses were taken into account so far and self-oscillation has to be avoided. Self-oscillation starts if the end-faces of the crystal are not completely antireflection (AR) coated and have remaining reflection coefficients R_1, R_2.

$$R_1 R_2 e^{2(g_0 - a)l} = 1. \tag{1.32}$$

This has to be taken into consideration for high amplification materials such as Nd:YAG.

1.3 Oscillator in the Stationary Case

First, the oscillator in the stationary case is discussed. Figure 1.4 shows the principle setup and the intensity distribution inside the oscillator [1.18].

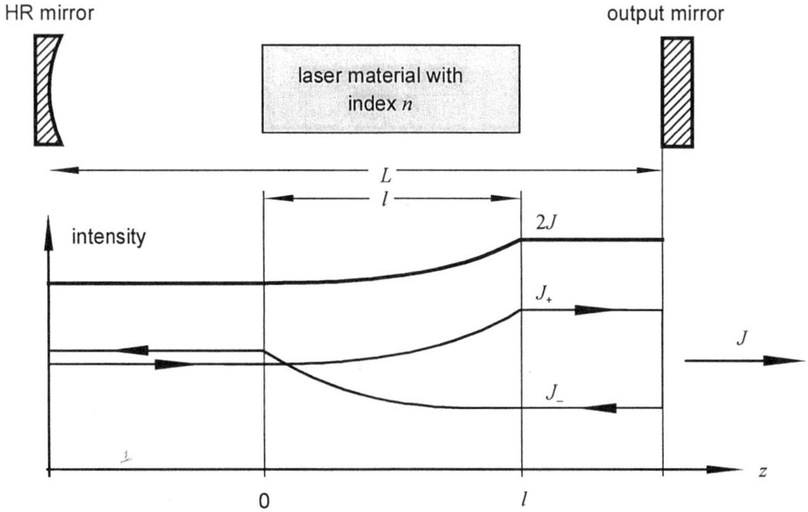

Fig. 1.4. Intensity distribution in an oscillator

The z dependence of the two intensities J_+ and J_- is described by (1.22)

$$\frac{\partial J_+(z)}{\partial z} = J_+(z)(g_s - \alpha)\,, \tag{1.33}$$

$$\frac{\partial J_-(z)}{\partial z} = -J_-(z)(g_s - \alpha)\,. \tag{1.34}$$

In Fig. 1.4 z increases from left to right. The gradient $\partial J_- / \partial z$ of (1.34) is therefore negative. The sum of the two equations yields

$$\frac{1}{J_-(z)}\frac{\partial J_-(z)}{\partial z} + \frac{1}{J_+(z)}\frac{\partial J_+(z)}{\partial z} = 0\,. \tag{1.35}$$

This is equivalent to

$$\frac{\partial\left(J_+(z) \times J_-(z)\right)}{\partial z} = 0 \tag{1.36}$$

or

$$J_+(z) \times J_-(z) = \text{const} = J^2\,. \tag{1.37}$$

The product of the intensities for the back- and forward-running wave inside the resonator is constant and independent of the position z.

With the boundary conditions

$$J_-(0) = J_+(0) \text{ and } J_-(l) = RJ_+(l) \tag{1.38}$$

it follows that

$$J_-(0) + J_+(0) = 2J \text{ and } J_-(l) + J_+(l) = \frac{1+R}{\sqrt{R}} J . \tag{1.39}$$

At the output mirror, with reflection R, there is, in general, the higher intensity. The difference of the average intensity $J_-(z) + J_+(z)$ from $2J$ is important only for low reflections. With $R = 0.55$ it is approximately 10%. Therefore, in the following only the average intensity $2J$ is considered. Exact solutions can be found for example in [1.2, 1.3].

With this, the rate equations, (1.6) and (1.7), become

$$\frac{\mathrm{d}N}{\mathrm{d}t} = WN_0 - \frac{N}{\tau}\left(1 + \frac{2J}{J_s}\right) \tag{1.40}$$

$$\frac{\mathrm{d}J}{\mathrm{d}t} = \frac{\sigma c l J\left(N - N_{\text{th}}\right)}{L + l(n-1)} , \tag{1.41}$$

L = resonator length,
l = length of the laser medium.

The change of the inversion is caused by the intensity $2J$ inside the laser material. The increased intensity of the radiation field caused by stimulated emission spreads out over the whole resonator length. The part that remains for stimulated emission in the medium is therefore only $nl/[L + l(n - l)]$. For the stationary case the rate equations become

$$0 = Wt - \frac{N}{N_0}\left(1 + \frac{2J}{J_s}\right) , \tag{1.42}$$

$$0 = \frac{l}{L + l(n-1)} \sigma c J\left(N - N_{\text{th}}\right) . \tag{1.43}$$

For $J > 0$, the last equation delivers

$$N = N_{\text{th}} = \text{const.} \tag{1.44}$$

The inversion above threshold remains constant, independent of the excitation power, as soon as threshold inversion is reached. The excitation power above threshold increases the intensity in the resonator and the output power.

Oscillator Condition

For steady-state oscillation, an amplifier must fulfill the oscillator condition. The intensity of the wave J_+, starting at point 0, must be reproduced after a full round-trip. All losses such as absorption, scattering and laser output must be fully compensated by the amplification

$$\left(VG_0RVG_0\right)J_+ = J_+ \,, \tag{1.45}$$

$R = R_1R_2$ = average reflection coefficient,
V = loss factor for internal losses,
$G_0 = \exp(g_0 l)$: small signal amplification factor ($J << J_s$),
$g_0 = \sigma N$: small signal gain coefficient,

$$RV^2G_0^2 = 1\,, \qquad \textit{oscillator condition.} \tag{1.46}$$

The inversion at laser threshold must not only compensate the internal losses in the active medium, but also the losses due to the output coupling. The threshold inversion follows from the oscillator condition as

$$\ln\left(RV^2\right) + 2\sigma N_{\text{th}} l = 0\,, \qquad N_{\text{th}} = -\frac{\ln\left(RV^2\right)}{2\sigma l} = -\frac{\ln\left(V\sqrt{R}\right)}{\sigma l}\,. \tag{1.47}$$

Equations (1.42) and (1.44) deliver the average intensity in the resonator:

$$J = \frac{J_s}{2}\left(\frac{WN_0\tau}{N_{\text{th}}} - 1\right). \tag{1.48}$$

Figure 1.5 shows schematically the inversion N and the intensity J inside the resonator dependent on the pump rate W. Introducing the threshold pump rate W_{th} for $J = 0$,

$$W_{\text{th}} = N_{\text{th}} / \tau N_0\,, \tag{1.49}$$

one obtains

$$J = \frac{J_s}{2}\left(\frac{W}{W_{\text{th}}} - 1\right). \tag{1.50}$$

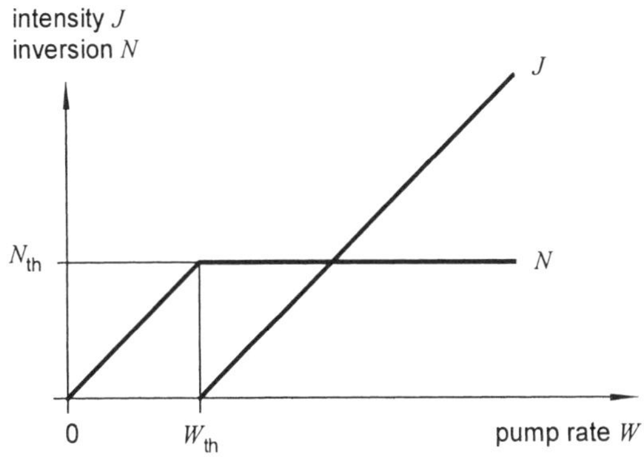

Fig. 1.5. Inversion N and intensity J in a 4-level laser

To excite N_0 ions per volume to the upper level an excitation power of $VN_0h\nu W$ for a volume V is required. This will be delivered by the pump power P_p with an excitation efficiency $\eta_{excit,}$ assuming that the pump light homogeneously excites the whole volume.

$$VN_0h\nu W = \eta_{excit} P_p \,, \tag{1.51}$$

η_{excit} = excitation efficiency,
$h\nu$ = photon energy,
V = pumped volume.

The pump power P_{th} at threshold is analogously defined as

$$P_{th} = VN_0 h\nu W_{th} / \eta_{excit} \,. \tag{1.52}$$

The average intensity in the resonator as a function of the electrical pump power P_p becomes

$$J = \frac{J_s}{2}\left(\frac{P_p}{P_{th}} - 1\right), \tag{1.53}$$

and the electrical pump power of the lamp at threshold is

$$P_{th} = \frac{VN_{th}h\nu}{\eta_{excit}\tau} = -\frac{fJ_s \ln\left(V\sqrt{R}\right)}{\eta_{excit}} = -P_s \ln\left(V\sqrt{R}\right), \tag{1.54}$$

$f = V/l$ = cross-sectional area of the active medium.

P_s is defined as a system constant since it depends only on parameters inherent to the system, such as pump lamp or cavity. It has the dimension of power:

$$P_s = \frac{fJ_s}{\eta_{excit}} \tag{1.55}$$

P_s = system constant.

In the stationary case the output laser intensity J_{cw} behind the mirror with the reflection R becomes

$$J_{cw}(l) = (1 - R)J_+(l) \tag{1.56}$$

and the reflected part is

$$J_-(l) = RJ_+(l)\,. \tag{1.57}$$

With $J^2 = J_+ J_-$ it follows that $J_+(l) = J/\sqrt{R}$ and

$$J_{cw} = \frac{1-R}{\sqrt{R}} J\,. \tag{1.58}$$

Therefore the output power P_{cw} is

$$P_{cw} = fJ_{cw} = f\frac{1-R}{\sqrt{R}} J\,. \tag{1.59}$$

With (1.52), the laser power behind the mirror can be expressed as

$$P_{cw} = fJ_s \frac{1-R}{2\sqrt{R}}\left(\frac{P_p}{P_{th}} - 1\right), \tag{1.60}$$

which holds for reflectivities which are not too low. With the approximation

$$-\ln\sqrt{R} \approx \frac{1-R}{2\sqrt{R}}, \tag{1.61}$$

for $0.4 < R \leq 1$ the error is less than 5%, one obtains

$$P_{cw} = \eta_{excit}\frac{\ln\sqrt{R}}{\ln\left(V\sqrt{R}\right)}\left(P_p - P_{th}\right). \tag{1.62}$$

This solution for the laser power is derived in [1.2] under much stricter conditions.

The usual form for the dependence of the laser power on the pump power is

$$P_{cw} = \eta_{slope}\left(P_p - P_{th}\right), \tag{1.63}$$

$$\text{with } \eta_{\text{slope}} = \eta_{\text{excit}} \frac{\ln\sqrt{R}}{\ln\left(V\sqrt{R}\right)} \text{ and } P_{\text{th}} = -P_{\text{s}} \ln\left(V\sqrt{R}\right).$$

This linear dependence is shown in Fig. 1.6. The slope efficiency η_{slope} and the threshold P_{th} determine the efficiency of a laser. Both, however, are dependent on the reflection coefficient R of the output mirror.

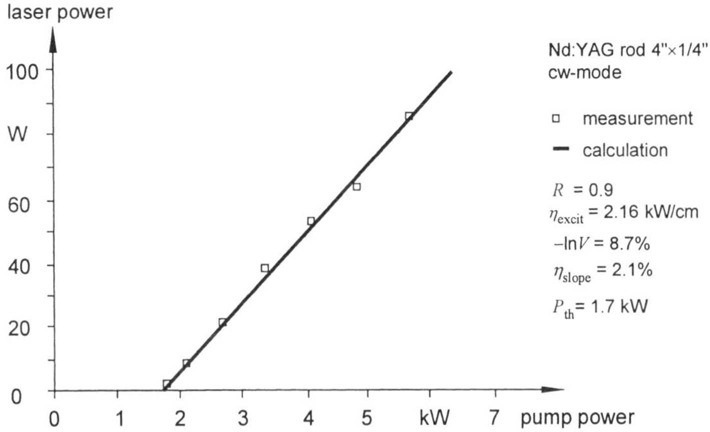

Fig. 1.6. Output power of a cw laser

Near the threshold, the measured output power or energy is not linear. A possible explanation for this is the inhomogeneous pump light distribution. The laser starts oscillating first in a small central area with low threshold where the pump light is highly concentrated. If the threshold condition over the whole area is reached, the laser characteristic depends linearly on the pump power.

Equation (1.62) also describes the dependence of the laser power on the reflection coefficient of the output coupler, as shown for the cw case in Fig. 1.7. The optimal reflection coefficient decreases with increasing pump power and should therefore be optimized for the maximum output power. From (1.62) and the constant P_{s} follows the laser output power

$$P_{\text{cw}} = \eta_{\text{excit}} \ln\sqrt{R} \left(\frac{P_{\text{p}}}{\ln\left(V\sqrt{R}\right)} + P_{\text{s}} \right), \quad \textit{laser output power}. \tag{1.64}$$

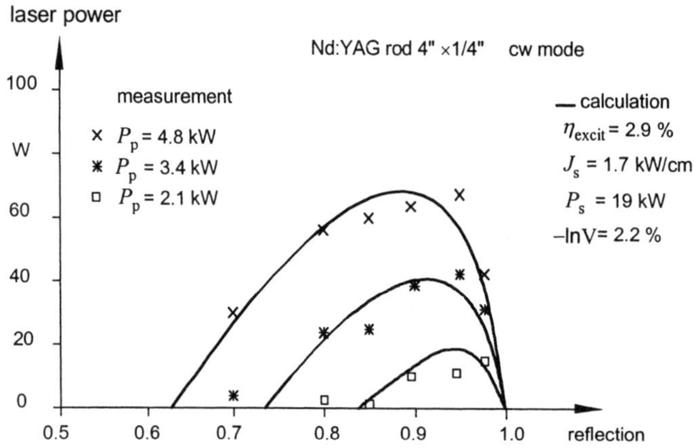

Fig. 1.7. Calculation of the laser power as a function of mirror reflection

Optimal Reflection Coefficient

The optimal mirror reflection R_{opt} can be derived from (1.64). This yields the maximum laser power for a given input power P_p:

$$-\ln\sqrt{R_{opt}} = \sqrt{\alpha l \frac{P_p}{P_s}} - \alpha l = \left(\sqrt{\alpha g_0} - \alpha\right) l, \tag{1.65}$$

$$R_{opt} = e^{-\sqrt{-\ln(V) P_p / P_s} + 2\ln(V)}, \qquad \text{with } \ln V = -\alpha l.$$

With the optimal reflection coefficient R_{opt} and the corresponding pump power P_p, the maximal laser output power P_{max} is

$$P_{max} = \frac{f J_s}{\alpha l}\left(\ln\sqrt{R_{opt}}\right)^2 = f J_s\left(\sqrt{g_0} - \sqrt{\alpha}\right)^2 l, \tag{1.66}$$

$$P_{max} = \eta_{excit}\left(\sqrt{P_p} - \sqrt{-P_s \ln(V)}\right)^2, \qquad \textit{optimized laser power.} \tag{1.67}$$

One selects the reflection coefficient of the output mirror with a safety margin towards larger values such that the efficiency is highest for the desired average power or energy. It should be noticed that with lower reflection coefficients, the resonator will be disturbed by back reflection from the workpiece, particularly during drilling and cutting.

1.4 Threshold and Losses

The oscillator threshold according to (1.54) is

$$-\ln\sqrt{R}=\frac{P_{\text{th}}}{P_{\text{s}}}+\ln V. \tag{1.68}$$

Plotting $-\ln\sqrt{R}$ as a function of the threshold value P_{th} [1.4], as in Fig. 1.8, one can extract from the measurements conclusions concerning the losses $-\ln V$ and the system performance P_{s} and thus also information about the small signal gain coefficient g_0.

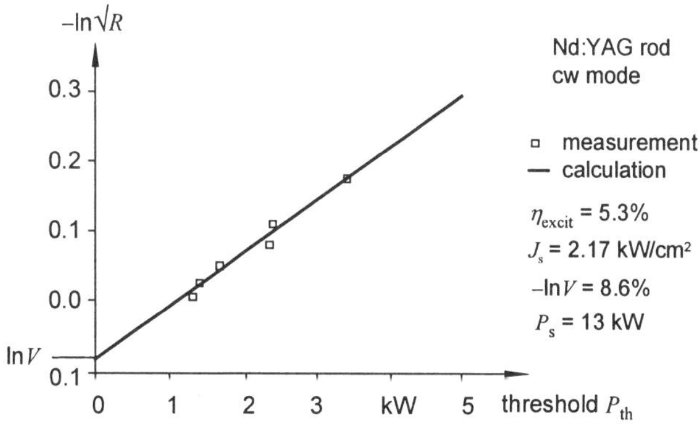

Fig. 1.8. Losses and threshold power of a Nd:YAG laser in the cw laser mode

Comparing (1.46) with (1.68) gives

$$g_0 l=\frac{P_{\text{p}}}{P_{\text{s}}} \qquad \text{with } g_0=\sigma N=\sigma N_{\text{th}}. \tag{1.69}$$

Measurements of the slope η_{slope} of the laser power as a function of the pump power at various reflection coefficients allows one to determine the excitation efficiency:

$$\eta_{\text{slope}}=\eta_{\text{excit}}\frac{\ln\sqrt{R}}{\ln(V\sqrt{R})}, \qquad \textit{slope efficiency}. \tag{1.70}$$

When determining the laser threshold one has to take into account that in many cases the excitation source has a threshold itself. The pump power versus electrical power has an offset. The losses are typically in the percentage range and are, among other things, due to absorption (α_{abs}) and scattering (α_{sc}) at the laser wavelength. They are related to the loss factor V by: $-\ln V=\alpha l$, $\alpha=\alpha_{\text{abs}}+....+\alpha_{\text{sc}}$.

In [1.5] a method for the measurement of scattering and absorption is described. Results are given in Table 1.1.

Table 1.1 Typical results for absorption and scattering, according to [1.5]

Material	Wavelength λ/nm	Absorption $\alpha_{abs}/10^{-3}$ cm^{-1}	Scattering $\alpha_{sc}/10^{-3}$ cm^{-1}	Losses $-\ln V/l/10^{-3}$ cm^{-1}
Nd:YAG	1064	0.6 – 3.8	2.0 – 2.9	10
Nd:Cr:GGG:		1.1 – 2.9		

The variations in the scatter coefficient are caused by material impurities or end-face deficiencies. The different absorption coefficients can be explained by the history of the laser rods. New rods have a value between 0.6×10^{-3} cm^{-1} and 0.8×10^{-3}cm^{-1}. Laser material pumped with short-wave radiation (λ < 500 nm) may result in short- and long-living color centers that considerably increase the absorption. Therefore the UV parts of a Xe flash lamp have to be blocked by a filter. The influence of Fe^{2+} ions on the 1 µm absorption is investigated in more detail in [1.6]. With a pump light between 230 and 360 nm the Fe^{2+} ions build short-living color centers, which absorb the laser light. Transferring the Fe^{2+} into Fe^{3+} ions by tempering the material for several hours at 1200°C reduces the amount of Fe^{2+}. This results in smaller losses at 1 µm and increased absorption by the Fe^{3+} ions at 260 nm.

Excitation caused by the laser wavelength can also start from the upper laser energy level, which is described in the literature as "excited state absorption" (ESA). Typical results for loss parameters are given in Table 1.2.

Table 1.2 Losses and excitation power in the cw mode

	f cm^2	$-\ln V$ %	P_s kW	η_{excit} %	J_s kW/cm^2
Nd:YAG					
Rod 4"×1/4"	0.32	8.6	11.6	6	2.18
Nd:GGG					
Rod 3"×1/4"	0.32	10	5	11	1.9

1.5 Influence of the Temperature

The temperature of the coolant and the heating due to the pump power determine the average temperature of the laser material. A higher average temperature in some cases is necessary to increase the efficiency, e.g. for alexandrite. In most other cases the material has to be operated at the lowest temperature possible.

The temperature in the laser material has several effects:

- the lifetime of the upper and lower level can change,
- a noticeable increase in the population of the lower laser energy level takes place,
- the cross section changes,
- the center frequency shifts.

Since the influence of the thermal population in a 4-level system is significant, this will be discussed in more detail. The inversion is reduced by the thermal population N_T and the assumption $N_1 \approx 0$ (1.4) is no longer valid. For a constant average temperature of the laser medium the thermal population is given by:

$$N_1 = N_T = \frac{N_0 e^{-\Delta E / k_B T}}{\sum_{i=0}^{n} e^{-\Delta E_i / k_B T}}, \tag{1.71}$$

k_B = Boltzmann constant,
ΔE = energy difference between ground state and lower laser energy level.

Example 1.1 Thermal population of the lower laser level in YAG

With $N_0 = 1.4\times10^{20}/\text{cm}^3$ and $T = 300$ K one finds:
the thermal population $N_T = 1.91\times10^{15}/\text{cm}^3$ for Nd with $\Delta E = 2110\ \text{cm}^{-1}$ and
the thermal population $N_T = 1.68\times10^{19}/\text{cm}^3$ for Yb with $\Delta E = 612\ \text{cm}^{-1}$.

With a rod length of 10 cm and a mirror reflectivity of about $R = 0.85$ the threshold inversion is typically $N_{th} = 5\times10^{16}/\text{cm}^3$ for Nd and $N_{th} = 5\times10^{17}/\text{cm}^3$ for Yb.

From (1.1) and (1.3), the rate equations for the oscillator in the stationary case therefore become

$$0 = W\tau N_0 - N_2 - \frac{J}{J_s}(N_2 - N_T), \tag{1.72}$$

$$0 = \frac{\sigma c J (N_2 - N_T - N_{th})}{L + l(n-1)}, \tag{1.73}$$

from which follows

$$N_2 = N_{th} + N_T. \tag{1.74}$$

With (1.48) the intensity in the resonator becomes

$$J = \frac{J_s}{2}\frac{W\tau N_0 - N_{th} - N_T}{N_{th}} = \frac{J_s}{2}\frac{-W\tau N_0 \ln(V\sqrt{R})/\sigma l - N_T}{-\ln(V\sqrt{R})/\sigma l} \tag{1.75}$$

and with (1.59), (1.61) and (1.62) the temperature-dependent output power is given as

$$P_{\text{cw}} = \frac{\ln\sqrt{R}}{\ln(V\sqrt{R})}\left(\eta_{\text{excit}} P_{\text{p}} + fJ_{\text{s}} \ln(V\sqrt{R}) - \sigma l f J_{\text{s}} N_{\text{T}}\right), \tag{1.76}$$

$$P_{\text{cw}} = \eta_{\text{slope}}\left(P_{\text{p}} - P_{\text{th}} - P_{\text{T}}\right).$$

Increasing the temperature will, depending on the pump power, decrease the laser power. The threshold power P_{th} is increased by

$$P_{\text{T}} = P_{\text{s}} \sigma l N_{\text{T}}. \tag{1.77}$$

In Fig. 1.9 the result of such a measurement is represented. The laser output power was measured as a function of the coolant temperature. The average crystal temperature is increased due to the pump-light absorption in the material and by the temperature jump between the coolant and the laser material. This increase of approximately 2°C per kW pump power was taken into account in the theoretical curves in Fig. 1.9.

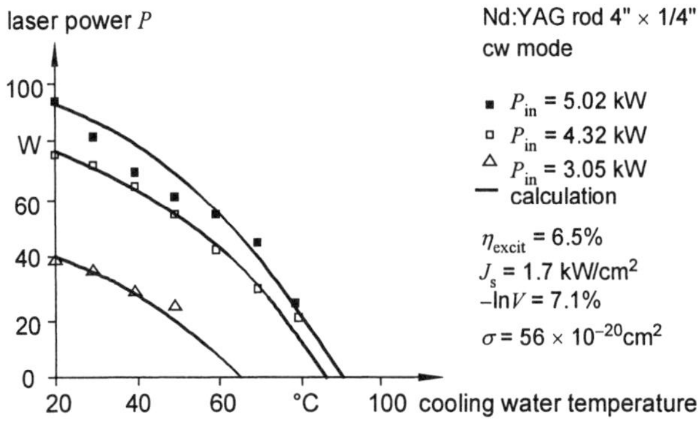

Fig. 1.9. Temperature dependence of a Nd:YAG laser in cw-mode

1.6 Oscillator in Pulsed Operation

Apart from some exceptions, most solid-state lasers for material processing up to an average power of about 1 kW are operated in the pulsed mode. The pulse length varies from the ns region in Q-switched mode (see Sect. 1.7) up to approximately 10 ms for pulse welding applications. When the pulse length is large compared to the upper laser lifetime, the system can be treated as in the stationary case.

Threshold

The pump light is switched on and the intensity in the resonator is very low up to the time T_{th}. It consists only of spontaneous emission which is emitted into the laser modes. The intensity J can be considered as nearly zero up to the time T_{th} (see Fig. 1.10). For $0 < t < T_{th}$ with $J = 0$, (1.40) becomes

$$\frac{\mathrm{d}N(t)}{\mathrm{d}t} = WN_0 - \frac{N(t)}{\tau}, \tag{1.78}$$

and at a constant pump rate W this may be integrated to

$$N(t) = W\tau N_0(1 - \mathrm{e}^{-t/\tau}). \tag{1.79}$$

Threshold inversion is reached at $t = T_{th}$, which delivers with (1.49)

$$W_{th} = W(1 - \mathrm{e}^{-T_{th}/\tau}), \tag{1.80}$$

and for the threshold itself with (1.51) and (1.52)

$$P_{th} = P_p(1 - \mathrm{e}^{-T_{th}/\tau}) \tag{1.81}$$

with P_p = const.

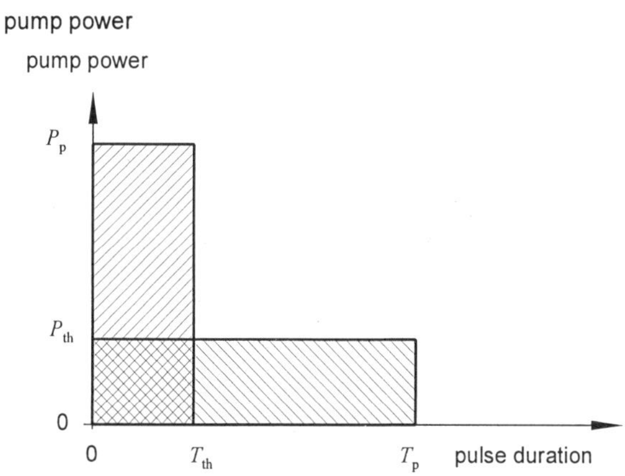

Fig. 1.10. Energy for reaching the laser threshold

Observing the onset of laser radiation (spiking), the time to reach threshold T_{th} can be measured and with $P_{th} = -P_s(\ln V + \ln \sqrt{R})$ the losses $\ln V$ for various reflection coefficients and the system constant P_S can be determined. Equation (1.81) only holds for rectangular pump light. A universal method is described in [4.9]. The time T_{th} to reach the threshold inversion is

$$T_{\text{th}} = \tau \ln\left(\frac{P_{\text{p}}}{P_{\text{p}} - P_{\text{th}}}\right) = \tau \ln\left(\frac{E_0}{E_0 - P_{\text{th}} T_{\text{th}}}\right). \tag{1.82}$$

E_0 is the pump energy, which is needed up to the onset of lasing,

$$E_0 = P_{\text{p}} T_{\text{th}}. \tag{1.83}$$

The total threshold energy E_{th} during the pump pulse T_{p} consists of the energy to reach the inversion and the energy to cover the losses in this quasistationary regime.

$$E_{\text{th}} = P_{\text{p}} T_{\text{th}} + P_{\text{th}}(T_{\text{p}} - T_{\text{th}}) = P_{\text{th}} T_{\text{p}} - (P_{\text{p}} - P_{\text{th}})\tau \ln\left(1 - P_{\text{th}} / P_{\text{p}}\right), \tag{1.84}$$

threshold energy in the pulsed operation.

This relation for constant pump power is sketched in Fig. 1.10.

More detailed information about absorption and losses of different laser systems as Nd:YAG and Cr:GGG may be found in [1.5].

High Pump Power $P_{\text{p}} >> P_{\text{th}}$

At high pump power (1.82) and (1.84) may be approximated by

$$T_{\text{th}} \approx \tau P_{\text{th}} / P_{\text{p}} \tag{1.85}$$

and

$$E_{\text{th}} \approx P_{\text{th}} T_{\text{p}} + \tau P_{\text{th}} - \tau P_{\text{th}}^2 / P_{\text{p}} \text{ with } P_{\text{th}} << P_{\text{p}}. \tag{1.86}$$

The pulse energy E of a laser pulse averaged over the spikes and for a constant pump pulse power during the pulse length T_{p} then follows from (1.63)

$$E = P_{\text{cw}}(T_{\text{p}} - T_{\text{th}}) = \eta_{\text{slope}}(P_{\text{p}} - P_{\text{th}})\left(T_{\text{p}} - \tau \frac{P_{\text{th}}}{P_{\text{p}}}\right), \tag{1.87}$$

$$E = \eta_{\text{slope}}\left(E_{\text{p}} - P_{\text{th}} T_{\text{p}} - \tau P_{\text{th}} - \tau \frac{P_{\text{th}}^2}{P_{\text{p}}}\right), \text{ \textit{energy in the pulsed operation.}} \tag{1.88}$$

With short pulses a good efficiency can only be reached at high pump power.

Relaxation Oscillation

After the threshold time T_{th} a radiation field can establish itself. The time-dependent differential equations (1.40) and (1.41) are now applicable.

$$\frac{dN}{dt} = WN_0 - \frac{N}{\tau}\left(1 + \frac{2J}{J_s}\right), \tag{1.89}$$

$$\frac{dJ}{dt} = \frac{\sigma d J\left(N - N_{th}\right)}{\tau_R}, \tag{1.90}$$

$\tau_R = [L + l(n - 1)]/c$ = resonator decay time.

In the following, small deviations from the stationary values ΔN and ΔJ are considered. Inversion and radiation field are assumed to be

$$N = N_{th} + \Delta N , \tag{1.91}$$

$J = J_\infty + \Delta J$, with $J_\infty = J_s\,(P_p/P_{th} - 1)$, according to (1.53). (1.92)

Inserting these into the rate equations (1.89) and (1.90), eliminating ΔJ and neglecting terms of second order the differential equation for the damped harmonic oscillator follows

$$0 = \tau\frac{d^2\Delta N}{dt^2} + \frac{P_p}{P_{th}}\frac{d\Delta N}{dt} + \frac{P_p - P_{th}}{\tau_R P_s}\Delta N , \tag{1.93}$$

with the solution

$$N = N_{th} + \Delta N\cos(2\pi f t)\exp(-t/\tau_D) \tag{1.94}$$

The damping time τ_D for the relaxation oscillation is

$$\tau_D = \frac{P_{th}}{P_P}\tau , \qquad (\tau_D = \tau_{th} \text{ according to (1.85)}), \tag{1.95}$$

and the frequency f

$$4\pi^2 f^2 = \frac{P_p - P_{th}}{\tau_R \tau P_s} - \frac{P_p^2}{4P_{th}^2\tau^2} \approx \frac{P_p - P_{th}}{\tau_R \tau P_s} . \tag{1.96}$$

The condition for damped oscillations is

$$\left(\frac{P_\mathrm{p}}{P_\mathrm{th}}\right)^2 < 4\frac{\tau}{\tau_\mathrm{R}}\frac{P_\mathrm{p}-P_\mathrm{th}}{P_\mathrm{s}} . \tag{1.97}$$

If the oscillation condition is fulfilled, and for small disturbances, the intensity and the inversion will perform damped oscillations around the stationary value. With Nd:YAG the damping time is in the 100 μs range and the period is about 10 μs.

Spiking

The linear approximative equations do not hold for transient effects and are also not valid for strong disturbances, in which case (1.89) as well as (1.90) must be solved numerically.

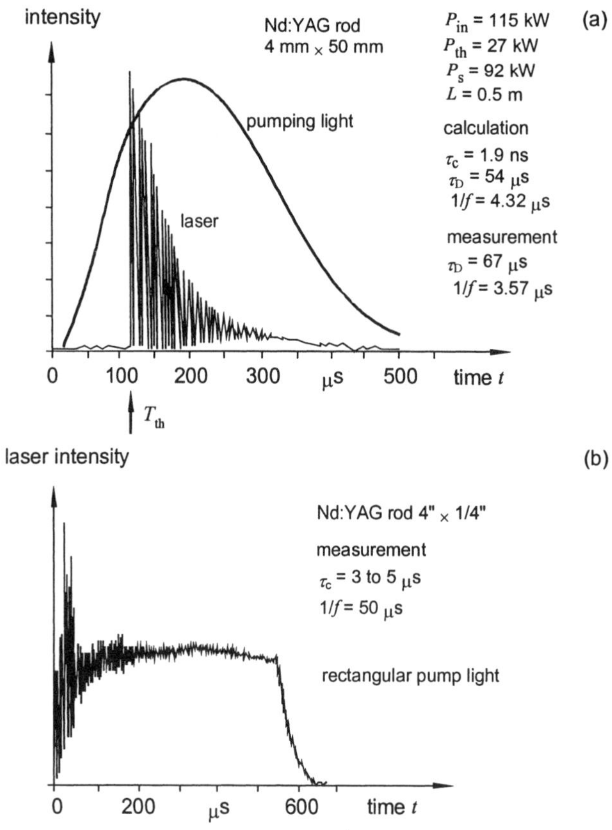

Fig. 1.11 (**a**) Spiking in monomode operation [1.8], (**b**) Spiking in multimode operation

Anharmonic oscillations appear, which are called spikes. When switching on a high pump power, the inversion quickly exceeds the threshold, without a sufficiently strong radiation field having been built up. The intensity then increases rapidly and reaches a value much higher than in the stationary case. The inversion is reduced below threshold and the radiation field drops again to a small value. As in the case of small perturbations, the stationary case is reached.

The temporal behavior of the laser radiation and the flash lamp light after switch-on for nearly monomode is shown in Fig. 1.11a, [1.8] and for multimode in Fig. 1.11b. Second harmonic generation inside the resonator may suppress or damp the spiking [1.20].

Long Pulse Regime

For long pulses with $T_p >> T_{th}$ and at high excitation the threshold energy becomes

$$E_{th} = P_{th} T_p , \qquad \textit{threshold energy.} \qquad (1.98)$$

The laser pulse duration T is equal to the pump pulse time T_p and the laser energy is given to a good approximation by

$$E = P_{cw} T = \frac{\eta_{excit} \ln \sqrt{R}}{\ln(V\sqrt{R})} (P_p - P_{th}) T = \eta_{slope} (E_p - E_{th}) . \qquad (1.99)$$

The amount of energy until the onset of laser activity is negligible with respect to the total threshold energy E_{th}. Analogously to Fig. 1.12 the threshold power can be determined from measurements.

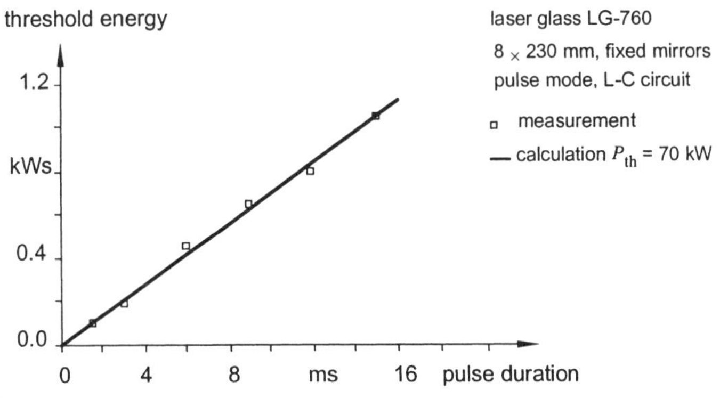

Fig. 1.12. Threshold energy of a pulsed glass laser

The temperature dependence is described analogously to the cw case, see (1.76).

$$E = \eta_{\text{excit}} \frac{\ln\sqrt{R}}{\ln(V\sqrt{R})}(P_{\text{p}} - P_{\text{th}} - P_{\text{T}})T \tag{1.100}$$

with $P_{\text{th}} = -P_{\text{s}} \ln(V\sqrt{R})$, $P_{\text{T}} = P_{\text{s}}\sigma l N_{\text{T}}$ and T = pulse duration.

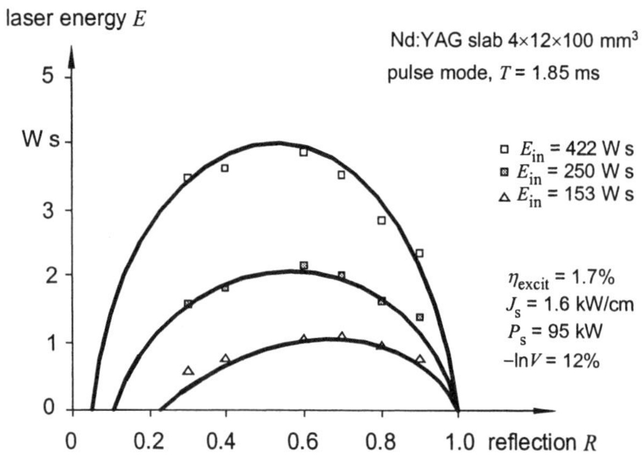

Fig. 1.13. Laser energy as a function of the mirror reflection

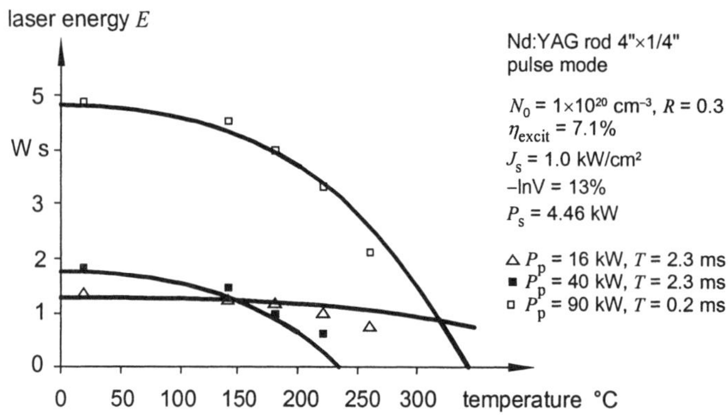

Fig. 1.14. Temperature dependence in the pulsed operation

The equation describing the laser energy for long pulses as function of the reflection factor and the temperature (1.98), is the same as in the cw case (Figs. 1.13 and 1.14).

1.7 Q-switch Operation

For material surface treatments such as depainting, scribing and marking, short, or very short laser pulses with high peak power are required to ablate the surface layers. The thermal impact on the bulk material is low. One method which is employed is focusing and scanning the laser beam in the so-called Q-switch mode [1.9], in which the quality of the resonator alternates rapidly between two extreme values with the help of a suitable optical switch.

In the ideal case the light path is interrupted in the resonator for a certain time and at a constant pump rate a maximal inversion is reached, limited only by the spontaneous emission. If the light path is then restored, a fast rising pulse is built up from the spontaneous emission and the inversion is reduced, creating a short, intense light pulse. Figure 1.15 shows the principle temporal behavior.

Two cases can be discussed in an approximate way. On the one hand the single pulse operation, in which the time between two pulses is long compared to the upper laser level lifetime and allows the maximal inversion. On the other hand the pulse operation with a frequency so high that the inversion varies only slightly around a stationary value before and after the Q-switch pulse. The range between these two extremes can be described only numerically. The rate equations were solved numerically in [1.10] and the theoretical results for pulse energy and average laser power were compared with the experimental results.

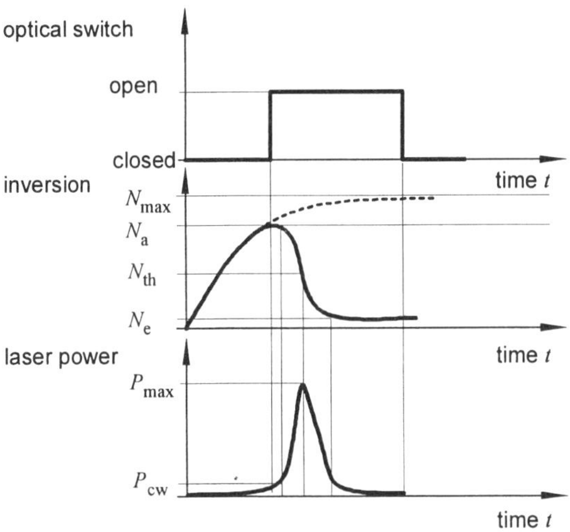

Fig. 1.15. Principle behavior of optical switch, inversion, and output power in the Q-switch mode

Single Q-switch Pulse

Equation (1.79) describes the temporal buildup of the inversion without induced emission:

$$N(t) = W\tau N_0(1-\mathrm{e}^{-t/\tau}) \qquad \text{for } W = \text{const.} \tag{1.101}$$

The maximum inversion after a sufficiently long pumping time ($T_P >> \tau$) is for $J = 0$ with (1.51),

$$N_{\max} = N_0 W\tau = \eta_{\text{excit}} \frac{\tau}{h\nu V} P_\text{p}\,, \tag{1.102}$$

and the maximum energy $E_{\max}$ which can be stored is

$$E_{\max} = h\nu N_{\max} V = \eta_{\text{excit}} \tau P_\text{p}\,. \tag{1.103}$$

Once the inversion has attained $N_{\max}$ the optical feedback is opened by a fast shutter. The buildup time of the laser pulse in normal resonators is much shorter than the upper laser level lifetime τ and the pumping time constant $1/W$. The rate equations (1.40) and (1.41) then can be approximated as

$$\frac{\mathrm{d}N}{\mathrm{d}t} = -\frac{2NJ}{\tau J_\text{s}} = -\frac{2NJ}{H_\text{s}}\,, \qquad \textit{rate equations for a fast Q-switch,} \tag{1.104}$$

$$\frac{\mathrm{d}J}{\mathrm{d}t} = \frac{\sigma l}{\tau_\text{c}} J(N - N_{\text{th}})\,, \tag{1.105}$$

$$\tau_\text{c} = \frac{L + l(n-1)}{c}\,, \qquad \textit{resonator transit time.} \tag{1.106}$$

At the beginning of the pulse $N_\text{a} \approx N_{\max}$ is assumed to be constant for a short time. The intensity inside the resonator and the pulse power increase exponentially

$$J(t) = J_\text{a} \mathrm{e}^{\sigma l (N_{\max} - N_{\text{th}}) t/\tau_\text{c}} \qquad \text{for } J << J_\text{s} \tag{1.107}$$

J_a is the spontaneous (initial) emission in the laser mode. Dividing (1.105) by (1.104) and integrating with the initial conditions $J(0) = J_\text{a}$ and $N(0) = N_\text{a}$ for $t = 0$ gives

$$J = \frac{lh\nu}{2\tau_\text{c}} \left[N_\text{a} - N + N_{\text{th}} \ln(N / N_\text{a}) \right] \,. \tag{1.108}$$

The maximum of the intensity in the resonator is reached for $N = N_{\text{th}}$ and (1.108) becomes

$$J_{\mathrm{max}} = \frac{lh\nu}{2\tau_{\mathrm{c}}}\left[N_{\mathrm{a}} - N_{\mathrm{th}} + N_{\mathrm{th}} \ln(N_{\mathrm{th}} / N_{\mathrm{a}})\right] \text{ neglecting } J_{\mathrm{a}}. \tag{1.109}$$

However, since the intensity is considerably higher than in the stationary case, the inversion is further reduced and the output power falls back to zero. The optical shutter is closed rapidly. After a certain time the inversion again reaches values larger than N_{th}, the switch can be opened again and the cycle starts once more.

With (1.59) and (1.109) the maximum of the available laser pulse power P_{max} can be estimated, to be

$$P_{\mathrm{max}} = -\ln(\sqrt{R}) fJ_{\mathrm{max}} \,. \tag{1.110}$$

At sufficiently long pump duration, and with $N_{\mathrm{a}} = N_{\mathrm{max}}$,the laser power is

$$P_{\mathrm{max}} = -\ln(\sqrt{R}) f \frac{lh\nu}{\tau_{\mathrm{c}}} N_{\mathrm{max}} \left(1 - \frac{N_{\mathrm{th}}}{N_{\mathrm{max}}} + \frac{N_{\mathrm{th}}}{N_{\mathrm{max}}} \ln\left(\frac{N_{\mathrm{th}}}{N_{\mathrm{max}}}\right)\right) \tag{1.111}$$

and with (1.103) it follows that

$$P_{\mathrm{max}} = -\ln(\sqrt{R}) f \frac{\tau}{\tau_{\mathrm{c}}} \eta_{\mathrm{excit}} P_p \left(1 - \frac{P_{\mathrm{th}}}{P_{\mathrm{p}}} + \frac{P_{\mathrm{th}}}{P_{\mathrm{p}}} \ln\left(\frac{P_{\mathrm{th}}}{P_{\mathrm{p}}}\right)\right), \tag{1.112}$$

which is the maximum attainable peak power of a Q-switch pulse.

The second zero of (1.108) gives the final inversion N_{f}.

$$0 = N_{\mathrm{a}} - N_{\mathrm{f}} + N_{\mathrm{th}} \ln(N_{\mathrm{f}} / N_{\mathrm{a}}) \,. \tag{1.113}$$

The output energy results from

$$E = \int_0^\infty P \mathrm{d}t = - f \ln(\sqrt{R}) \int_0^\infty J \mathrm{d}t = fH_{\mathrm{s}} \ln(\sqrt{R}) \int_{N_{\mathrm{a}}}^{N_{\mathrm{f}}} \frac{\mathrm{d}N}{N} = fH_{\mathrm{s}} \ln(\sqrt{R}) \ln \frac{N_{\mathrm{f}}}{N_{\mathrm{a}}} \tag{1.114}$$

and reads with (1.113)

$$E = -f \ln(\sqrt{R}) \frac{N_{\mathrm{a}} - N_{\mathrm{f}}}{N_{\mathrm{th}}} = \frac{Vh\nu}{-\ln(V\sqrt{R})} (N_{\mathrm{max}} - N_{\mathrm{f}}) \text{ for } N_{\mathrm{a}} \approx N_{\mathrm{max}}. \tag{1.115}$$

At high pump power N_{f} can be neglected compared to N_{max} and (1.115) with (1.112) therefore gives

$$E_{\mathrm{max}} = \eta_{\mathrm{excit}} \frac{\ln\sqrt{R}}{\ln\left(V\sqrt{R}\right)} \tau P_{\mathrm{p}} = \eta_{\mathrm{slope}} \tau P_{\mathrm{p}} \,, \tag{1.116}$$

the maximum energy of a single Q-switch pulse. The peak power P_{max} and the pulse energy E can be used to define the laser pulse duration T:

$$T = \frac{E}{P_{max}} = \tau_c \frac{P_p}{-\ln\left(V\sqrt{R}\right)\left[P_p - P_{th} - P_{th}\ln(P_p / P_{th})\right]} . \quad (1.117)$$

Periodical Q-switch for High Pulse Repetition Rate ($r\tau >> 1$)

In the periodical Q-switch mode the laser is continuously pumped and the Q-switch is switched on and off rectangularly with the pulse repetition rate r. The inversion varies slightly around the threshold inversion N_{th} between N_a and N_f (Fig. 1.16). In the interval T_p where $J \approx 0$ the evolution of the inversion is with (1.79)

$$N(t) = N_f + (N_{max} - N_f)\left(1 - e^{-t/\tau}\right). \quad (1.118)$$

With the approximation,

$$T_p \approx 1/r , \quad (1.119)$$

the inversion at the end of a pump cycle is

$$N_a = N_f + (N_{max} - N_f)\left(1 - e^{-1/(r\tau)}\right). \quad (1.120)$$

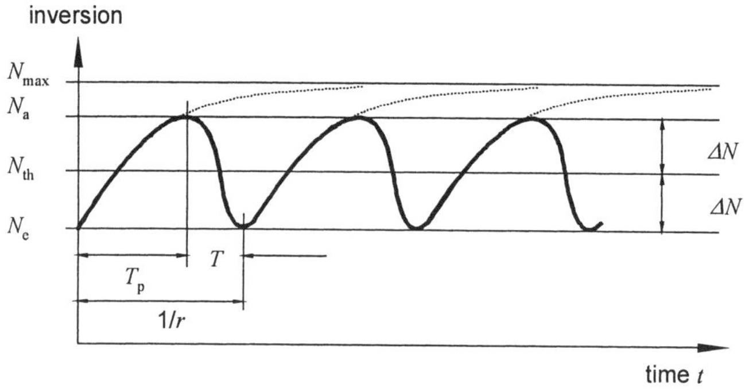

Fig. 1.16. Inversion for the periodical Q-switch versus time ($N_i - N_{th} << N_{th}$)

For small values of the inversion oscillations, $\Delta N << N_{th}$, (1.113) delivers in second order

$$\Delta N = N_a - N_{th} = N_{th} - N_e \tag{1.121}$$

the amplitude of N can then be described as

$$\Delta N = \left(\frac{1-e^{-1/(r\tau)}}{1+e^{-1/(r\tau)}}\right)(N_{max} - N_{th}) \,. \tag{1.122}$$

With (1.115) the energy of a pulse in the periodical Q-switching mode becomes

$$E = -fH_s \ln\sqrt{R}\,\frac{2\Delta N}{N_{th}} \tag{1.123}$$

and the peak power with (1.111) and (1.103)

$$P_{max} = -V \ln\sqrt{R}\,\frac{h\nu}{\tau_R}\left[\Delta N - N_{th}\ln\left(1+\frac{\Delta N}{N_{th}}\right)\right] \approx \frac{-fh\nu}{2}\ln\sqrt{R}\,\frac{H_s}{\tau_R}\left(\frac{\Delta N}{N_{th}}\right)^2 \tag{1.124}$$

with the $\tau_R = \tau_c/(-\ln V\sqrt{R})$ resonator decay time. The pulse duration T can be estimated to be

$$T = \frac{E}{P_{max}} = \frac{4\tau_R}{l\sigma\Delta N} \,. \tag{1.125}$$

The product of pulse energy and pulse duration in the periodical Q-switch is a system constant

$$E \times T = 8\eta_{slope} P_s \tau\tau_R = C_Q \,. \tag{1.126}$$

Example 1.2 Nd:YAG Q-switch parameter

With the following experimental parameters

$\tau = 200$ µs,	$\eta_{excit} = 5\%$,	$\eta_{slope} = 2\%$,	$L = 1$ m,
$P_p = 5$ kW,	$P_{th} = 1$ kW,	$R = 0.9$,	$l = 10$ cm.

and the above equations one obtains:

stored energy	$E_{max} = 50$ mW s,	maximum laser energy	$E = 20$ mW s,
resonator transit time	$\tau_c = 3.3$ ns,	peak pulse power	$P_{max} = 0.4$ MW,
pulse duration	$T = 50$ ns.		

An increase of the pulse energy results in a shorter pulse duration. Equation (1.121) inserted into (1.122) gives for the energy with N_{max}, according to (1.102) and with N_{th}, according to (1.54)

$$E = 2\left(\frac{1-e^{-1/(r\tau)}}{1+e^{-1/(r\tau)}}\right)\tau P_{cw} \,, \tag{1.127}$$

and for the average power

$$P_{av} = Er, \qquad T,\ T_p \text{ and } 1/r \text{ according to Fig. 1.16.} \tag{1.128}$$

At high pulse repetition rate r the average power in Q-switch mode approaches the cw laser power. With the help of the system constant C_Q the pulse duration and peak power follow as

$$T = \frac{(E \times T)}{E} = \frac{C_Q}{E} \tag{1.129}$$

and

$$P_{max} = \frac{E}{T} = \frac{E^2}{(E \times T)} = \frac{E^2}{C_Q}. \tag{1.130}$$

Figure 1.17 shows experimental results [1.10] for the average power and for the pulse energy.

The relation between pump duration, laser pulse duration and Q-switch repetition rate determine how regular the pulse train is. If the Q-switch is opened quickly after reaching the threshold, a small pulse arises, changing the inversion insignificantly. In the next pump cycle a high inversion can then be reached, which leads to a large pulse and low initial inversion. Small pulses alternate therefore with large ones. Pulse patterns are possible that recur more or less regularly. Since the thresholds are different for the various modes, a mode hopping can be observed [1.11].

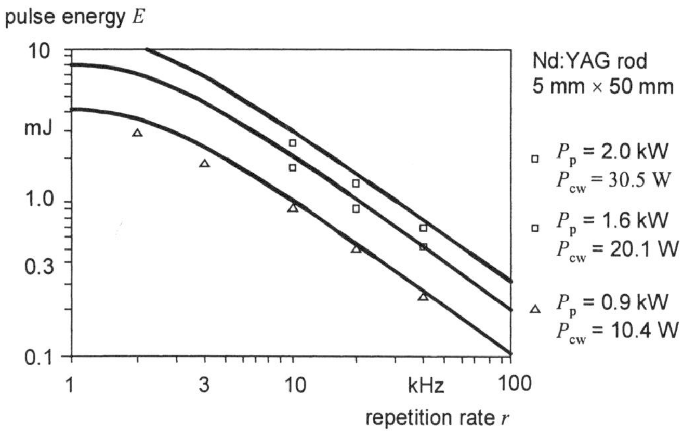

Fig. 1.17. Pulse energy versus the repetition rate in the Q-switch mode, [1.10]

1.8 The 3-Level System

Using a simplified energy diagram of a 3-level system as in Fig. 1.18, the differences with respect to the 4-level system will be derived.

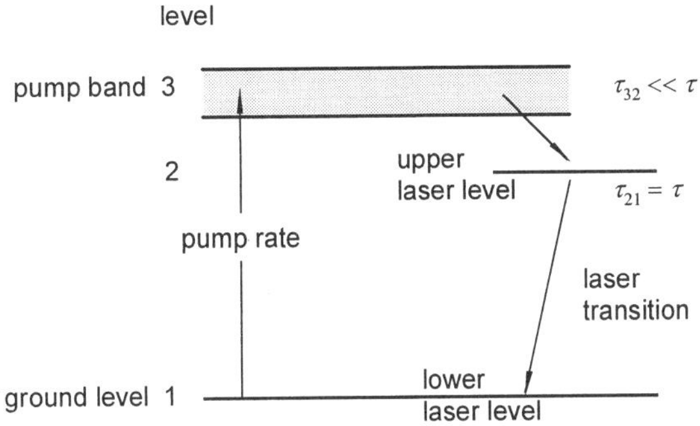

Fig. 1.18. Simplified energy diagram of a 3-level system

The rate equations for the upper laser energy level and the intensity are

$$\frac{\mathrm{d}N_2}{\mathrm{d}t} = WN_1 - \frac{N_2}{\tau} - \frac{\sigma J}{h\nu}\left(N_2 - N_1\right), \tag{1.131}$$

$$\frac{\mathrm{d}J}{\mathrm{d}t} = \frac{\sigma c J}{n}\left(N_2 - N_1\right) - \frac{\sigma c J}{n} N_{\mathrm{th}} \,. \tag{1.132}$$

With the help of the following definitions

$N = N_2 - N_1$ *the inversion,*

$N_0 = N_2 + N_1$ *the total density of the active ions and*

$J_{\mathrm{s}} = \dfrac{h\nu}{2\sigma\tau}$ *the saturation intensity,*

the densities N_1 and N_2 in (1.131) and (1.132) can be eliminated and one obtains the rate equations for the 3-level system

$$\frac{\mathrm{d}N}{\mathrm{d}t} = W\left(N_0 - N\right) - \frac{1}{\tau}\left(N_0 - N\right) - \frac{J}{J_{\mathrm{s}}} N \,, \tag{1.133}$$

$$\frac{\mathrm{d}J}{\mathrm{d}t} = \frac{\sigma c J}{n}\left(N - N_{\mathrm{th}}\right). \tag{1.134}$$

Inversion requires $N \geq 0$, i.e. half of the ions must be excited into the upper laser level, $N_2 \geq N_0/2$.

Stationary Case with d*N*/d*t* = 0 and d*J*/d*t* = 0

The stationary case is the same as for the 4-level system

$$N = N_{\mathrm{th}}. \tag{1.135}$$

The steady-state inversion equals the threshold inversion. The intensity in the resonator follows with (1.133)

$$J = J_{\mathrm{s}}\left[W\tau\left(\frac{N_0}{N_{\mathrm{th}}} - 1\right) - \left(\frac{N_0}{N_{\mathrm{th}}} + 1\right)\right]. \tag{1.136}$$

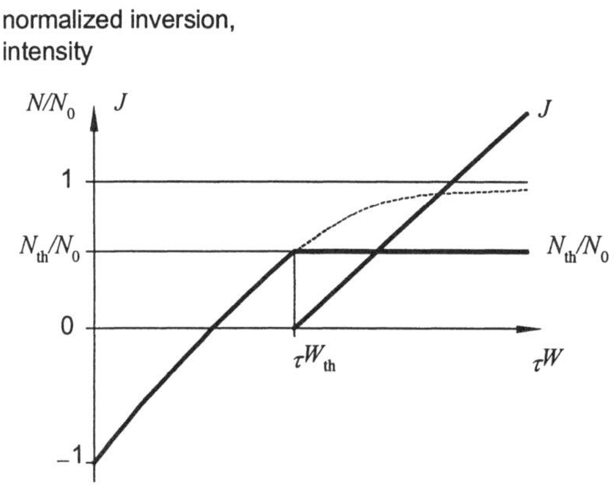

Fig. 1.19. Schematic diagram of inversion and intensity in the 3-level system

The threshold pump rate reads:

$$W_{\mathrm{th}} = \frac{1}{\tau}\frac{N_0 + N_{\mathrm{th}}}{N_0 - N_{\mathrm{th}}}. \tag{1.137}$$

Figure 1.19 shows schematically the intensity J and the inversion N as a function of the normalized pump rate τW. The pump rate is, via (1.51)

$$W = \eta_{\mathrm{excit}}\frac{2P_{\mathrm{p}}}{V(N_0 - N_{\mathrm{th}})h\nu}. \tag{1.138}$$

For the intensity in the resonator it follows that

$$J = \eta_{\text{excit}} \frac{1}{\sigma V N_{\text{th}}} \left(P_{\text{p}} - \frac{\sigma V J_s (N_0 + N_{\text{th}})}{\eta_{\text{excit}}} \right), \tag{1.139}$$

and the stationary laser power P_{cw} becomes with (1.47), (1.58) and (1.61)

$$P_{\text{cw}} = \eta_{\text{slope}} \left(P_{\text{p}} - P_{\text{s}} \left(\sigma N_0 l - \ln(V\sqrt{R}) \right) \right), \tag{1.140}$$

with $P_{\text{s}} = \dfrac{f J_{\text{s}}}{\eta_{\text{excit}}}$ and the slope-efficiency $\eta_{\text{slope}} = \eta_{\text{excit}} \dfrac{\ln\sqrt{R}}{\ln(V\sqrt{R})}$.

In contrast to the 4-level-system, the threshold is increased by $\sigma N_0 l$. The pump power P_{p} has to be larger by the amount $\sigma N_0 l P_{\text{s}}$ than in the 4-level system. Because of the larger thermal load 3-level lasers are mainly used in pulsed operation.

The laser´s pulse energy is

$$E = \eta_{\text{slope}} \left(P_{\text{p}} - P_{\text{s}} \left(\sigma N_0 l - \ln(V\sqrt{R}) \right) \right) T, \tag{1.141}$$

with the threshold energy

$$E_{\text{th}} = P_{\text{s}} \left(\sigma N_0 l - \ln(V\sqrt{R}) \right) T, \qquad T = \text{pulse duration } (T >> \tau). \tag{1.142}$$

Example 1.3 Ruby rod with 1 cm diameter and 10 cm length with a 0.05 wt% Cr doping.

With the supplier data

$\sigma = 2\times10^{-20}$ cm^2,	$N_0 = 0.16\times10^{20}$ cm^{-3},	$l = 10$ cm,	$r_0 = 0.5$ cm,
$R = 0.5$,	$V = 0.9$,	$\eta_{\text{excit}} = 0.8\%$	

the parameters are calculated to be

$\sigma N_0 l = 3.2$,	$-\ln(V\sqrt{R}) = 0.45$,	$J_{\text{s}} = 1.9$ kW/cm^2,	$f \approx 0.8$ cm^2,
$P_{\text{s}} \approx 190$ kW,	$E_{\text{th}} \approx 3.4$ kWs,	$\eta_{\text{slope}} \approx 0.7\%$,	

and yield a laser energy of approximately 5 Ws at 10 kW s pump energy.

By measuring E_{th} for the corresponding reflection R of the output mirror losses $(\ln V - \sigma l N_0)$ can be determined. The influence of the reflection coefficient R on the threshold compared to the absorption $\sigma l N_0$, e.g. for ruby, is insignificant. The optimal reflection R_{opt} of the output mirror, for which the maximum laser pulse energy $E_{\max}$ can be extracted, is, analogous to (1.65),

$$\ln\sqrt{R_{\text{opt}}} = -2\sqrt{\left(P_{\text{p}} / P_{\text{s}} - \sigma N_0 l \right) \ln V} + 2\ln V \tag{1.143}$$

and the maximum laser energy is

$$E_{\max} = fJ_s T\left(\sqrt{P_p / P_s - \sigma N_0 l} - \sqrt{-\ln V}\right)^2 . \tag{1.144}$$

All equations for the 3- and 4-level systems were derived under simplified assumptions and can basically be validated by measurements. Some restrictions, however, have to be discussed in more detail:

- The inversion is not created and reduced homogeneously over the material cross section. The radiation field generally has a transverse structure. There is a mutual influence of the radiation field and the inversion. The longitudinal radiation field does not necessarily agree with the geometry of the laser medium. This is taken into account basically by the extraction efficiency η_{extr} [1.13].

- As the laser preferably oscillates near threshold in the regions of highest inversion, the laser energy or power as a function of excitation (at low pump power) is nonlinear.

- For a high reflection coefficient and high excitation power the linear dependence of the laser energy or power on the excitation no longer holds. A saturation effect occurs.

Table 1.3 Typical results for losses and excitation power for pulsed operation

4-level	Dimensions mm or "	f cm^2	$-\ln V$ %	P_s kW	η_{excit} %	J_s kW/ cm^2	Ref.
Nd glass LG 760							
Slab	6×20×300	1.8	10	307	4	6.8	[1.14]
Nd:YAG							
Rod [a]	5×65	0.2		33	0.9	1.6	
Rod	3"×1/4"	0.32			3.5		[1.15]
Rod	4 " × 1/4 "	0.32	12	4.53	7	1	
Rod	4×12×100	0.87	13	108	1.7	2.1	
Nd:GGG							
Rod [b]	5×59	0.2	10	53	1.8	4.9	
Rod	3"×1/4"					7.7	[1.15]
Cr:Alexandrite							
Rod	4"×1/4"	0.32	7	2200	2.2	157	[1.14]

3-level	Dimensions mm	f cm^2	σN_0 % cm^{-1}	T ms	η_{slope} %	E_{th} kW s	Ref.
Cr:ruby							
Rod	104×9.5	2.84	0.2	0.8	0.75	2.2	[1.1]
Rod	150×9.5	2.84	0.2		0.82	1.3	[1.16]

[a)] Pump length 43 mm, cavity not optimized
[b)] Pump length not optimized to the cavity, GGG without AR coating

All of these effects cause losses and heating of the laser material. However, they have not been examined in the equations explicitly but are taken into account by the loss terms $\ln V$ and P_s. They are discussed in detail in [1.3, 1.7]. Typical parameters for cw and pulsed operation are given in Tables 1.2 and 1.3. All essential equations for 4- and 3-level systems are compiled in Table 1.4.

Table 1.4 Essential equations

	4-level (Nd:YAG)	**3-level** (Cr:ruby)
saturation intensity	$J_s = \frac{h\nu}{\sigma\tau}$	$J_s = \frac{h\nu}{2\sigma\tau}$
saturation energy density	$H_s = \frac{h\nu}{\sigma}$	$H_s = \frac{h\nu}{2\sigma}$
small-signal gain factor	$G_0 = \exp(\sigma N l)$	$G_0 = \exp(\sigma N l)$
inversion	$N = N_2$	$N = N_2 - N_1$
saturated gain coefficient	$g_s = g_0/(1 + J/J_s)$	$g_s = g_0/(1 + J/J_s)$
loss factor per round trip	$V = \exp(-\alpha l)$	$V = \exp(-\alpha l)$
system constant	$P_s = fJ_s/\eta_{excit}$	$P_s = fJ_s/\eta_{excit}$
threshold (inversion)	$N_{th} = -\frac{\ln\left(V\sqrt{R}\right)}{\sigma l}$	$N_{th} = -\frac{\ln\left(V\sqrt{R}\right)}{\sigma l}$
threshold (pump power)	$P_{th} = -P_s \ln(V\sqrt{R})$	$P_{th} = P_s(\sigma l N_0 - \ln(V\sqrt{R}))$
threshold (temperature)		$P_T = P_s \sigma l N_T$
slope efficiency	$\eta_{slope} = \frac{\eta_{excit} \ln\sqrt{R}}{\ln(V\sqrt{R})}$	$\eta_{slope} = \frac{\eta_{excit} \ln\sqrt{R}}{\ln(V\sqrt{R})}$
laser power	$P_{cw} = \eta_{slope}(P_p - P_{th})$	$P_{cw} = \eta_{slope}(P_p - P_{th})$
laser energy (rectangular pulse)	$E = P_{cw}T$, $\tau << T$	$E = \eta_{slope}(E_p - E_{th})$

f = beam cross section
J_s = saturation intensity
α = loss coefficient
P_p = pump power
τ = upper level lifetime
J = intensity of the laser beam
l = length of the laser medium
N_T = thermal population
σ = cross section
V = loss factor
h = Planck's constant
N = inversion density
η_{excit} = excitation efficiency
R = reflection coefficient
T = pulse duration

The following loss mechanisms for pump light and laser radiation field in 3 and 4-level systems exist:

- amplification of spontaneously emitted ASE light outside the utizable area,
- inversion reduction by thermal population of the lower laser level,
- radiation field absorption in the medium by the upper laser energy level and excitation of higher energy states, excited state absorption (= ESA),
- scattering of the radiation field in the material and at the mirrors,
- diffraction of the radiation field by the finite aperture of the laser medium,
- radiation field and inversion are not adapted to each other.

1.9 Efficiency and Optimization

The different efficiencies from the power supply to the useable process laser power are explained and defined by the power transfer diagram, Fig. 1.21. An analogous definition of the power efficiency is valid for pulsed lasers if the pulse duration is large compared to the lifetime of the upper laser level and the laser is pumped far above the threshold. The efficiency determines the size of the power supply and the cooling unit and thus also fundamentally the price of the complete laser equipment. Typical values for the various efficiencies are given below [1.17].

Power Supply

The electrical power is provided by a power supply in a suitable form for the excitation source. In the case of pulsed operation the necessary energy is stored at first in capacitors (except Q-switch). Power for control circuits as well as for the cooling unit are not taken into account initially.

$$P_{in} = \eta_{plug} P_{plug}, \qquad \eta_{plug} \approx 0.8 - 0.9, \qquad (1.145)$$

η_{plug} = efficiency of the power supply.

Excitation Source

The excitation sources convert the electrical power, both into radiation within the pump levels, and also to nonutilizable radiation and to heat

$$P_{light} = \eta_{lamp}(P_{in} - P_{pth}), \qquad \eta_{lamp} \approx 0.5 - 0.6. \qquad (1.146)$$

P_{pth} takes into account the power that must be supplied to the lamp up to the first light emission.

Cavity

The pump light in the cavity is concentrated by a reflector onto the laser medium. Geometric and absorption losses by reflector, flow tubes and coolants occur. The radiation inside the laser material can then be utilized for optical pumping. The pump light power P_p is given by

$$P_p = \eta_{cav} P_{light}, \qquad \eta_{cav} \approx 0.5. \tag{1.147}$$

Upper Laser Level

Losses in the energy level diagram, Fig. 1.20, that can occur in the pump levels and in the upper laser energy level are:

- nonradiative transitions (quantum efficiency, η_Q),
- multiphoton excitation activation directly from the already excited upper laser level (excited state absorption, ESA),
- spontaneous and amplified spontaneous emission (ASE)

$$P_{niv} = \eta_{med} \eta_{niv} P_p, \qquad \eta_{niv} = \eta_{QD}(1 - \eta_{ESA})(1 - \eta_{ASE}), \tag{1.148}$$

$$P_{niv} - VN_0 h\nu W.$$

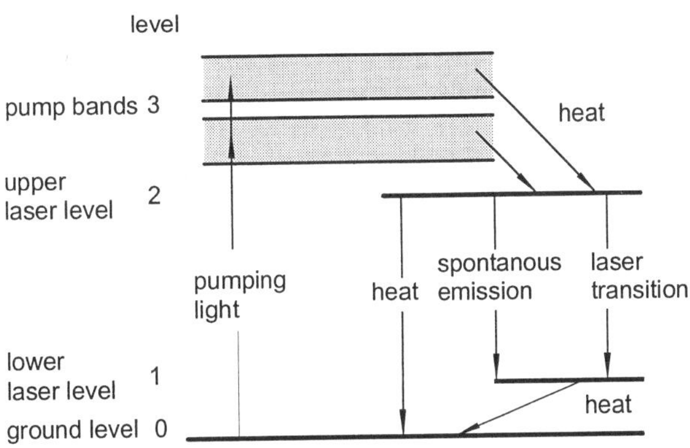

Fig. 1.20. Losses in the various levels

In [1.19] the efficiency for the absorption into excited levels is defined as $\eta_{ESA} = N_3\, \sigma_{ESA} d$, with d as the material thickness. Measured results can also be found there.

Laser Material

The pump light, absorbed by the laser material into the pump levels excites the upper laser level. All effects that cause losses of the pump radiation are contained in the efficiency η_{mat} of the laser material.

One can distinguish the following single loss mechanisms:

- absorption of the pump light by color centers,
- nonradiative transitions from the upper levels, (quantum efficiency η_Q)
- parts of the pump light not being absorbed,
- the quantum defect η_{QD}, i.e. the energy difference between pump and laser light.

The energy losses of the above effects heat the laser material. The first two are combined in the quantum efficiency η_Q and the quantum defect η_{QD} is described through

$$\eta_{QD} = \frac{h\nu}{h\nu_p}, \tag{1.149}$$

η_{QD} = quantum defect.

The efficiency in the medium is made up of

$$\eta_{mat} = \eta_Q \eta_{QD}\, f(\alpha), \tag{1.150}$$

$f(\alpha)$ = function that describes the absorption,
$\eta_{mat} \approx 0.4 - 0.5$,
$\eta_Q \approx 0.8$

Resonator (Extraction Efficiency)

The excitation energy available in the upper laser level cannot be fully coupled out. Usually this efficiency is described as extraction efficiency [1.13]. The resonator prescribes a radiation structure longitudinally and radially. Parts of the laser medium that are not exposed to the radiation field of the resonator cannot contribute to the energy from the upper laser level by stimulated emission. In addition scattering, diffraction and absorption also occur in the resonator. These losses are included in the resonator efficiency as extraction efficiency η_{extr} and the laser radiation behind the output mirror is therefore given by

$$P = \eta_{extr} P_{niv}, \qquad \eta_{extr} = -\ln\sqrt{R}\left(\frac{1}{-\ln(V\sqrt{R})} - \frac{1}{gl}\right). \tag{1.151}$$

With $V = 1$ and rectangular mode profile (large mode order, mode volume ≈ 1) η_{extr} approaches 1

$$\eta_{extr} = \left(1 - \frac{\ln\sqrt{R}}{gl}\right), \qquad \eta_{extr} \approx 0.5.$$

For $R = R_{opt}$, η_{extr} becomes

$$\eta_{extr} = \left(\frac{\sqrt{gl} - \sqrt{\alpha l}}{gl}\right)^2 .$$

Optics

The optical system transforms the laser beam for the application as far as focus diameter and focus depth are concerned. There are losses due to reflection, e.g. at the lens surfaces or fiber ends, or by geometric restrictions.

$$P_{use} = \eta_{opt} P, \qquad \eta_{opt} \approx 0.85 - 0.95. \tag{1.152}$$

Workpiece

The laser light is partially reflected by the worpiece, is absorbed by the plasma, disappears through joint gaps or is conducted as heat into the workpiece

$$P_{proc} = \eta_{proc} P_{use} , \tag{1.153}$$

η_{proc} = process efficiency.

For measurement reasons the various efficiencies are partially combined.

Excitation Efficiency, Storage Efficiency

For pulsed operation with $T >> \tau$ or in the cw mode one has

$$P_{niv} = \eta_{excit} P_p , \tag{1.154}$$

η_{excit} = excitation efficiency, according to (1.50),

$\eta_{excit} = \eta_{mat} \eta_{cav}$, $\qquad \eta_{excit}$ » 0.07 – 0.12.

For pulsed operation with short pulses $T < \tau$,

$$E_{\text{niv}} = \eta_{\text{stor}} P_{\text{p}} , \tag{1.155}$$

η_{stor} = storage efficiency.

Total Efficiency

The total efficiency is defined as the ratio of average laser output power to electrical input power.

$$P = \eta_{\text{tot}} P_{\text{in}} , \tag{1.156}$$

$$\eta_{\text{tot}} = \eta_{\text{excit}}\, \eta_{\text{cav}}\, \eta_{\text{mat}}\, \eta_{\text{niv}}\, \eta_{\text{res}} ,$$

$$\eta_{\text{tot}} = \eta_{\text{excit}}\, \eta_{\text{cav}}\, \eta_{\text{extr}} , \qquad \eta_{\text{tot}} \approx 0.04.$$

Slope Efficiency

Since the total efficiency depends on the operating point, one uses the differential efficiency η_{slope}

$$\mathrm{d}P_{\text{L}} = \eta_{\text{slope}} \mathrm{d}P_{\text{in}} , \qquad \eta_{\text{slope}} \approx 0.05 \tag{1.157}$$

$$\mathrm{d}E_{\text{L}} = \eta_{\text{slope}} \mathrm{d}E_{\text{in}} ,$$

$$\eta_{\text{slope}} = \eta_{\text{excit}} \frac{-\ln\sqrt{R}}{-\ln(V\sqrt{R})} \quad . \tag{1.158}$$

Total Efficiency of the Laser Unit

The utilizable average laser power is related to the total mains power as follows

$$P = \eta_{\text{plug}} P_{\text{plug}} , \tag{1.159}$$

$$\eta_{\text{plug}} = \eta_{\text{opt}}\, \eta_{\text{tot}}\, \eta_{\text{plug}} , \qquad \eta_{\text{plug}} \approx 0.03.$$

Laser System Efficiency

Important, however, is the overall energy of a processing task. Therefore it is necessary to take into account the energy for auxiliary circuits such as cooling, control units and also the process efficiency. Under certain circumstances it is more favorable to work with an inferior efficiency but to obtain better processing results with a high laser beam quality.

In the following, some suggestions are given on how to improve the overall efficiency:

- The power for the control, the fans, the auxiliary circuits and for the cooling unit is up to 30% of the total power supply power and these losses must be extracted as heat. The power supply efficiency can be improved with loss-reduced power mains (5 to 10%).
- The connection cross section must be sufficiently large to avoid inadmissible heating of the power transmission cable.

Example 1.4 Power losses in the connection cable

The heat dissipation P from the connection cable in the pulsed operation is $P = I^2 R f T$ with a resistance R for rectangular pulses T. Operating two lamps in parallel at $I = 200$ A each having two leads of R = 5 mΩ (corresponds to 3 m length with a cross-sectional area of 6 mm^2) at a pulse frequency f = 100 Hz and a pulse duration T = 1 ms the cable is heated with $P = 4 \times 200^2\ \text{A}^2 \times 5\ \text{m}\Omega \times 100\ \text{Hz} \times 1\ \text{ms} = 80\ \text{W}$.

- The choice of the pump gases can influence the lamp efficiency. For pump Nd lasers the efficiency of Kr lamps is generally higher, the overall light efficiency of Xe lamps, e.g. the ratio of light to electrical power, however, is better.
- Due to the better spectral adaptation to the laser medium, preference is given to Kr lamps rather than to the higher light efficiency of the Xe lamps. Optical feedback into the lamp through selectively coated flow tubes around the lamps recycles the energy outside the absorption levels.
  The reflector has to be adapted as closely as possible to the laser medium and the lamps. The pump light should not see any "gaps", by which it may escape. Boundary surfaces (flow tube of the rod) must be antireflective-coated. The wall material of the lamp should be absorption-free.
- Water has a strong absorption at a wavelength of 700 nm, increasing with wavelength. The pump light path through the water should therefore be as short as possible. Additives in the coolant, for example fluorescence converters, must be spectrally adapted. Silver reflectors are spectrally more favorable than gold reflectors, however, they must be protected against oxidation.
- The contribution to the excitation of short-wavelength ($\lambda < 550$ nm) light is generally small and leads to an additional heating of the laser material.
- The ratio of polished surface areas to ground surface areas of the laser medium should be small to avoid losses by spontaneous amplified emission and whisper modes. Good material quality is necessary to keep scattering losses small.
- The radial and longitudinal mode volume can be maximized for the application by optimizing the resonator.
- The laser optics must be AR-coated for the laser wavelength. The diameters must be chosen to be sufficiently large.

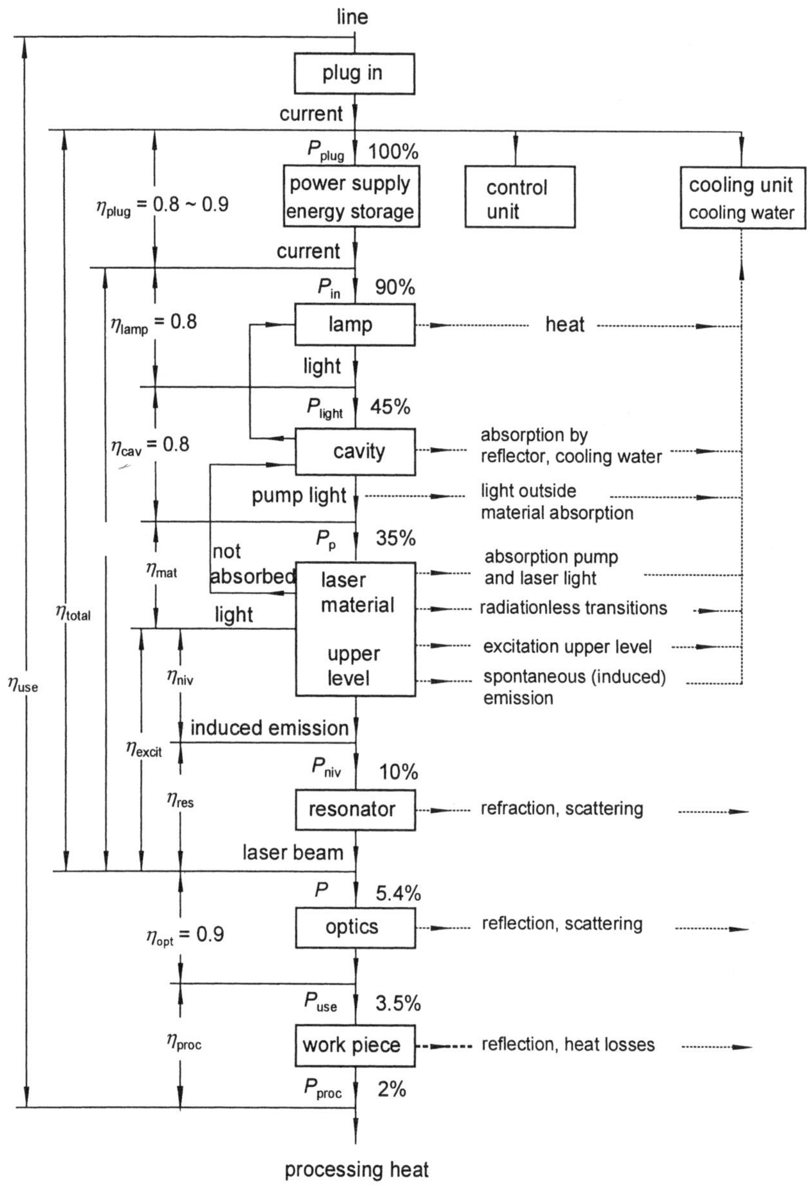

Fig. 1.21. Numerical values of efficiencies, [1.12]

2 Gaussian Optics

The interaction of the radiation field with the excited atoms in the laser material, the stimulated emission, creates the utilizable laser power. The quality of the radiation field, particularly the intensity distribution, determines, next to the laser power, the field of laser applications. Therefore the laser beam must be characterized with respect to its optical qualities. In this chapter Gaussian optics is applied, which means:

- Paraxial beams with a small slope α to the axis of the optical system. The trigonometric functions can be approximated by their argument, $\sin\alpha \cong \tan\alpha \cong \alpha$.
- Only spherical or cylindrical lenses and mirrors are used, which in the paraxial approximation can be replaced by parabolic optics.

2.1 Gaussian Beam

As an approximation the light wave in the resonator is considered as a scalar wave field. The diffraction due to the finite transversal dimensions, e.g. of the laser crystal, is taken into account only in a first order approximation. The amplitude A of the electrical field then fulfills the time-independent wave equation [2.1] in its SVE-approximation (slowly varying envelope). With the formulation for the electric field $E = A(x,y,z)\exp(-\mathrm{i}kz)$ the SVE-wave equation reads

$$\frac{\partial^2 A}{\partial x^2} + \frac{\partial^2 A}{\partial y^2} - 2\mathrm{i}k\frac{\partial A}{\partial z} = 0 \tag{2.1}$$

with

$k = 2\pi / \lambda$, *wave number.*

The simplest solution for rotational symmetry [2.2] is the Gaussian beam or fundamental mode TEM_{00} with an amplitude distribution $A\,(r, z)$ as:

$$A(r,z) = A_0 \frac{w_{T0}}{w_0} \exp\left(-r^2 \left(\frac{1}{w_0^2} + \frac{\mathrm{i}k}{2R} \right) + \mathrm{i}(kz - \varphi) \right) \tag{2.2}$$

with

w_{T0}	beam waist of 00-mode,
$w_0 = w_0(z)$	beam radius of 00-mode,
$\varphi = \arctan(z/z_{\mathrm{R}})$	longitudinal phase,
$R = R(z)$	curvature of the wavefront.
$z_{\mathrm{R}} = \pi w_{\mathrm{T0}}^2 / \lambda$	the Rayleigh length, focal depth or confocal parameter.

The beam radius w_0 and the curvature R are given by (see Sect. 2.6)

$$w_0^2(z) = w_{\mathrm{T0}}^2\left(1 + (z / z_{\mathrm{R}})^2\right), \qquad \textit{beam radius,} \tag{2.3}$$

$$R(z) = z_{\mathrm{R}}\left(z_{\mathrm{R}} / z + z / z_{\mathrm{R}}\right), \qquad \textit{curvature,} \tag{2.4}$$

provided the origin of the z coordinate is in the waist.

The Gaussian beam is fixed by two parameters: the beam waist w_{T0} and the Rayleigh length z_{R}. A beam has a waist with radius w_{T0} when the radius of curvature is infinite, i.e. for $R = \infty$. The divergence θ_0 of the fundamental beam TEM_{00} is defined as

$$\theta_0 = \lim_{z\to\infty} \frac{w_0(z)}{z} = \frac{\lambda}{\pi}, \qquad \textit{divergence.} \tag{2.5}$$

Defining the complex beam parameter $q = q(z)$,

$$\frac{1}{q} = \frac{1}{R} - \frac{\mathrm{i}\lambda}{\pi w_0^2}, \quad q = \frac{R\pi w_0^2}{\pi w_0^2 - \mathrm{i}\lambda R} \quad \text{or} \quad q = z + \mathrm{i}z_{\mathrm{R}}, \tag{2.6}$$

Equation (2.2) becomes

$$A(r,z) = A_0 \frac{w_{\mathrm{T0}}}{w_0} \exp\left(-\frac{kr^2}{2q} + \mathrm{i}(kz - \varphi) \right). \tag{2.7}$$

Radius of curvature and beam radius are related to q according to Eq. (2.6)

$$R = (\mathrm{Re}\, 1/q)^{-1} \tag{2.8}$$

$$w_0^2 = -\frac{\lambda}{\pi} (\mathrm{Im}\, 1/q)^{-1}. \tag{2.9}$$

The complex beam parameter permits, in a simple way, in conjunction with the ABCD law, calculation of the beam propagation to a first order approximation in parabolic optical systems. This will be shown in the following. The intensity $J = J(r,z)$ follows from (2.2) as the square of the amplitude $A(r,z)$

$$J(r,z) = J_0 \frac{w_{\mathrm{T0}}^2}{w_0^2} \mathrm{e}^{-2r^2/w_0^2}, \qquad \textit{intensity of a Gaussian beam.} \tag{2.10}$$

J_0 is the peak intensity of the Gaussian beam in the waist w_{T0}. Figure 2.1 shows the profile of the beam along the z direction.

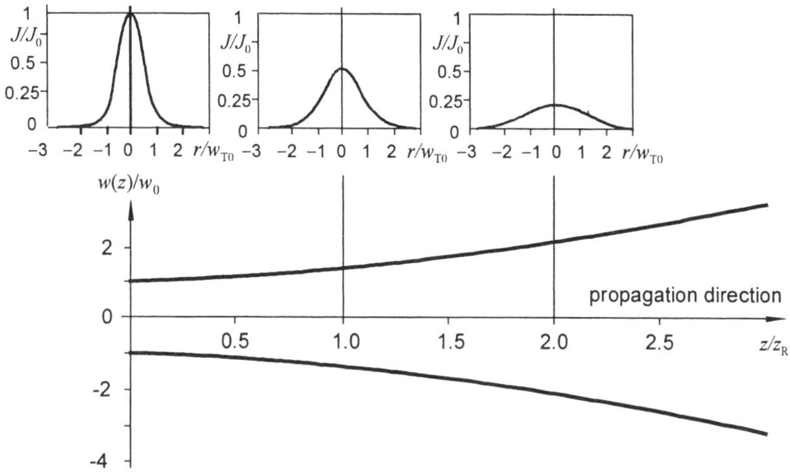

Fig. 2.1. Beam profile and normalized beam radius along the propagation direction z

The radial intensity distribution in the waist for $z = 0$ is shown in Fig. 2.2:

$$J(r,0) = J_0 e^{-2r^2/w_{T0}^2} . \tag{2.11}$$

The power P in the beam cross-sectional area with radius r is given by

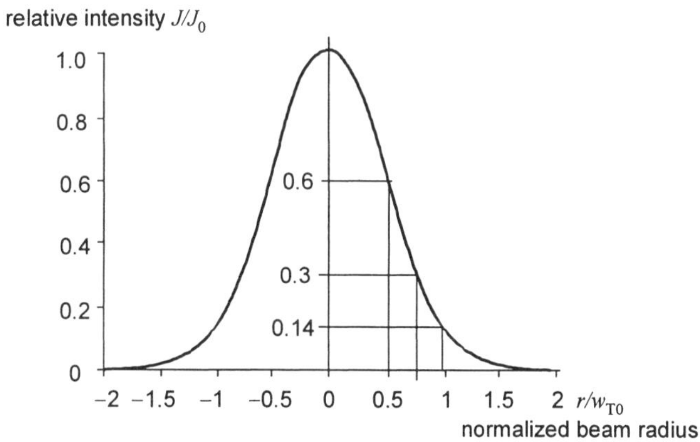

Fig. 2.2. Radial intensity distribution of a Gaussian beam

$$P(r,z) = 2\pi \int_0^r J(r,z) r \mathrm{d}r = P_0\left(1 - \mathrm{e}^{-2r^2/w_0^2}\right) \tag{2.12}$$

with $P_0 = 2\pi \int_0^\infty J(r,0) r \mathrm{d}r = \frac{\pi}{2} w_{T0}^2 J_0$.

This Gaussian power distribution is shown in Fig. 2.3.

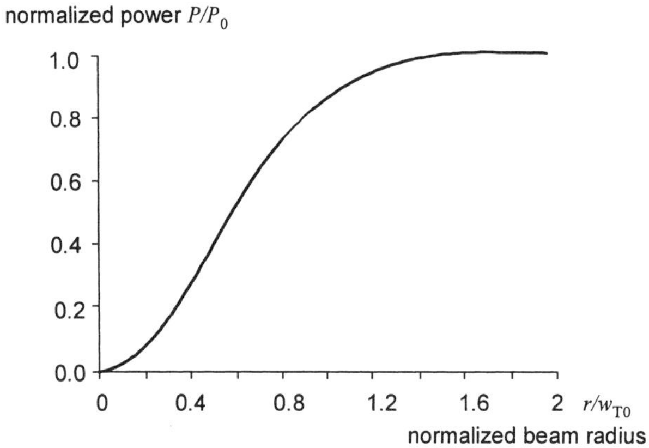

Fig. 2.3. Power distribution of a Gaussian beam within a radius r

Choosing the lenses and apertures traversed by the laser beam to have a diameter $D = 2 \times 1.5 w_0$, the power loss due to the geometric shading per element is about 1%. Table 2.1 gives the corresponding values.

Table 2.1 Intensity and power content of a Gaussian beam

r/w_{T0} or r/w_0	0.23	0.59	1.0	1.25	1.5	2.0
$J(r/w_{T0})/J_0$	0.9	0.5	0.135	0.044	0.011	0.0003
$P(r/w_0)/P_0$	0.1	0.5	0.865	0.956	0.989	0.9997

2.2 Higher Order Modes

Equation (2.1) has an infinite number of solutions, the eigenfunctions, which propagate in free space with increasing or decreasing diameter, but without changing their structure, but their transversal dimensions. The type of eigenfunction depends on the symmetry of the resonator. Only the most important results will be presented here, details can be taken from the textbooks [2.13].

The frequency of the higher order modes is determined mainly by the resonator length and characterized by the mode order q. These longitudinal modes will not be considered here any further, because the mode order q is of no importance for material processing so far. Only the transverse modes are considered. Power and the focus diameter of the beam are of importance. The transverse modes are abbreviated by TEM_{mn} (transverse electromagnetic wave with the mode orders m and n).

Rectangular Symmetry

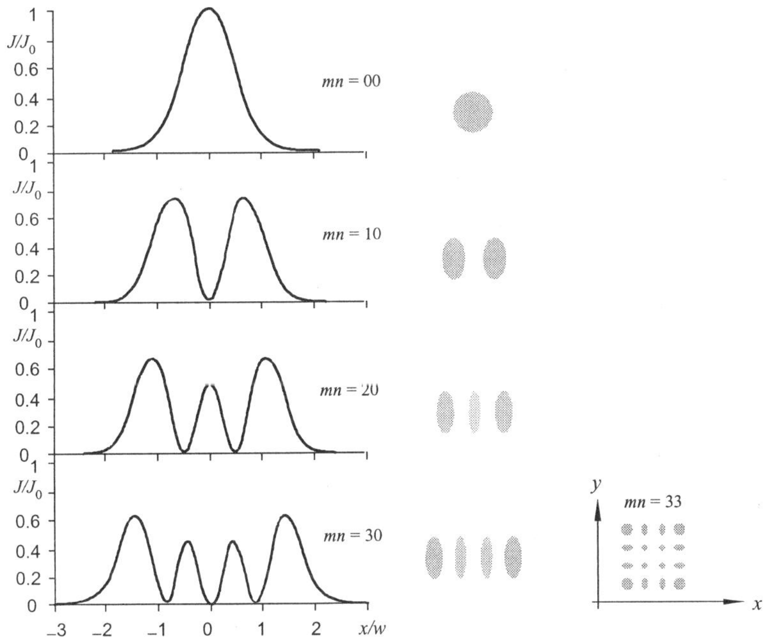

Fig. 2.4. Transverse beam structure for rectangular symmetry at low mode order *mn*

For rectangular geometry like a slab or a rod with a rectangular aperture the normalized intensity of a transverse mode is given, in the general case, as follows:

$$J_{mn}(x,y) = \frac{P_{mn}}{\pi w_0^2/2} \frac{1}{n!m!2^{n+m}} \mathrm{e}^{-2\left(x^2+y^2/w_0^2\right)} \left(H_m(\sqrt{2}\,x/w_0)\right)^2 \left(H_n(\sqrt{2}\,y/w_0)\right)^2$$

intensity distribution for rectangular geometry, (2.13)

with w_0 depending on z according to (2.3)

$P_{mn} = \iint J_{mn} \mathrm{d}x\mathrm{d}y$ is the total power of the beam,

$H_i(\Omega)$ = Hermite polynomials with m = mode order in x direction, n = mode order in y direction,

$H_0(\Omega) = 1$, $H_1(\Omega) = \Omega$, $H_2(\Omega) = 4\,\Omega^2 - 2$,

$H_3(\Omega) = 8\,\Omega^3 - 12\,\Omega$, $H_4(\Omega) = 16\,\Omega^4 - 48\,\Omega + 12$.

In Fig. 2.4 the intensity distributions of low order modes are shown. The mode order indicates the number of the intensity zeros within the beam cross section, for the corresponding coordinate. On the right side a two-dimensional contour plot of the intensity is shown.

Circular Symmetry

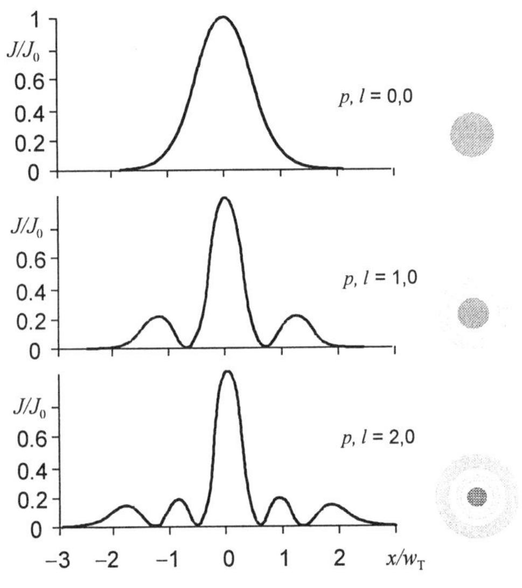

Fig. 2.5. Intensity distribution with circular symmetry at low mode orders TEM_{p0}

For rotational symmetry, like a rod or a tube, the normalized intensity distribution is given by

$$J_{p,l}(r,\delta) = \frac{P_{p,l}}{\pi / 2w_0^2} \frac{2p!}{(l+p)!} \left(\frac{2r^2}{w_0^2}\right)^2 \left(L_p^{(l)}\left(2r^2 / w_0^2\right)\right)^2 \mathrm{e}^{-2r^2/w_0^2} \mathrm{f}(\delta), \tag{2.14}$$

$P_{p,l} = \iint J_{p,l} r\mathrm{d}r\mathrm{d}\delta$ is the total power in the beam,

$L_p^{(l)}(\Omega)$ = Laguerre polynomial *with* $\Omega = 2r^2/w_0^2$,

$L_0^{(l)}(\Omega) = 1,\ L_1^{(l)}(\Omega) = l + 1 - \Omega,$ $\quad L_2^{(l)}(\Omega) = (l+1)(l+2)/2 - (l+2)\Omega + \Omega^2/2,$

l = azimuthal mode order, $\quad p$ = radial mode order,

δ = azimuthal angle, $\quad f(\delta) = \sin^2 l\delta,\ \cos^2 l\delta.$

For the mode of lowest order (fundamental mode), i.e. n, m, l and p equal to zero one has

$$J(x,y) = J_0(z)\mathrm{e}^{-2(x^2+y^2)/w_0^2} = J_0(z)\mathrm{e}^{-2r^2/w_0^2}\,, \tag{2.15}$$

i.e. a Gaussian beam with a beam radius w_0 at the position $z = 0$.

1. $l = 0$ and $f(\delta) = 1$

In Fig. 2.5 the intensity distribution for a circular symmetry for low-order modes is shown.

2. $l \neq 0$ Donut Modes

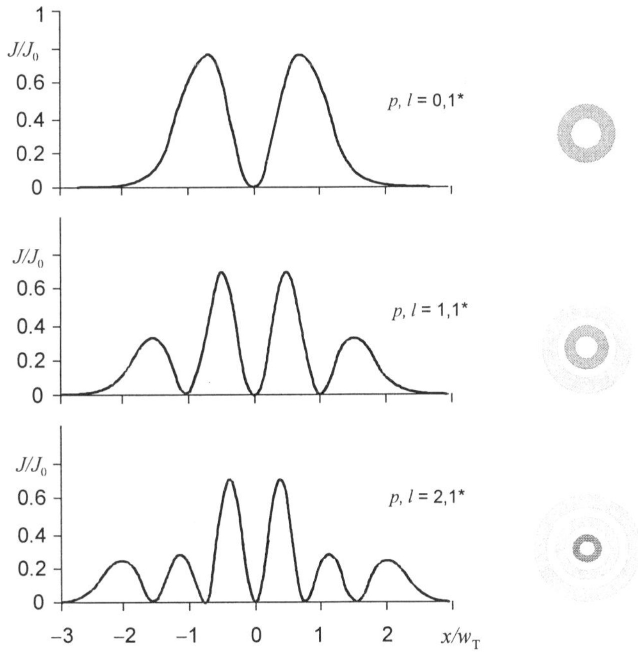

Fig. 2.6. Intensity distribution for hybrid modes TEM$^*_{pl}$

For $l \neq 0$ the intensity is always zero at $r = 0$, (2.13). Two intensity distributions $J_{p,l}$ with $(\sin^2 l\delta + \cos^2 l\delta) = 1$ can be superimposed to give rotational-symmetric distributions, so-called Donut modes, with a zero in the center, Fig. 2.6.

3. $p = 0$ and $f(\delta) = \sin^2 l\delta$ or $\cos^2 l\delta$

The distribution is characterized by an azimuthal structure (Fig. 2.7).

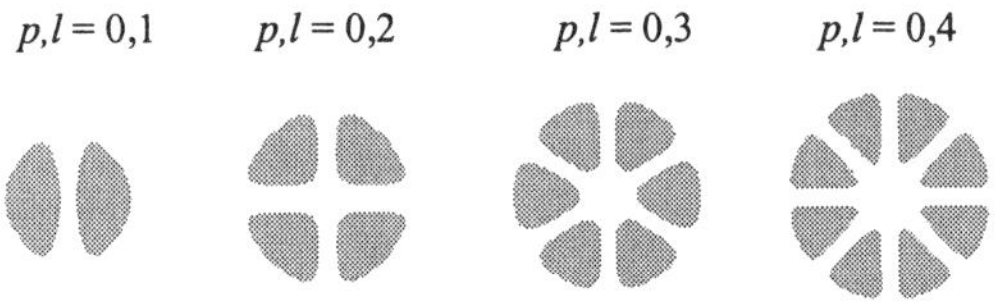

Fig. 2.7. Azimuthal mode structure

The mode, which is finally generated by the laser system is determined by the geometry of the limiting aperture, i.e. circular-(rod) or rectangular geometry (slab) and by the transverse gain structure. In most cases, especially in high power systems, several modes of different frequencies oscillate, which is called multimode operation.

Real Resonator

Figures 2.4 to 2.7 only apply to an empty resonator. In a resonator with amplifying medium the wings are raised [2.3] and the zeros become local minima (Fig. 2.8).

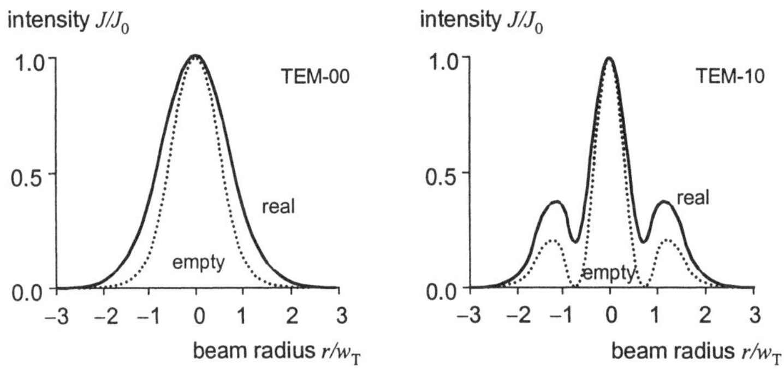

Fig. 2.8. Real intensity distribution, according to [2.4]

The burn pattern of a pulsed or switched laser beam on black photographic paper gives an inverse intensity distribution. This is a rapid but approximate method to check the laser alignment and the symmetry. In Fig. 2.9 such burn patterns for a circular rod and for a slab with rectangular symmetry are shown.

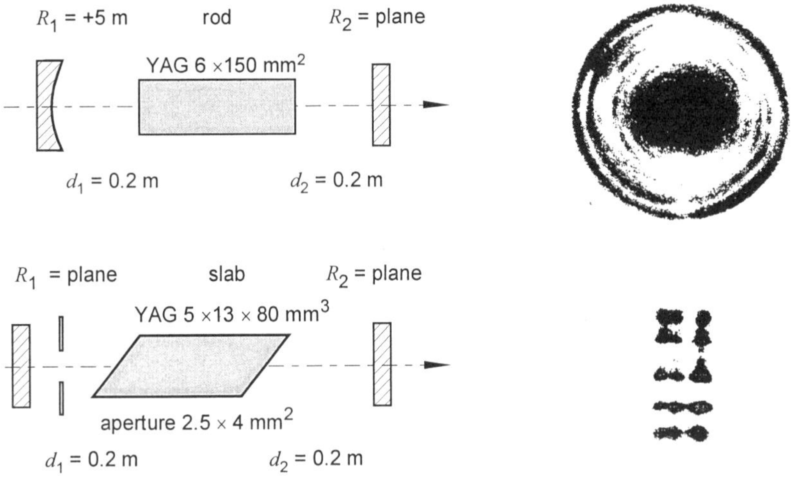

Fig. 2.9. Burn pattern in photographic paper for two different resonators

In [2.5, 2.6] it is shown that the beam radius is proportional to the square root of the mode orders and to the 00-mode radius. For rectangular symmetry it may be different in x, y directions

$$w_{mx,y} = M_{x,y}\, w_0 \qquad M_x = \sqrt{m+1} \qquad M_y = \sqrt{n+1} \tag{2.16}$$

$w_{mx,y}$ = beam radius for a mode of rectangular symmetry,

$$w_{p,l} = M w_0 \qquad M = \sqrt{2p+l+1}$$

$w_{p,l}$ = beam radius for a mode of circular symmetry.

Instead of the 1/e decrease the power content values or the second moments are used (see Sect. 10.5) is defined as the beam radius. In most cases, particularly with rods, the intensity distribution is of circular symmetry, i.e. $l = 0$. In the following the two different symmetries are no longer distinguished. From now on M is used as the mode order. Applying (2.3a) delivers the divergence of higher order modes

$$\theta_M = M\theta_0, \tag{2.17}$$

θ_M = divergence for a mode of order M
θ_0 = divergence for the fundamental mode,

$$z_R = \frac{w_{TM}}{\theta_M} = \frac{w_{T0}}{\theta_0} = \frac{\pi w_{T0}^2}{\lambda} = \frac{\pi w_{TM}^2}{\lambda M^2}, \tag{2.18}$$

z_R = focal depth (Rayleigh length),
w_{TM} = waist radius for mode of order M.

The focal depth is independent of the mode order [2.7].

Aperture

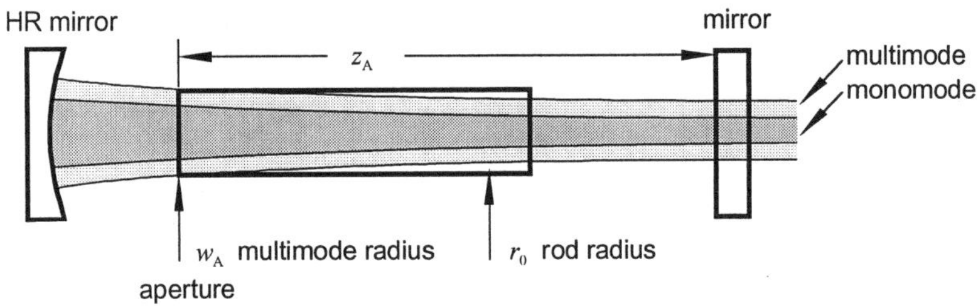

Fig. 2.10. Aperture formed by the laser material

The limiting aperture in the resonator determines the radius of the highest mode order. Normally the laser medium itself is the aperture, for example the rod diameter, and the M is

$$M = \frac{r_0(z_A)}{w_0(z_A)}, \tag{2.19}$$

with w_0 the 00-mode radius in the laser medium at the position of the aperture z_A. Figure 2.10 exemplifies this for the case of a laser rod.

On the other hand a smaller aperture radius a than the rod radius enforces a smaller mode than (2.18) but with a reduction in efficiency. The energy stored in the volume $\pi(r_0^2 - a^2)$ is not available for stimulated emission and the beam power is therefore reduced. At high average power (> 20 W) the aperture needs to be cooled.

2.3 ABCD Law

The description of the ray propagation for the 00-mode and higher modes is normally carried out by means of the matrix formalism [2.2, 2.8 – 2.10].

Geometric Optics

A beam can be characterized in every position z by its angle α_1 and its distance w_1 relative to a given axis (Fig. 2.11). This holds only for small angles, the so-called paraxial approximation.

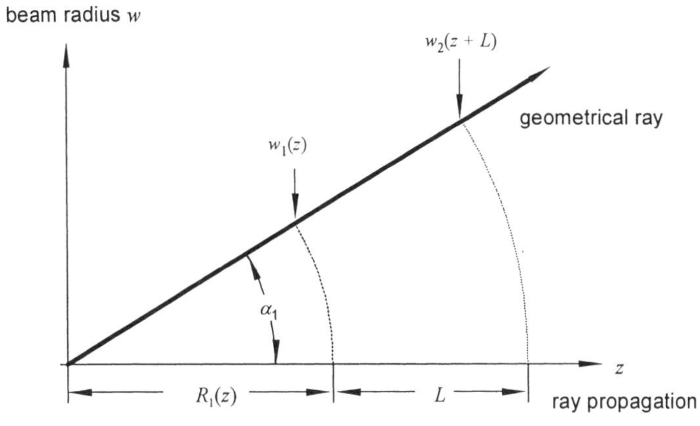

Fig. 2.11. Propagation of a geometric beam

If the beam starts at the origin, one can assign the curvature R_1 at the position z

$$R_1 = \frac{w_1}{\alpha_1}. \tag{2.20}$$

After a distance L the new parameters α_2, w_2 are

$$w_2 = (1 \times w_1) + (L \times \alpha_1), \tag{2.21}$$

$$\alpha_2 = (0 \times w_1) + (1 \times \alpha_1), \text{ or} \qquad \begin{vmatrix} w_2 \\ \alpha_2 \end{vmatrix} = \begin{vmatrix} 1 & L \\ 0 & 1 \end{vmatrix} \begin{vmatrix} w_1 \\ \alpha_1 \end{vmatrix},$$

or, more generally,

$$w_2 = (A \times w_1) + (B \times \alpha_1), \tag{2.22}$$

$$\alpha_2 = (C \times w_1) + (D \times \alpha_1), \qquad \begin{vmatrix} w_2 \\ \alpha_2 \end{vmatrix} = \begin{vmatrix} A & B \\ C & D \end{vmatrix} \begin{vmatrix} w_1 \\ \alpha_1 \end{vmatrix}.$$

The propagation in a homogeneous, isotropic material is described by a matrix M, or, even more general, every first order optical element is represented by its matrix M

$$M = \begin{vmatrix} A & B \\ C & D \end{vmatrix}, \qquad \textit{optical matrix.} \tag{2.23}$$

For the curvature R_2, with (2.20), by division of the two equations the geometric ABCD law is:

$$R_2 = \frac{AR_1 + B}{CR_1 + D}, \qquad \textit{geometrical ABCD law.} \tag{2.24}$$

For simple elements such as lenses or disks the optical matrices can be easily derived by means of geometric optics. The matrices for optical elements are listed in Sect. 2.5 and the most essential definitions are arranged in Tables 2.2 to 2.4.

ABCD Law

The complex beam parameter q according to (2.5)

$$\frac{1}{q} = \frac{1}{R} - \frac{i\lambda}{\pi w_m^2}, \tag{2.25}$$

is transformed by optical elements and a modified ABCD-law holds. [2.10, 2.11],

$$q_2 = \frac{Aq_1 + B}{Cq_1 + D}, \qquad \textit{ABCD law.} \tag{2.26}$$

With this transformation the optical beam parameters can be easily calculated. The most essential formulae for the transformation of one beam with the parameters θ_1 and w_1 into a new beam with the parameters θ_2 and w_2 (Fig. 2.12) are listed in the following. The matrix of the optical elements relates the beam parameters from one reference plane to the next reference plane.

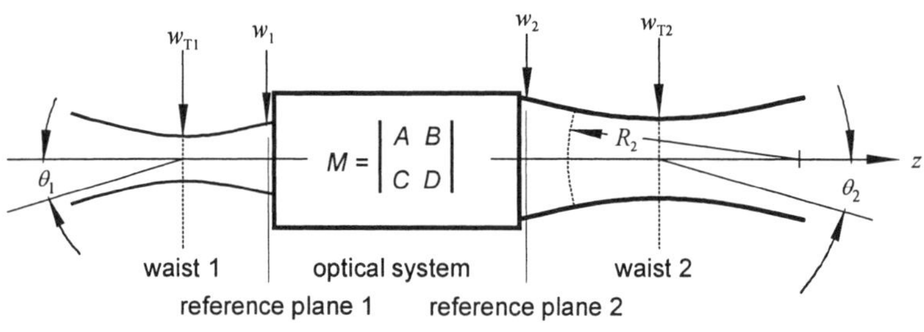

Fig 2.12. Transformation of the optical beam parameters

The equations to transform the beam parameters are

$$\frac{1}{R_2} = \frac{AC + (AD + BC)/R_1 + BD\left(1/R_1^2 + \lambda^2/(\pi w_1^2)^2\right)}{\left(A + B/R_1\right)^2 + \left(B\lambda/\pi w_1^2\right)^2}, \tag{2.27}$$

R_i = radius of curvature,

$$w_2^2 = w_1^2\left((A + B/R_1)^2 + (B\lambda/\pi w_1^2)^2\right), \qquad w_i = \text{beam radius}, \tag{2.28}$$

$$\theta_2^2 = w_1^2\left((C + D / R_1)^2 + (D\lambda / \pi w_1^2)^2\right), \qquad \theta_i = \text{divergence}. \tag{2.29}$$

The equations (2.26) to (2.28) apply to every mode order. Therefore from here onward no distinction will be made between fundamental mode and higher order modes, unless it is necessary for better understanding.

Waist Transformation

If a beam waist w_{T1} with $R_1 = \infty$ is transformed by an optical system one has from (2.27) and (2.28) the following relations [2.7]

$$w_2^2 = A^2 w_{T1}^2 + B^2 \theta_1^2, \qquad w_{T1} = \text{waist radius}, \tag{2.30}$$

$$\theta_2^2 = C^2 w_{T1}^2 + D^2 \theta_1^2, \tag{2.31}$$

$$\frac{1}{R_2} = \frac{ACz_{R_1}^2 + BD}{A^2 z_{R_1}^2 + B^2}, \qquad z_{R_1} = \pi w_{T1}^2 / \lambda. \tag{2.32}$$

Waist-to-Waist Transformation

Transforming a beam waist of w_{T1} into a new waist w_{T2}, one can show that for the product of waist and divergence one has

$$w_{T2}\theta_2 = w_{T1}\theta_1 = \text{const}, \qquad w_{T2}, w_{T1} = \text{waist radius}. \tag{2.33}$$

This product is constant in ideal first-order optical systems (Abbe's sine law). For the Rayleigh length holds

$$z_R = \frac{w_T^2}{M^2 \lambda / \pi} \qquad M = \text{mode order, (2.16)}. \tag{2.34}$$

This parameter of the laser beam cannot be reduced by passive optics without any losses. The product of the full divergence angle $2\times\theta$ and the waist diameter $2w_T$ is therefore a convenient measure for the optical beam quality. The smaller this product or the mode order, the higher is the quality of the laser beam.

The matrix relates the beam parameters of two subsequent reference planes. It is convention to define as the beam parameter product the full width of the waist times the full far field divergence:

$$(2w_T)(2\theta) = 4M^2\lambda/\pi, \qquad \textit{beam parameter product}. \tag{2.35}$$

Example 2.1 Mode order of a Nd:YAG laser

A diffraction limited Nd:YAG lasers ($M = 1$ and $\lambda = 1$ µm) has a beam parameter product of $2w_T \times 2\theta \approx 1.3$ mm mrad. A laser rod with 10 mm diameter in the range of 100 W average laser power produces a divergence of about 10 mrad. The beam parameter product is therefore about 100 mm mrad and the mode order $M^2 \approx 80$.

2.4 Sign Conventions

For the radii of curvature of the optical elements in Table 2.3 the correct signs have to be used [2.10]. In particular, an output mirror has a different sign from a reflective mirror or a refractive lens. It is therefore recommended to state whether the radii are convex or concave.

Propagation in z Direction

The direction of propagation of the beam is chosen as the positive z axis. Through this convention the matrix of a plane mirror is the identity matrix.

Boundary Surface Radius *R*

The radius R of a refracting element is positive, if the beam encounters a convex surface in its direction of propagation.

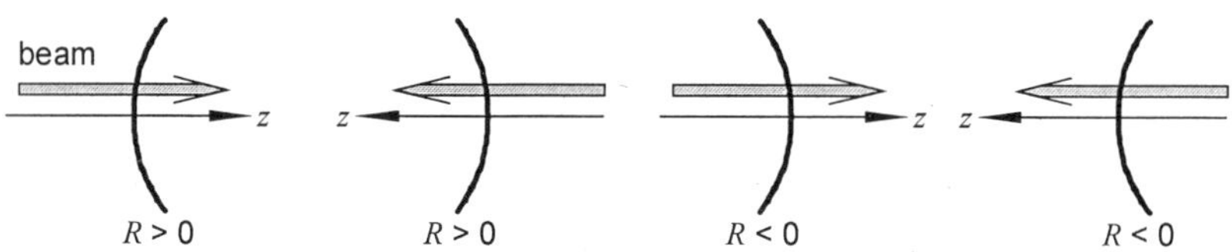

Mirror Radius *R*

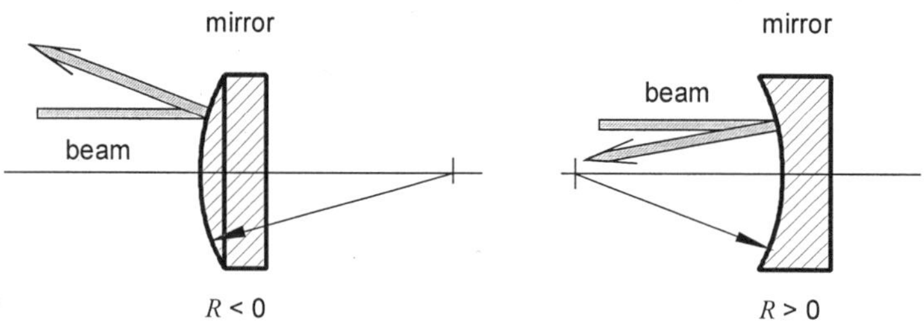

The mirror radius R is positive if the beam encounters a concave surface, just the opposite of the previous situation. This has particularly to be taken into account for a semitransparent mirror, which may act as a mirror and a lens simultaneously.

Curvature R of a Beam

The phase surface has a positive curvature if it has a concave curvature in the direction of propagation.

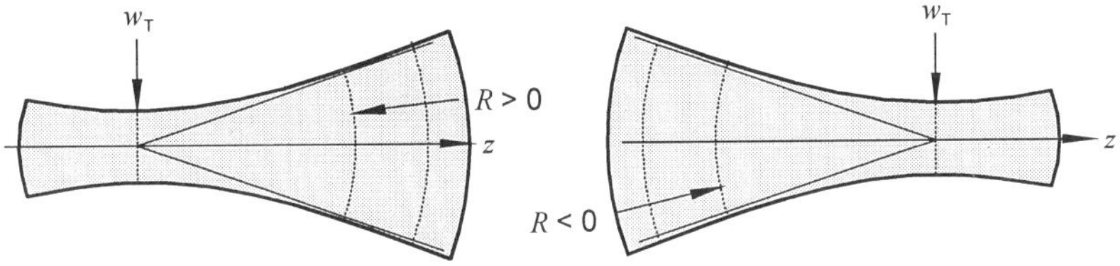

Focal Length f

A converging lens has a positive, a diverging lens a negative, focal length.

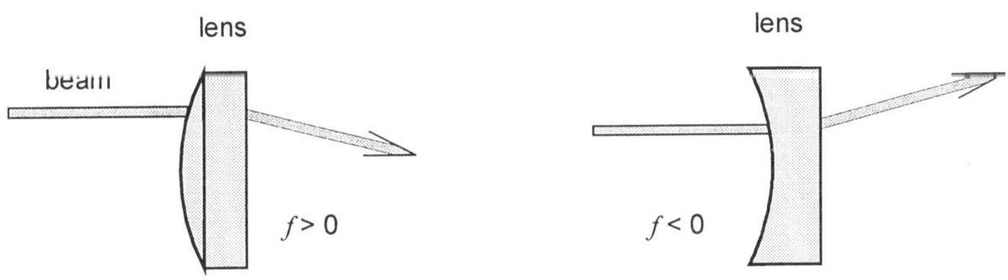

Reference Plane Distance h, Intercept Length s

The sum of the reference plane distance and the corresponding intercept length gives the focal length

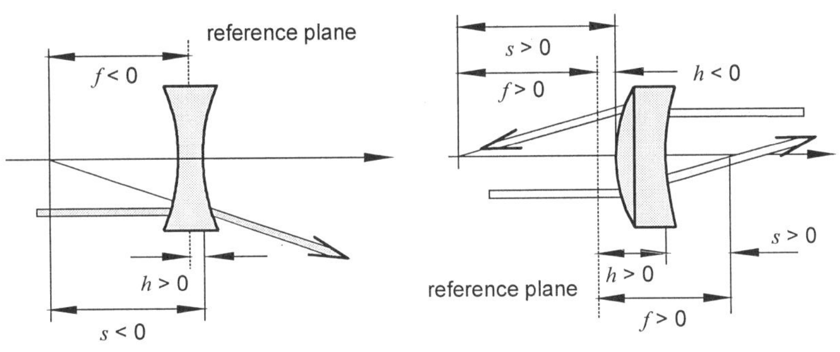

2.5 Matrices of Optical Elements

The matrices of simple optical elements are listed in Tables 2.2 to 2.4 [2.1, 2.9, 2.10, 2.11]. The elements must be centered and not tilted, the beam must propagate paraxially. The propagation is only in the positive z- direction. The z axis is also reflected by a mirror, and therefore the reflected beam propagates again in the positive z direction.

Table 2.2 Optical matrices of simple elements

Free propagation

$$M = \begin{vmatrix} 1 & z \\ 0 & 1 \end{vmatrix}$$

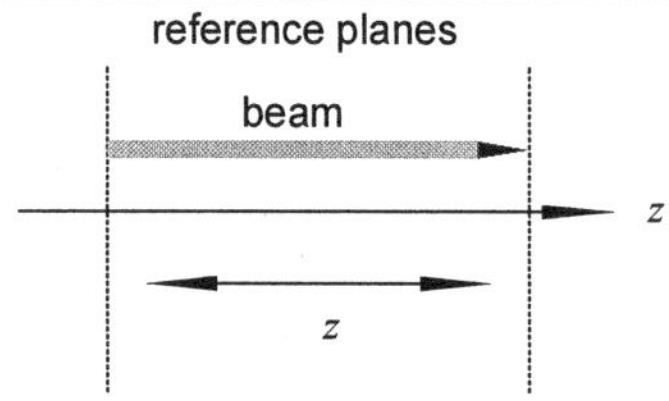

Refraction

$$M = \begin{vmatrix} 1 & 0 \\ \dfrac{n_1 - n_2}{n_2 R} & \dfrac{n_1}{n_2} \end{vmatrix}$$

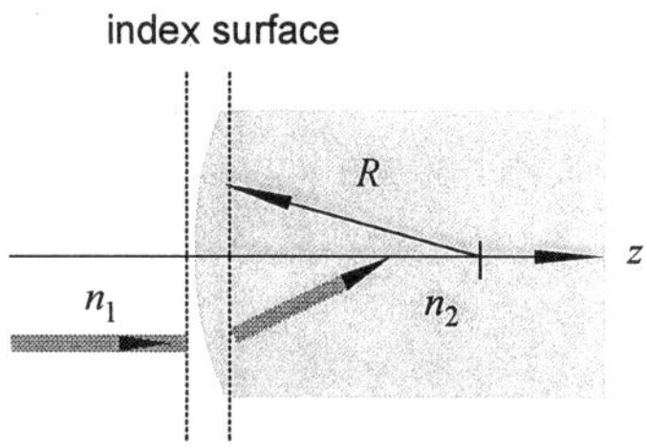

Thin lens

$$M = \begin{vmatrix} 1 & 0 \\ -1/f & 1 \end{vmatrix}$$

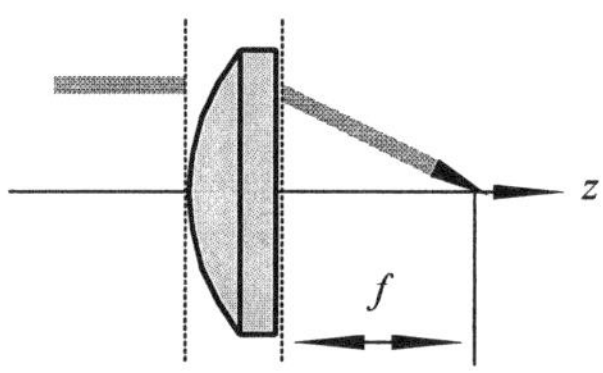

Mirror

$$M = \begin{vmatrix} 1 & 0 \\ -1/R & 1 \end{vmatrix}$$

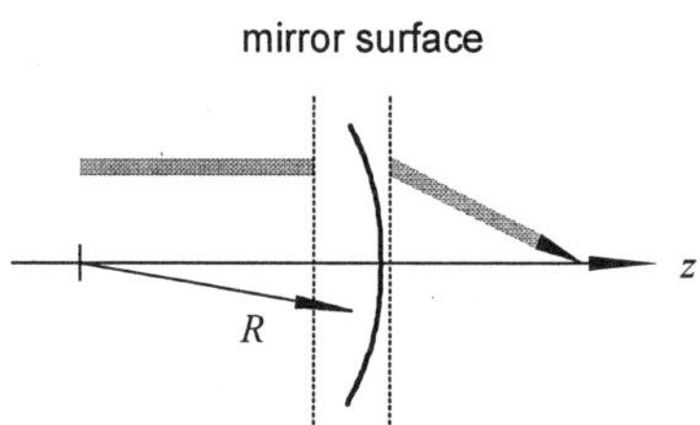

Table 2.3 Optical matrices of combined elements

Telescope with $f_1 + f_2 = l$

$$M = \begin{vmatrix} -f_2/f_1 & f_1+f_2 \\ 0 & -f_1/f_2 \end{vmatrix}$$

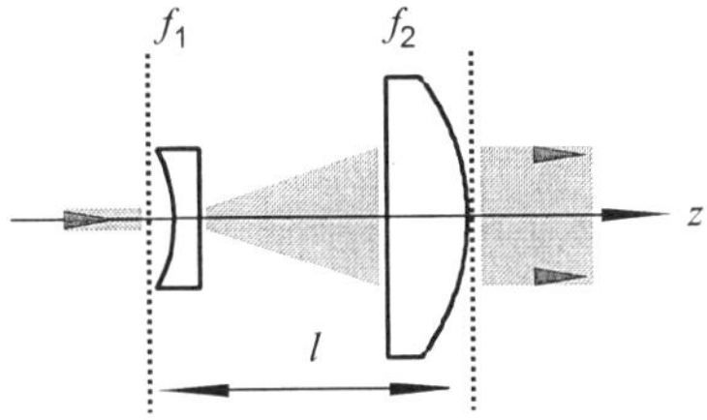

Flat plate

$$M = \begin{vmatrix} 1 & dn_1/n_2 \\ 0 & 1 \end{vmatrix}$$

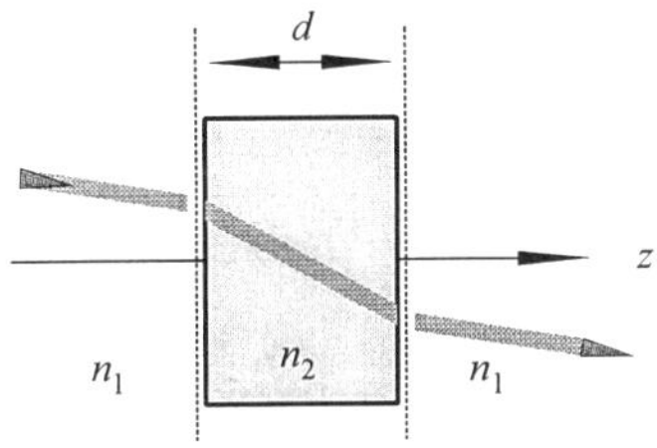

Lens ($d << R$)

$$M = \begin{vmatrix} 1 & 0 \\ -(n-1)\left(\frac{1}{R_1} - \frac{1}{R_2}\right) & 1 \end{vmatrix}$$

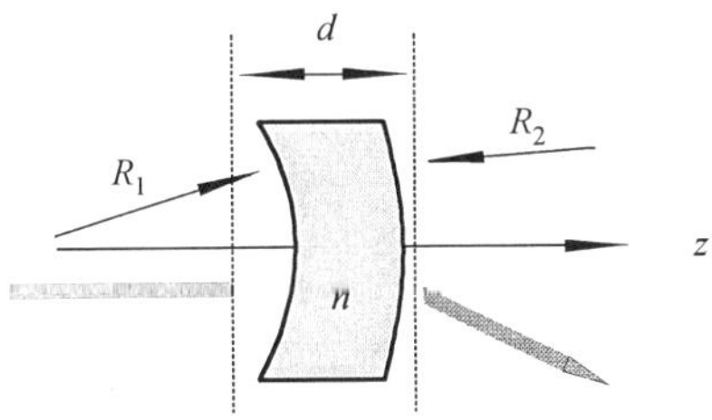

Thick lens

$$M = \begin{vmatrix} 1-h_2/f & h_1+h_2-h_1h_2/f \\ -1/f & 1-h_1/f \end{vmatrix} = $$

$$= \begin{vmatrix} 1 & h_2 \\ 0 & 1 \end{vmatrix} \begin{vmatrix} 1 & 0 \\ -1/f & 1 \end{vmatrix} \begin{vmatrix} 1 & h_1 \\ 0 & 1 \end{vmatrix}$$

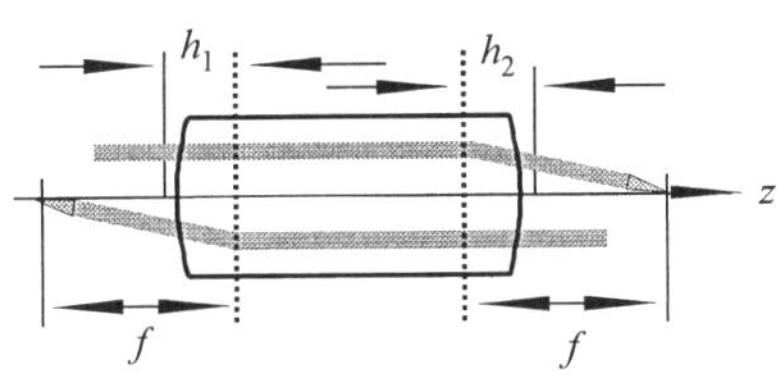

Adapted mirror, [3.16]

$$M = \begin{vmatrix} 1 & 0 \\ \frac{n}{R} & 1 \end{vmatrix} \begin{vmatrix} 1 & d \\ 0 & 1 \end{vmatrix} \begin{vmatrix} 1 & 0 \\ \frac{1-n}{nR} & \frac{1}{n} \end{vmatrix}$$

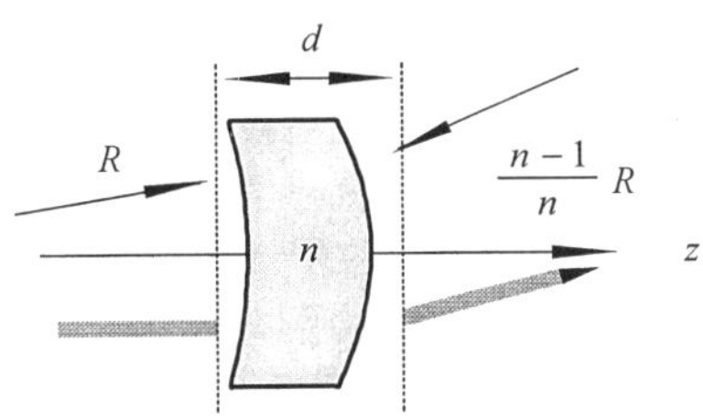

Table 2.4 Optical matrices of the thermal lens with radial refractive index distribution

Thermal lens, focusing, $n(r) = n\left(1 - r^2/2\beta^2\right)$

β = characteristic length of the thermal lens

Reference planes inside [2.1]

$$M = \begin{vmatrix} \cos(l/\beta) & \beta\sin(l/\beta) \\ \dfrac{-\sin(l/\beta)}{\beta} & \cos(l/\beta) \end{vmatrix}$$

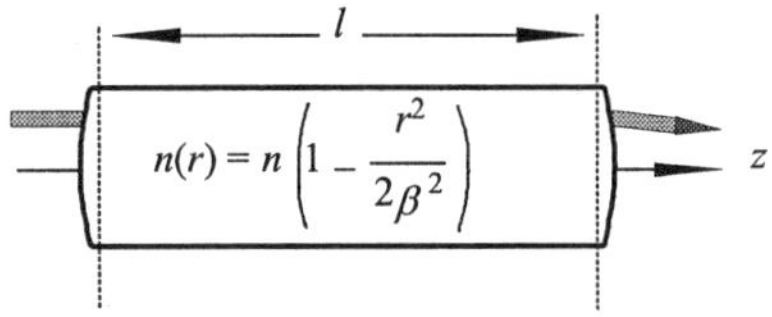

Reference planes outside [2.1]

$$M = \begin{vmatrix} \cos(l/\beta) & \dfrac{\beta\sin(l/\beta)}{n} \\ \dfrac{-n\sin(l/\beta)}{\beta} & \cos(l/\beta) \end{vmatrix}$$

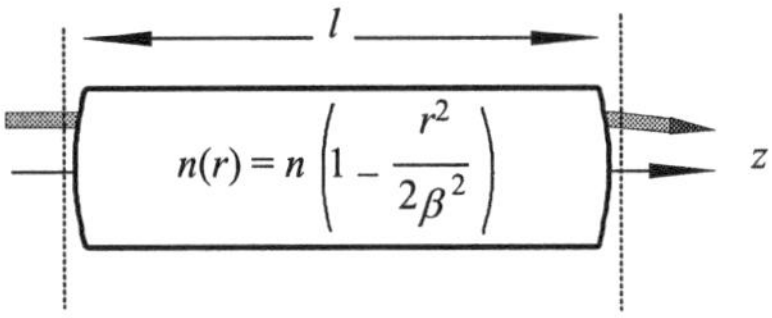

For $\beta >> l$ the thermal lens can be approximated as a thick lens.

$$M = \begin{vmatrix} 1 - l^2/2\beta^2 & l(1 - l^2/4\beta^2)/n \\ -nl/\beta^2 & 1 - l^2/2\beta^2 \end{vmatrix}$$

In comparison with the thick lens one has $1/f = nl/\beta^2$ and $h = l/2n$

Thermal lens, divergent $n(r) = (1 + r^2/2\beta^2)$

Reference planes inside [2.1]

$$M = \begin{vmatrix} \cosh(l/\beta) & \beta\sinh(l/\beta) \\ \dfrac{-\sinh(l/\beta)}{\beta} & \cosh(l/\beta) \end{vmatrix}$$

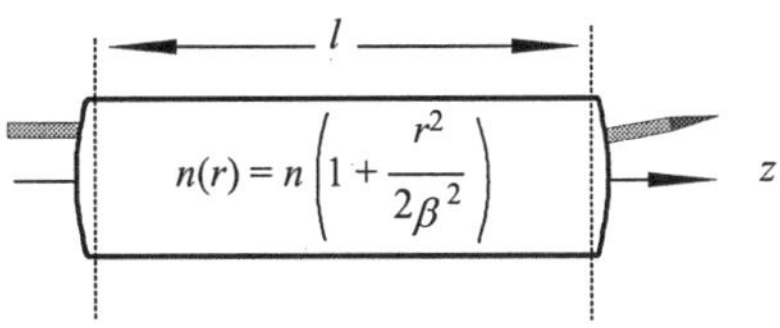

Reference planes outside [2.1]

$$M = \begin{vmatrix} \cosh(l/\beta) & \dfrac{\beta\sinh(l/\beta)}{n} \\ \dfrac{-n\sinh(l/\beta)}{\beta} & \cosh(l/\beta) \end{vmatrix}$$

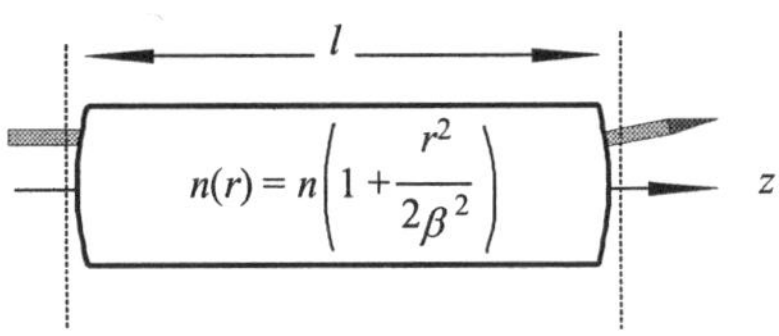

2.6 Application of the ABCD Law

In the following some special applications of the ABCD law will be considered in more detail.

Free Propagation in *z* Direction

The matrix for free propagation [2.12], according to Table 2.2, yields the z dependence of beam radius and curvature, (2.26) to (2.28),

$$w_2^2(z) = w_1^2(1 + z / R_1)^2 + z^2\theta_1^2, \qquad \textit{beam radius,} \qquad (2.36)$$

$$\frac{1}{R_2(z)} = \frac{1/R_1 + z\left(1/R_1^2 + \lambda^2/(\pi w_1^2)^2\right)}{(1 + z/R_1)^2 + (z/\pi w_1^2)^2}, \qquad \textit{curvature,} \qquad (2.37)$$

$$\theta_2^2 = \left(\frac{w_1}{R_1}\right)^2 + \theta_1^2, \qquad \textit{divergence.} \qquad (2.38)$$

The divergence at position z can be calculated from the beam parameters $w_1(0)$, θ_1 and $R_1(0)$. The first term gives the geometric aperture angle, the second the diffraction part of the divergence.

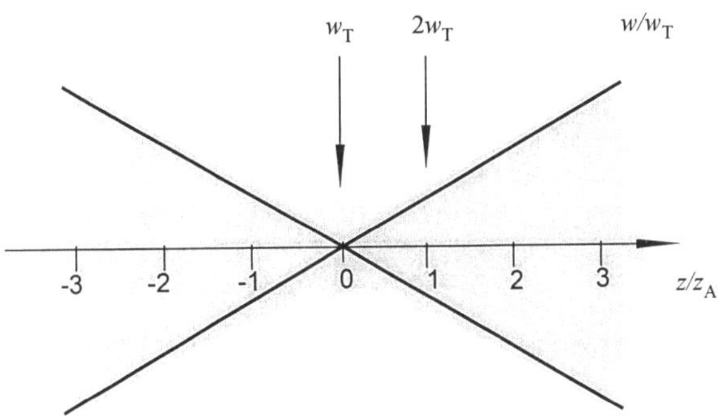

Fig 2.13. Free beam propagation

If the origin w_1 is a waist ($w_1 = w_T$, $R_1 = \infty$, Fig. 2.13), then it follows

$$w^2(z) = w_T^2\left(1 + (z/z_R)^2\right) \qquad \textit{beam radius,} \qquad (2.39)$$

$$R(z) = z_R\left(z_R / z + z / z_R\right) \qquad \textit{curvature,} \qquad (2.40)$$

$$\theta = \frac{w_T}{z_R} \qquad \textit{divergence.} \tag{2.41}$$

The cross section of the beam doubles for the distance $z = z_R$. Therefore the Rayleigh length z_R can be considered as the focal depth. Simultaneously, the curvature $R(z_R) = 2z_R$ has a minimum in z_R. In practice, intensity changes of around 10% can be tolerated, $(w(z)^2/w_T^2 = 1.1)$. This intensity change occurs approximately at $z/z_R \approx 0.33$.

Imaging with a Thin Lens

By a lens, or an arrangement of lenses, the waist of the laser beam is focused on the workpiece, or on the end of a fiber (Fig. 2.14).

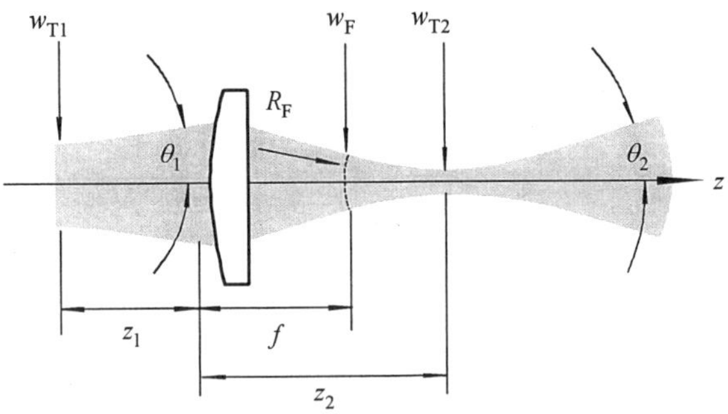

Fig 2.14. Imaging with a thin lens

The beam parameter product remains constant $w_{T2}\theta_2 = w_T\theta$ and cannot be reduced by lenses. In the focal plane $z_2 = f$ the following relations hold

$$2w_F = 2f\theta \qquad \text{for the diameter and} \tag{2.42}$$

$$\frac{1}{R_F} = \frac{1}{f} - \frac{z_1}{f^2} \qquad \text{for the curvature.} \tag{2.43}$$

Since the beam diameter in the focal plane is proportional to the divergence this offers a method to measure the divergence. The imaging equations for the new waist take the form

$$\frac{1}{z_2} = \frac{1}{f} - \frac{1}{z_1 + z_R^2/(z_1 - f)} \quad \text{or} \quad z_2 = f + \frac{f^2(z_1 - f)}{(z_1 - f)^2 + z_R^2} \tag{2.44}$$

for the distance, and

$$w_{T2}^2 = \frac{\theta^2 f^2}{1+(z_1 - f)^2 / z_R^2} \quad \text{or} \quad w_{T2}^2 = \frac{w_T^2 f^2}{(z_1 - f)^2 + z_R^2} \tag{2.45}$$

for the waist radius.

These equations differ from the standard imaging equations of geometrical optics by incorporating the focal depth z_R, which may depend on the operating conditions of the laser. The aperture angle θ_2 behind the lens is (2.30)

$$\theta_2^2 = \frac{w_T^2}{f^2} + \frac{(z_1 - f)^2}{f^2} . \tag{2.46}$$

For $z_1 = f$ one obtains

$$z_2 = z_1, \qquad \text{(1:1 imaging)} \tag{2.47}$$

$$w_2 = w_F = f\theta, \tag{2.48}$$

$$\theta_F = w_0/f \qquad \text{and} \tag{2.49}$$

$$z_{R1} = \frac{f_2^2\theta^2}{\theta w_0} = \frac{w_F^2}{\theta w_0} . \tag{2.50}$$

Imaging with a Telescope

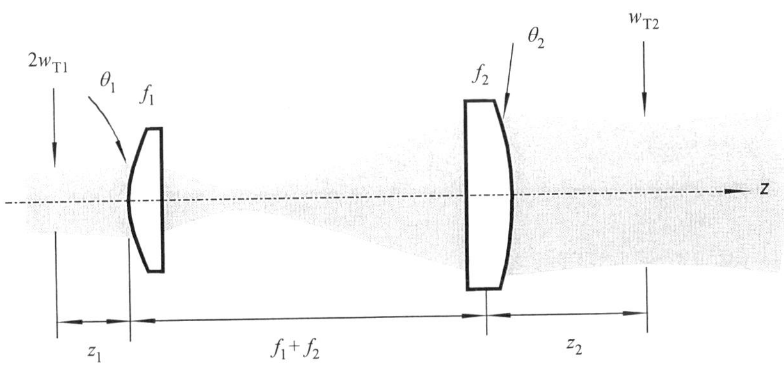

Fig. 2.15. Imaging by means of a telescope

Let the telescope (in Fig. 2.15) be focused, i.e. $l = f_1 + f_2$. The zoom factor m is defined as

$$m = \frac{|f_2|}{|f_1|}. \tag{2.51}$$

Application of the ABCD law with the corresponding matrix for the telescope gives the following beam parameters

$$w_{T2} = m w_{T1} \quad \textit{waist radius,} \tag{2.52}$$

$$\theta_2 = \theta_1 / m \quad \textit{divergence,} \tag{2.53}$$

$$z_{R2} = w_{T2}/\theta_2 = m^2 z_{R1} \quad \textit{focal depth and} \tag{2.54}$$

$$z_2 = m(f_1 + f_2) - m^2 z_1 \quad \textit{waist position.} \tag{2.55}$$

The new beam waist w_{T2} is magnified by the factor m, the divergence is reduced by the factor m. The waist position for a Galilei telescope ($f_1 < 0$) appears to be at a distance z_2 in front of the telescope, for a Kepler telescope it is behind the telescope at the real distance z_2. The use of a telescope in front of the focusing lens allows the reduction of the focus diameter by the zoom factor m, at the expense of the focus depth and the aperture angle in the focus.

Transformation of the Beam Parameters

The equations (2.29) and (2.30) describe the beam parameters of an optical arrangement that is represented by the corresponding matrix M. By a suitable choice of the optics one can transform the beam parameters, such as beam radius and beam divergence,

$$w_2^2 = A^2 w_T^2 + B^2 \theta_2, \quad w_T = \text{waist of the incident beam,} \tag{2.56}$$

$$\theta_2^2 = C^2 w_T^2 + D^2 \theta^2, \quad \theta = \text{divergence of the incident beam.} \tag{2.57}$$

$$M = \begin{vmatrix} A & B \\ C & D \end{vmatrix}, \quad \textit{optical matrix.}$$

In the case of the same refractive index in front of and behind the lens, all optical matrices are subject to [2.10]

$$AD - BC = 1. \tag{2.58}$$

The condition for the existence of a new waist is [2.7]

$$AC - BD/z_{R1}^2 = 0. \tag{2.59}$$

With the help of these four equations the imaging properties of special optics will be discussed further. In each case the general matrix and a specific example are given.

1. $A = 0$ $\quad w_2 = |B|\theta \quad B = -1/C$

Example 2.2 Imaging with a lens

The beam radius in the focal plane is proportional to the divergence, independent of the waist distance to the lens. This arrangement can be used with a diode array to measure the divergence.

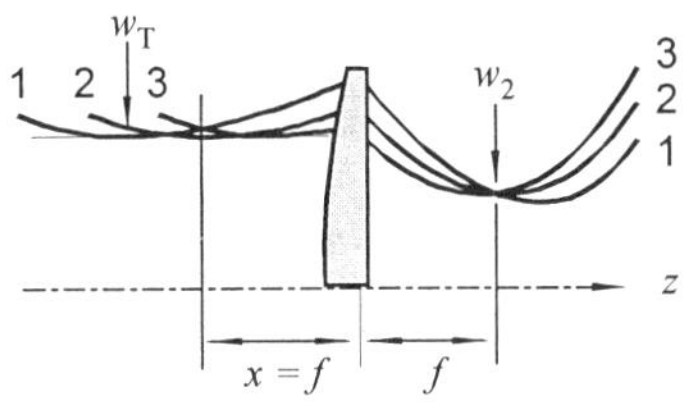

$$\begin{vmatrix} 0 & B \\ -\frac{1}{B} & D \end{vmatrix} = \begin{vmatrix} 0 & f \\ -\frac{1}{f} & \frac{f-x}{f} \end{vmatrix}$$

$$w_2 = f\theta$$

2. $B = 0$ $\quad w_2 = |A|w_T \quad A = 1/D$

Example 2.3 Imaging with a lens

With a constant waist radius w_T a constant beam radius can be generated, independent of the changing divergence.

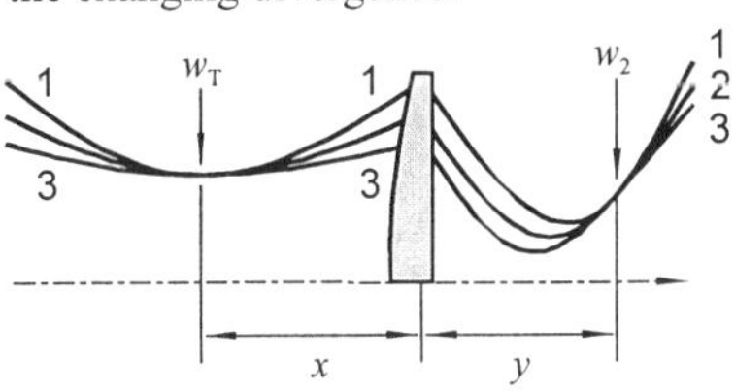

$$\begin{vmatrix} A & 0 \\ C & \frac{1}{A} \end{vmatrix} = \begin{vmatrix} \frac{f}{f-x} & 0 \\ \frac{-1}{f} & \frac{f-x}{f} \end{vmatrix}$$

$$y = \frac{xf}{x-f}$$

3. $C = 0$ $\quad \theta_2 = |D|\theta \quad A = 1/D$

Example 2.4 Telescope $y = f + g$

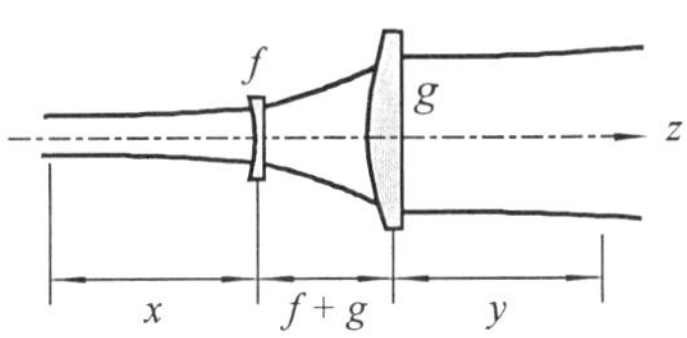

$$\begin{vmatrix} -\frac{g}{f} & -\frac{xg}{f} - \frac{zf}{g} + f + g \\ 0 & -\frac{g}{f} \end{vmatrix}$$

Example 2.5 Sequence of lenses

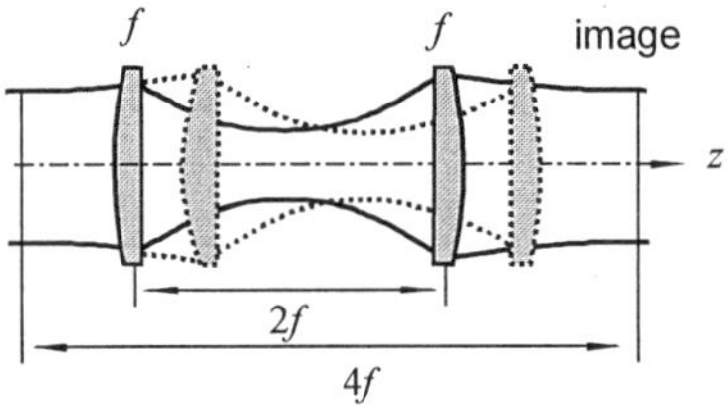

$C = 0$, $B = 0$, $A = -1$, $D = -1$

$w_{T2} = w_T$, $\theta_2 = \theta$

$$\begin{vmatrix} A & B \\ C & D \end{vmatrix} = \begin{vmatrix} -1 & 0 \\ 0 & -1 \end{vmatrix}$$

for $x + y = 2f$

The fixed coupled pair of lenses can be moved within the distance 4*f*, without changing the 1:1 image in the position and beam parameters.

4. $D = 0$ $\quad \theta_2 = |C| w_T \quad B = -1/C$

Example 2.6 Divergence proportional to the waist

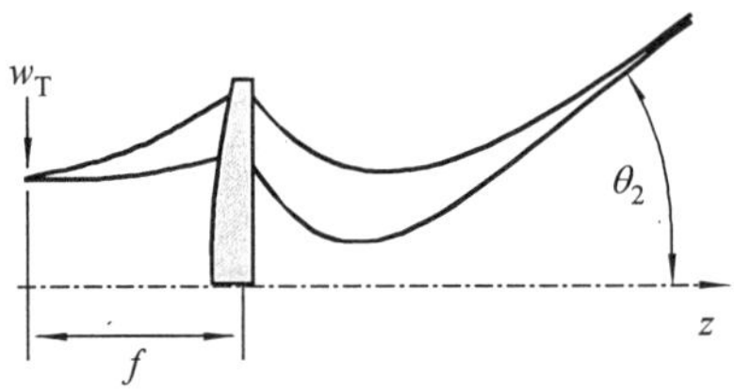

$$\begin{vmatrix} A & B \\ \dfrac{-1}{B} & 0 \end{vmatrix} = \begin{vmatrix} \dfrac{f-y}{f} & f \\ \dfrac{-1}{f} & 0 \end{vmatrix}$$

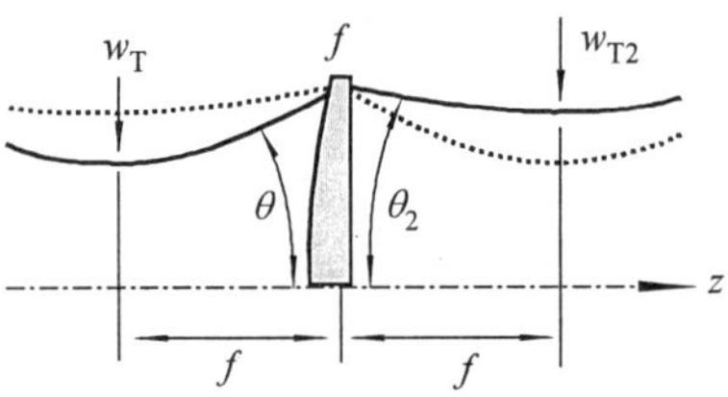

$A = 0$, $D = 0$

$w_{T2} = |B|\theta$, $\theta_2 = |1/B| w_T$

$$\begin{vmatrix} 0 & B \\ \dfrac{-1}{B} & 0 \end{vmatrix} = \begin{vmatrix} 0 & f \\ \dfrac{-1}{f} & 0 \end{vmatrix}$$

Once again the conservation of the beam parameter product is exemplified.

3 Resonators

The beam parameters of a laser system for material processing are supposed to remain constant at the workpiece surface under all operating conditions. In particular this means: a constant focus diameter and focal depth, constant focus position and constant laser energy and laser power. For economic reasons the efficiency and the beam quality should be high, which guarantees a wide field of applications. Asymmetry in the processing direction must be avoided, the beam cross section, except in special cases, should be circular and the beam polarization circular or random.

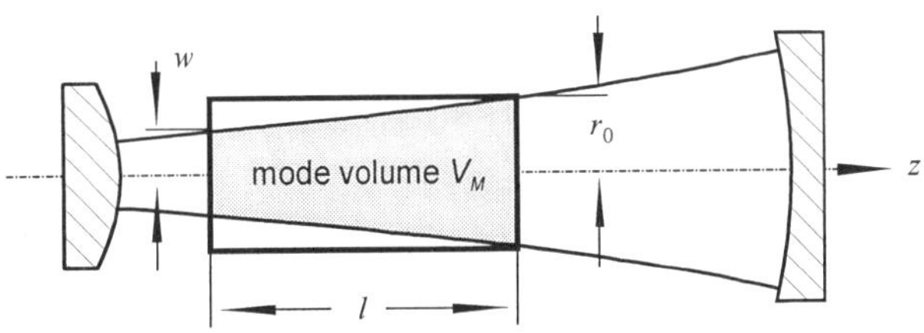

Fig. 3.1. Mode volume

Assuming stable initial parameters, such as pumping power and cooling conditions, the above requirements have the following implications for the resonator.

- *Constant optical parameters*
  This means constant and low mode order and, in particular, constant beam radius and, simultaneously, constant divergence. Thereby the focus position is also stable:

$$w\theta = M^2\lambda/\pi = \text{const.} \tag{3.1}$$

- *High efficiency*
  The mode volume (Fig. 3.1) must be adapted to the geometry of the laser medium. For rods the mode volume V_M can be estimated for a linearized beam as a truncated cone [3.1]:

$$\frac{V_M}{\pi r_0^2 l} = \frac{1}{3} + \frac{w}{3r_0} + \frac{w^2}{3r_0^2}, \tag{3.2}$$

- *Constant laser parameters, such as energy and power*
  The mode volume must remain constant during all changes in the processing parameters, such as pulse frequency or laser power.

- *Rotational symmetry for the beam geometry and its polarization*
  Apertures and mirrors must be circular, no polarizing elements, such as beam splitters and Brewster windows, should be inside the resonator.

3.1 Stable Resonators

To date stable resonators are almost exclusively used in material processing with solid-state lasers. The resonator is defined as stable if the mode diameter of the TEM-00 mode is confined and remains finite for unlimited mirrors and unlimited medium size. The stationary intensity distribution can be described within the paraxial approximation and the matrix formalism can be applied.

Eigenvalues for TEM-00 Mode

The beam is described by the beam parameter q, as given in (2.24). For stationary conditions q must be reproduced after a round-trip of the wave (Fig. 3.2):

$$q = q_1. \tag{3.3}$$

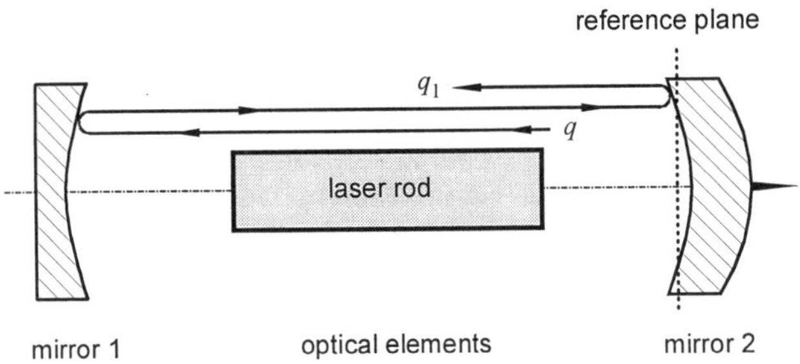

Fig. 3.2. Stable resonator

The ABCD law for a round-trip,

$$q = \frac{Aq + B}{Cq + D}, \qquad \begin{vmatrix} A & B \\ C & D \end{vmatrix} = \textit{matrix for a full round-trip}, \tag{3.4}$$

leads with (2.58) to the solution

$$\frac{1}{q} = \frac{D-A}{2B} \pm \frac{\mathrm{i}}{2B}\sqrt{4-(A+D)^2}\,. \tag{3.5}$$

The expression under the square root must be positive in order for the beam radius to remain real. Therefore the stationary value of the beam parameter in a stable resonator requires

$$2 < A + D < 2, \qquad \textit{stability criterion.} \tag{3.6}$$

Such a resonator is called stable. Choosing the position of the reference plane, such that there is a flat wavefront (plane mirror), we have

$$\frac{1}{R} = \mathrm{Re}\left(\frac{1}{q}\right) = \frac{D-A}{2B} = 0\,. \tag{3.7}$$

The beam waist and the Rayleigh length at mirror 2 follow from (2.9) and (2.18) as

$$w_{\mathrm{T}2}^2 = \frac{\lambda}{\pi}\sqrt{-\frac{B}{C}}\,, \tag{3.8}$$

$$z_{\mathrm{R}2}^2 = -\frac{B}{C}\,. \tag{3.9}$$

In Fig. 3.3 M_{12} is defined as the beam matrix for a single transition from reference plane 1 to plane 2. M_{12} contains all optical elements including the plane mirror surfaces.

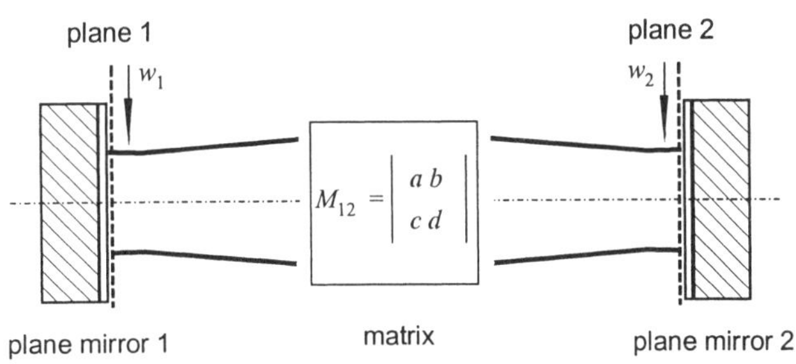

Fig. 3.3. Modified matrix representation

The matrix for a round-trip from plane 2 to mirror 2 through the resonator to mirror 1 and back to plane 2 reads

$$\begin{vmatrix} A & B \\ C & D \end{vmatrix} = \begin{vmatrix} a & b \\ c & d \end{vmatrix}\begin{vmatrix} 1 & 0 \\ 0 & 1 \end{vmatrix}\begin{vmatrix} d & b \\ c & a \end{vmatrix}\begin{vmatrix} 1 & 0 \\ 0 & 1 \end{vmatrix} = \begin{vmatrix} ad+bc & 2ab \\ 2cd & ad+bc \end{vmatrix}. \tag{3.10}$$

Since $A = D$, (3.9) and (3.10) give the following expressions for the Rayleigh length and waist radii:

$$z_{R2}^2 = -\frac{ab}{cd}, \quad z_{R1}^2 = -\frac{db}{ca}, \qquad \textit{Rayleigh length } [3.2], \tag{3.11}$$

$$w_{T1} = \sqrt{\frac{\lambda z_{R1}}{\pi}}, \qquad w_{T2} = \sqrt{\frac{\lambda z_{R2}}{\pi}}, \quad \textit{waist radii.}$$

The stability condition (3.8) with (3.11) delivers

$$abcd < 0, \qquad \text{the } \textit{stability criterion.} \tag{3.12}$$

As soon as one of the parameters a to d is zero, the resonator becomes unstable. This simplified description of the full round-trip with flat wavefronts in the reference planes will be used in the following to consider various resonators.

Introducing a Virtual Reference Plane

The position of the reference plane can be chosen such that a flat wavefront apparently lies inside it. Since a curved mirror can be represented by a combination of a lens and a flat mirror (Fig 3.4), the mirror matrix can be written in the form

$$\begin{vmatrix} 1 & 0 \\ -2/R & 1 \end{vmatrix} = \begin{vmatrix} 1 & 0 \\ -1/f & 1 \end{vmatrix} \begin{vmatrix} 1 & 0 \\ 0 & 1 \end{vmatrix} \begin{vmatrix} 1 & 0 \\ -1/f & 1 \end{vmatrix} = \begin{vmatrix} 1 & 0 \\ -2/f & 1 \end{vmatrix}, \tag{3.13}$$

with $R = f$. The virtual plane mirror is located inside the real mirror.

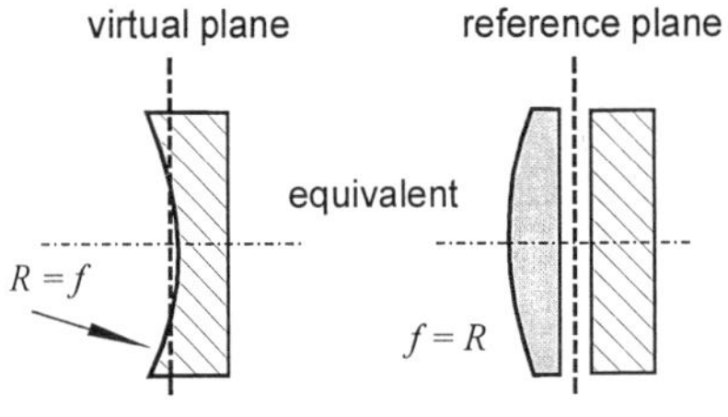

Fig. 3.4. Equivalence of a curved mirror and lens with a plane mirror

For the beam radius w_2 at the mirror it follows analogously to (3.12) that

$$w_2 = \sqrt{\frac{\lambda p_2}{\pi}} \text{ with } p_2^2 = -\frac{ab}{cd}, \qquad \textit{resonator parameter.} \tag{3.14}$$

For plane mirrors, the resonator parameter p is the Rayleigh length z_R. The part of the beam transmitted by the partially reflecting mirror (Fig. 3.5), enters the mirror with the refractive index n, gets transformed with the corresponding mirror matrix and enters air with the curvature R_3.

The correct sign of the curvatures R_i has to be chosen according to whether R_i is regarded as a lens radius or a curvature for the wave front (Sect. 2.4):

$$M_{24} = \begin{vmatrix} 1 & 0 \\ \dfrac{n-1}{R_3} & n \end{vmatrix} \begin{vmatrix} 1 & 0 \\ \dfrac{-1}{R_2} & 1 \end{vmatrix} \begin{vmatrix} 1 & 0 \\ 0 & \dfrac{1}{n} \end{vmatrix} = \begin{vmatrix} 1 & 0 \\ \dfrac{n-1}{R_3} - \dfrac{n}{R_2} & 1 \end{vmatrix} . \tag{3.15}$$

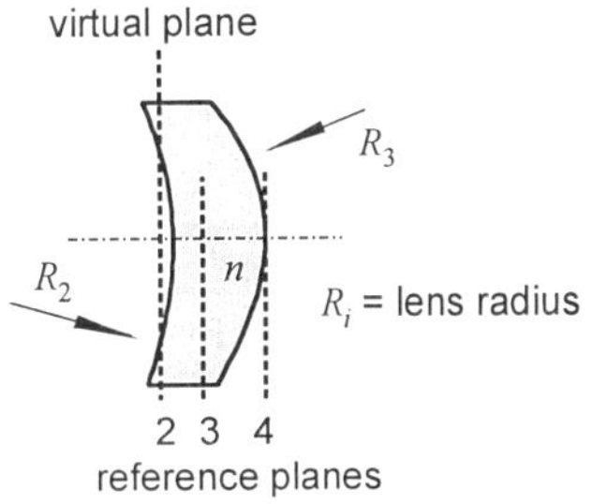

Fig. 3.5. Reference planes in the output mirror

For adapted mirrors, which also produce a flat wavefront in plane 3 [3.4], not only the matrix element B in R_2 of (3.16) has to vanish, but also the matrix element C (see (2.26)). Therefore the transition matrix of reference plane 2 to plane 4 with adapted mirrors requires $R_3 = R_2\,(1 - 1/n)$.

The matrix elements a to d result from the optical elements inside the resonator. In general, a laser crystal shows a lens effect due to the optical pumping. This is taken into account to a first approximation by the corresponding refracting power. The matrix for a thermal, thin lens is

$$M = \begin{vmatrix} 1 & 0 \\ -D & 1 \end{vmatrix}, \tag{3.16}$$

D = refractive power.

The matrix elements a to d for a single transit are dependent on the refracting power D. The values of D are called critical refracting powers D_i, whenever one of the corresponding matrix element a, b, c or d vanishes:

$$i(D_i) = 0, \tag{3.17}$$

D_i = critical refracting power, $\qquad i = a, b, c, d.$

In order to calculate the beam radius and the divergence of a given resonator, the following steps have to be performed:

- The resonator matrix from plane 1 to 2, with the elements a to d is determined from the elements of the corresponding resonator. The critical refracting power D_i is calculated.
- The resonator parameters p_i at the two mirrors are calculated with (3.15).
- The mode order is determined by calculating the beam radius at the position of the aperture (2.29). The ratio of the squares of the radii then yields the mode order, according to (2.18).
- Thus, with the mode order and the Rayleigh length, the multimode radius, (3.15) is known in the reference plane.
- The multimode divergence in the mirror plane is determined with (2.34), and for the output divergence the mirror itself has to be taken into account according to (2.30).

3.2 Empty Resonators

The simplest resonator, for which the range of stability can be discussed, is the empty resonator (Fig. 3.6). The mirrors are assumed to be infinitely thin, providing the curvature of the beam.

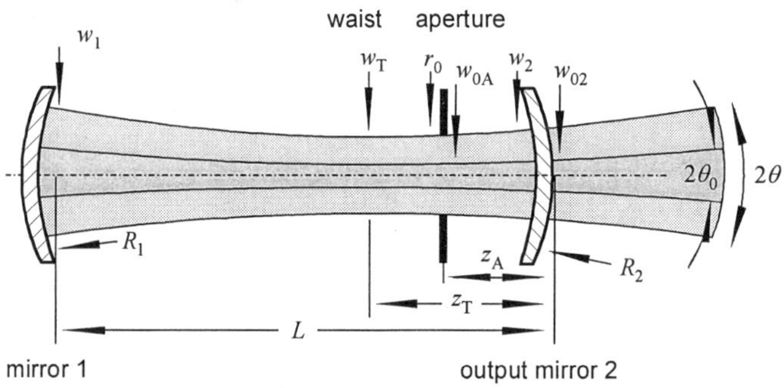

Fig. 3.6. Empty resonator

With given aperture, length and mirror radii the beam parameters can be calculated according to the above steps.

Resonator matrix:

$$M = \begin{vmatrix} a & b \\ c & d \end{vmatrix} = \begin{vmatrix} 1 & 0 \\ -1/R_2 & 1 \end{vmatrix} \begin{vmatrix} 1 & L \\ 0 & 1 \end{vmatrix} \begin{vmatrix} 1 & 0 \\ -1/R_1 & 1 \end{vmatrix} . \tag{3.18}$$

Resonator matrix elements:

$$a = 1 - \frac{L}{R_1}, \qquad b = L, \tag{3.19}$$

$$c = \frac{-1}{R_1} - \frac{1}{R_2} + \frac{L}{R_1 R_2}, \qquad d = 1 - \frac{L}{R_2}.$$

Resonator parameters and beam radius:

$$p_2^2 = -\frac{ab}{cd}, \qquad w_{02}^2 = \frac{\lambda p_2}{\pi}. \tag{3.20}$$

Transition matrix from w_{02} to $w(z)$:

$$\begin{vmatrix} 1 & z \\ 0 & 1 \end{vmatrix} \begin{vmatrix} 1 & 0 \\ -1/R_2 & 1 \end{vmatrix} = \begin{vmatrix} 1 - z/R_2 & z \\ -1/R_2 & 1 \end{vmatrix}. \tag{3.21}$$

Position of the waist (from R_2):

$$z_T = \frac{R_2 p_2^2}{R_2^2 + p_2^2}. \tag{3.22}$$

Fundamental-mode radius $w_0(z)$ at the position z (aperture z_A or waist z_1):

$$w_{0A}^2 = \left(1 - \frac{z}{R_2}\right)^2 w_{02}^2 + \frac{z^2 w_{02}^2}{p_2^2}, \qquad \text{according to (2.29).} \tag{3.23}$$

Multimode radius w_2 at the output mirror:

With $w_2^2 = r_0^2 w_{02}^2 / w_{0A}^2$, using (2.15) and (2.18), one obtains

$$\frac{w_2^2}{r_0^2} = \frac{1}{(R_2 - z_A)^2 / R_2^2 + z_A^2 / p_2^2}. \tag{3.24}$$

Multimode divergence θ_2 at the output coupler:

$$\frac{\theta_2^2}{r_0^2} = \frac{1}{p_2^2 (R_2 - z_A)^2 / R_2^2 + z_A^2}. \tag{3.25}$$

Table 3.1 indicates the beam parameters of the 00-modes in the empty resonator for several examples. The notations are chosen according to Fig. 3.6.

Table 3.1 Beam dimensions for the 00-mode in the empty resonator

Resonator			Matrix elements				Beam parameters					
L	R_1	R_2	a	b	c	d	z_{R2}	w_2	z_T	w_T	θ
m	m	m	-	m	1/m	-	m	mm	m	mm	mrad
1	5	5	0.8	1.0	−0.36	0.80	1.67	0.73	0.5	0.69	0.46
10	20	20	0.5	10	−0.08	0.50	11.55	1.92	5	1.66	0.19
0,5	1	1	0.5	0.5	−1.50	0.50	0.58	0.43	0.25	0.37	0.86
0.5	−3	3	1.17	0.5	−0.06	0.83	3.55	1.06	1.75	0.69	0.46
0.3	−0.5	0.5	1.60	0.3	−1.20	0.40	1.00	0.56	0.4	0.25	1.26
0.3	0.5	−0.5	0.40	0.3	−1.20	1.60	0.25	0.28	−0.1	0.25	1.26

w_2 = radius at the output coupler, z_T = position of the waist relative to the mirror R_2
w_T = waist radius, θ = divergence of the laser beam.

g-parameters

It is convenient to characterize a resonator by two parameters:

$$g_1 = 1 - L/R_1 \text{ and } g_2 = 1 - L/R_2. \tag{3.26}$$

The resonator matrix then reads

$$M = \begin{vmatrix} g_1 & L \\ -(1 - g_1 g_2)/L & g_2 \end{vmatrix}. \tag{3.27}$$

The stability criterion, $abcd = g_1 g_2(g_1 g_2 - 1) < 0$, is fulfilled for

$$0 < g_1 g_2 < 1. \tag{3.28}$$

The stability diagram is plotted in Fig. 3.7, with the boundaries between stable and unstable regions and some special resonators configurations named.

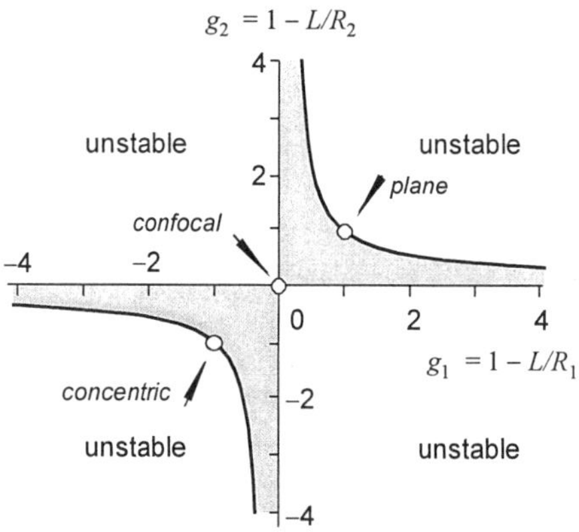

Fig. 3.7. Stability diagram

3.3 Conventional Stable Resonators

In the following, results are presented for beam divergence and beam radius for specific resonators. The details refer to the output coupler and for beams propagating from the left to the right.

Laser Rod with Spherical Endfaces as Mirrors

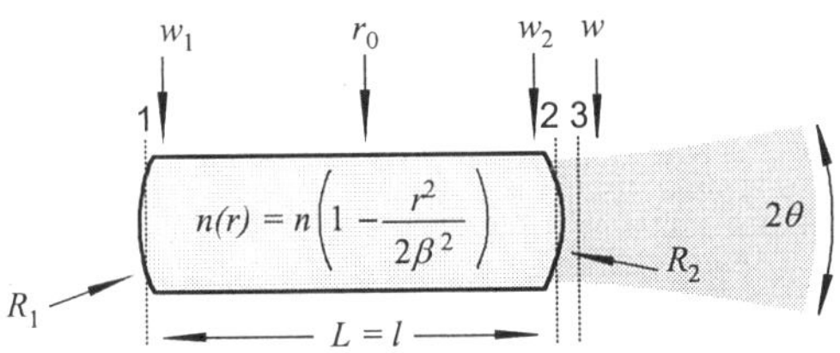

Fig. 3.8. Laser rod with fixed mirrors at the endfaces 1 and 2 [3.5]

The matrix M_{rod} for the laser medium (rod in Fig. 3.8) has to be chosen according to Table 2.4,

$$M_{12} = \begin{vmatrix} a & b \\ c & d \end{vmatrix} = \begin{vmatrix} 1 & 0 \\ -1/R_2 & 1 \end{vmatrix} M_{\text{rod}} \begin{vmatrix} 1 & 0 \\ -1/R_1 & 1 \end{vmatrix}, \textit{ resonator matrix,} \qquad (3.29)$$

$$p_2^2 = -\frac{ab}{cd}, \qquad \textit{resonator parameter,} \qquad (3.30)$$

$$M_{23} = \begin{vmatrix} 1 & 0 \\ -1/R_2 & n \end{vmatrix}, \qquad \textit{mirror transition matrix.} \qquad (3.31)$$

R_2 has the sign for a mirror, reference planes 1 and 2 are inside the mirror,

$$w_2 \approx r_0, \qquad \textit{multimode radius,} \qquad (3.32)$$

$$\frac{\theta_2^2}{r_0^2} = \frac{w_2^2}{w_A^2}\left(\frac{1}{R_2^2} + \frac{n^2}{p_2^2}\right), \qquad \textit{multimode divergence,} \qquad (3.33)$$

w_A = TEM $_{00}$ mode radius at the aperture position.

For a plane-plane resonator holds

$$w_2 \approx r_0, \qquad \theta_2 = nr_0/\beta, \qquad (3.34)$$

with $n(r) = n\left(1 \pm r^2/2\beta^2\right)$, the thermally induced refractive index. A plane-plane system is only stable if the refractive index decreases with r.

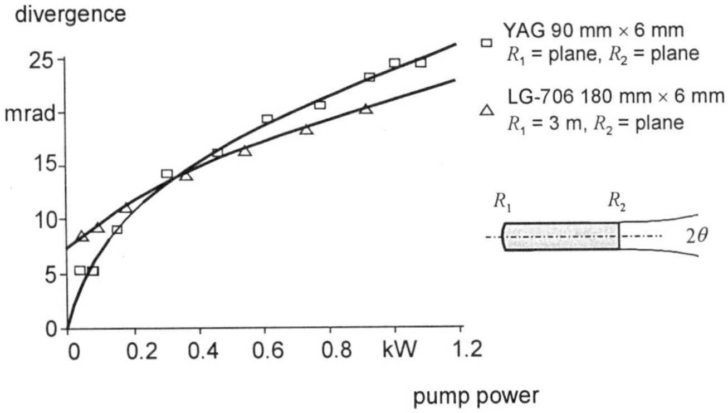

Fig. 3.9. Divergence of laser rods with fixed mirrors

For a plane spherical resonator one has

$$w_2 \approx r_0, \qquad \frac{\theta_2^2}{r_0^2} = \frac{n^2}{\beta^2} + \frac{n^2}{R_1 L}, \tag{3.35}$$

with $n(r) = n\left(1 \pm r^2/2\beta^2\right)$ and $R_2 = \infty$.

In Chap. 4 it is shown that the refracting power of a rod is proportional to the pump power P_p and proportional to $1/\beta^2$. The square of the divergence of a rod with fixed mirrors thus becomes proportional to the pump power, (3.36). In Fig. 3.9 measurements are presented for a plane-plane Nd:YAG rod and for a plane-spherical glass rod.

One Rod with a Fixed Mirror and Plane External Mirror

The refracting power of the rod in Fig. 3.10 is approximated by a thick lens with $D = nl/\beta^2$

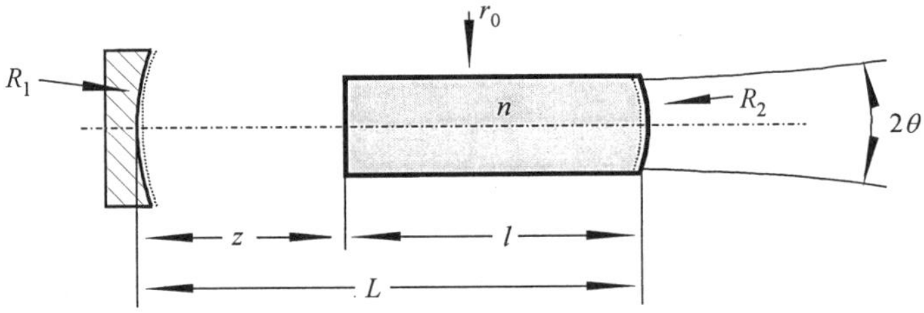

Fig. 3.10. Rod with one external mirror

Resonator matrix:

$$M_{12} = \begin{vmatrix} a & b \\ c & d \end{vmatrix} = \begin{vmatrix} 1 & 0 \\ -1/R_2 & 1 \end{vmatrix} \begin{vmatrix} 1 & 0 \\ -D & 1 \end{vmatrix} \begin{vmatrix} 1 & 0 \\ 0 & 1/n \end{vmatrix} \begin{vmatrix} 1 & z \\ 0 & 1 \end{vmatrix} \begin{vmatrix} 1 & 0 \\ -1/R_1 & 1 \end{vmatrix}. \tag{3.36}$$

Matrix elements:

$$a = 1 - \frac{z}{R_1}, \qquad c = -\frac{1}{R_2} - D + \frac{z}{R_1 R_2} + \frac{zD}{R_1} - \frac{1}{nR_1}, \tag{3.37}$$

$$b = z, \qquad d = -\frac{z}{R_2} - zD + \frac{1}{n_1}.$$

Critical refracting power:

$$D_c = -\frac{1}{R_2} - \frac{1}{n(R_1 - z)}, \qquad D_d = \frac{1}{zn} - \frac{1}{R_2}. \tag{3.38}$$

Stable range of operation: $\Delta D = D_d - D_c$. (3.39)

Matrix elements in the representation with critical refracting power:

$$a = \frac{R_1 - z}{R_1}, \qquad b = z, \tag{3.40}$$

$$c = \frac{(R_1 - z)(D_c - D)}{R_1}, \qquad d = z(D_d - D).$$

Resonator parameter:

$$p_2^2 = \frac{1}{(D - D_c)(D_d - D)}. \tag{3.41}$$

Mirror matrix:

$$M = \begin{vmatrix} 1 & 0 \\ 1/R_2 & n \end{vmatrix}. \tag{3.42}$$

Multimode divergence:

$$\theta_2^2 = r_0^2 \left(\frac{1}{R_2^2} + \frac{n^2}{p^2} \right). \tag{3.43}$$

Multimode radius: $w_2 = r_0$ (3.44)

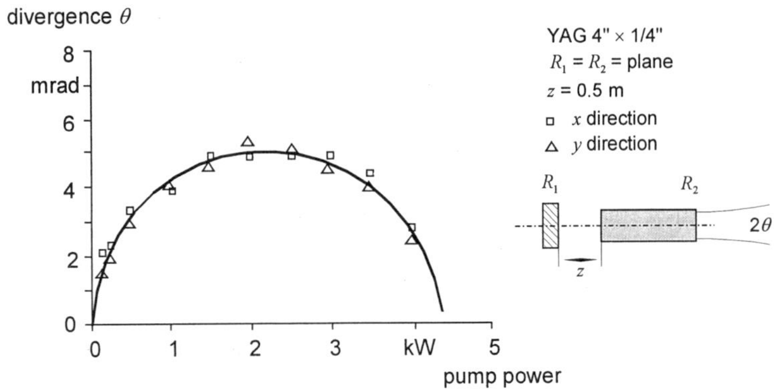

Fig. 3.11. Divergence of a laser rod with one fixed and one external plane mirror

For a plane-plane resonator holds

$$\theta_{max} = 0.5\ r_0/z. \tag{3.45}$$

In plane-plane resonators with one fixed and one external mirror two values for the critical refracting power exist and therefore two critical values for the pump power, at which the resonator becomes unstable. With (3.39) this is the case for $D_c = 0$ and for $D_d = 1/(zn)$. Figure 3.11 demonstrates that the divergence at these two critical values approaches the lower TEM-00-mode divergence.

Two External Mirrors, One Rod

Resonator matrix:

$$M = \begin{vmatrix} a & b \\ c & d \end{vmatrix} = \begin{vmatrix} 1 & 0 \\ -1/R_2 & 1 \end{vmatrix} \begin{vmatrix} 1 & d_2 \\ 0 & 1 \end{vmatrix} \begin{vmatrix} 1 & 0 \\ -D & 1 \end{vmatrix} \begin{vmatrix} 1 & d_1 \\ 0 & 1 \end{vmatrix} \begin{vmatrix} 1 & 0 \\ -1/R_1 & 1 \end{vmatrix}. \tag{3.46}$$

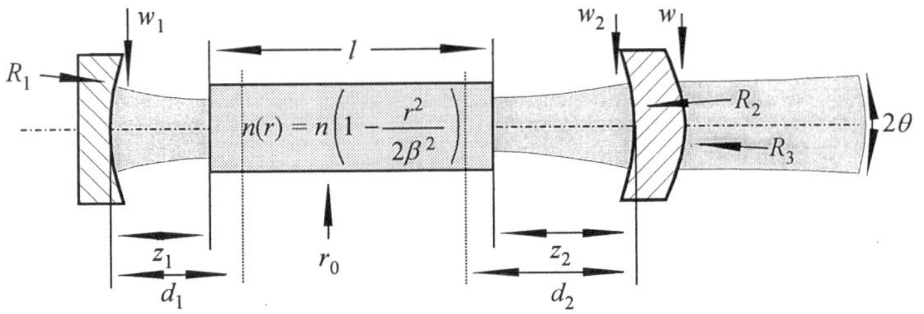

Fig. 3.12. Resonator with two external mirrors

The refracting power of the rod is approximated by a thick lens [3.6, 3.7, 3.8] (Fig. 3.12)

$$D = \frac{nl}{\beta^2}, \; d_i = z_i + \frac{l}{2n}, \qquad i = 1, 2 \qquad l = \textit{rod length.}$$

For numerical calculations the matrix from Table 2.4 has to be used.

Matrix elements:

$$a = \frac{1}{R_1}(R_1 - d_1 - d_2) + \frac{d_2}{R_1}(d_1 - R_1)D, \tag{3.47}$$

$$b = d_1 + d_2 - d_1 d_2 D,$$

$$c = \frac{1}{R_1 R_2}(d_1 + d_2 - R_1 - R_2) - \frac{1}{R_1 R_2}(d_1 - R_1)(d_2 - R_2)D,$$

$$d = \frac{1}{R_2}(R_2 - d_1 - d_2) + \frac{d_1}{R_2}(d_2 - R_2)D.$$

Critical refracting power:

$$D_a = \frac{1}{d_2} + \frac{1}{d_1 - R_1}, \qquad D_b = \frac{1}{d_2} + \frac{1}{d_1}, \tag{3.48}$$

$$D_c = \frac{1}{d_2 - R_2} + \frac{1}{d_1 - R_1}, \qquad D_d = \frac{1}{d_1} + \frac{1}{d_2 - R_2}.$$

Resonator parameter: $$p_2^2 = -\frac{ab}{cd} = \frac{(d_2 R_2)^2 (D_a - D)(D_b - D)}{(d_2 - R_2)^2 (D_c - D)(D - D_d)} \tag{3.49}$$

Mirror matrix for transmitted beam:

$$M_{23} = \begin{vmatrix} 1 & 0 \\ \frac{n-1}{R_3} - \frac{n}{R_2} & 1 \end{vmatrix} \qquad R_2, R_3 > 0 \text{ in Fig. 3.12} \tag{3.50}$$

Multimode divergence:

$$\frac{\theta_2^2}{r_0^2} = \frac{1}{p^2 (R_2 - d_2)^2 / R_2^2 + d_2^2}, \qquad \text{aperture at } d_2 \text{ and adapted mirror.} \tag{3.51}$$

Multimode radius:

$$\frac{w_2^2}{r_0^2} = \frac{1}{(R_2 - d_2)^2 / R_2^2 + d_2^2 / p^2}, \qquad \text{aperture at } d_2 \text{ and adapted mirror.} \tag{3.52}$$

Since up to now the resonator type with two external mirrors has found the widest use, some special systems will be discussed in more detail. The critical refracting powers D_i with $i = a$,..., d determine the limits of the two stability regions. At the stability limits the beam is focused on the mirrors or the beam radius becomes infinite. Since the refracting power of the rod is determined by the pump power it limits the stable range of laser operation.

For the critical refracting power the following relation holds

$$D_a + D_d = D_b + D_c \,, \tag{3.53}$$

and the matrix elements may be expressed as functions of the critical refracting power

$$a = \frac{d_2}{R_1}(d_1 - R_1)(D - D_a)\,, \qquad b = d_1 d_2 (D_b - D)\,, \tag{3.54}$$

$$c = \frac{(d_1 - R_1)(d_2 - R_2)}{R_1 R_2}(D_c - D)\,, \qquad d = \frac{d_1}{R_2}(d_2 - R_2)(D - D_d)\,.$$

The divergence versus the refracting power of the thermal lens and the dependence on the critical refracting power are shown in Fig. 3.13.

The two principle limitations for the beam parameter product of this resonator type are, according to [3.9] and [3.10],

$$\frac{D_i - D_j}{2} \geq \frac{\mathrm{Max}(w)\mathrm{Max}(\theta)}{r_0^2} \tag{3.55}$$

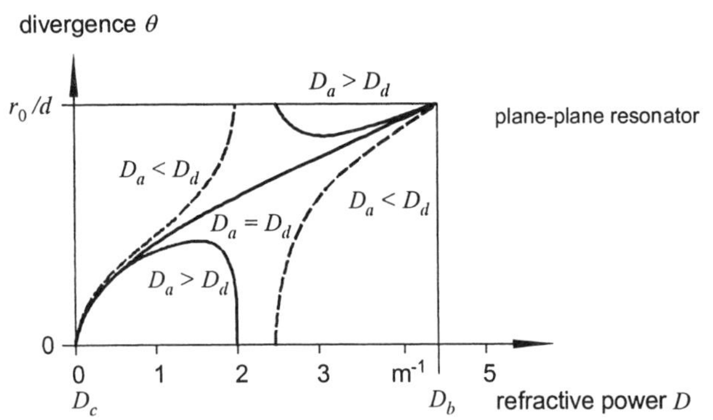

Fig. 3.13. Principle multimode divergence of a resonator with two external mirrors

with $D_i - D_j = \Delta D$ for the stable range of operation and

$$\frac{\mathrm{Max}(w\theta)}{r_0^2} \geq \frac{D_i - D_j}{4}\,. \tag{3.56}$$

A large stable range results also in a large beam parameter product. High beam quality is possible only with a reduced range of stability.

Beam Propagation in the Resonator

If the multimode radius and the Rayleigh length (e.g. for external mirrors, according to (3.50, 3.53)) are known, the beam propagation can be calculated in the resonator with the resonator matrices, (3.47) and (2.27). With varying pump power the refracting power in the rod changes and therefore the corresponding matrix. In Fig. 3.14 the propagation for two different, averaged pump powers is shown. The utilizable rod volume (mode volume) is reduced with increasing pumping power.

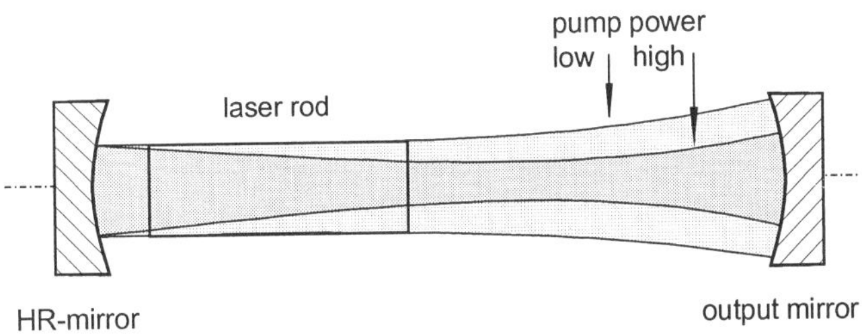

Fig. 3.14. Beam propagation in the resonator dependent on the refracting power

3.4 Special Resonators with Two External Mirrors

The aim for a resonator concept is, besides reaching constant parameters, to find a simple and reliable construction. In the following, some specific resonators are discussed, which stand out by having at least one constant parameter in the complete stable range or by a simple construction.

As stable range ΔD, a section of the stable range $D_d - D_c$ is chosen (Fig. 3.15). The divergence change is thereby minimized. $D_d < D_a$ is the condition that a maximum exists. If two of the possible four parameters are already determined by the design, then the other two parameters can be calculated.

Resonator with Approximately Constant Divergence

1. The distances d_1 and d_2 are given, the mirror radii are

$$R_1 = \frac{d_1^2}{d_1 - 1/(D_d - D_c)}, \qquad R_2 = \frac{d_1}{1 - d_1 D_d} + d_2. \tag{3.57}$$

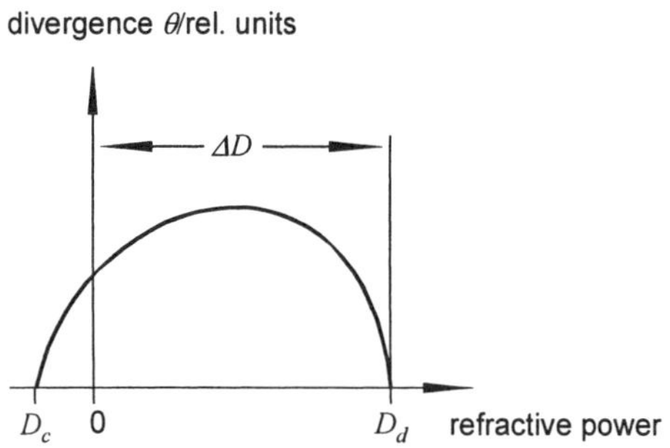

Fig. 3.15. Divergence for a nearly constant beam radius

2. The mirror radii R_1 and R_2 are given, the distances are

$$d_1 = \frac{R_1}{2} \pm \sqrt{\frac{R_1^2}{4} - \frac{R_1}{D_d - D_c}}, \qquad d_2 = \frac{d_1}{1 - d_1 D_d}. \tag{3.58}$$

Plane-plane resonator: (3.59)

$$R_1 = R_2 = \infty \qquad a = 1 - d_2 D, \qquad b = d_1 + d_2 - d_1 d_2 D,$$

$$c = -D, \qquad d = 1 - d_1 D,$$

$$p^2 = -\frac{ab}{cd} \qquad D_a = 1/d_2, \qquad D_b = 1/d_1 + 1/d_2,$$

$$D_c = 0, \qquad D_d = 1/d_1,$$

$$\frac{\theta^2}{r_0^2} = \frac{1}{p^2 + d_2^2} \qquad \frac{w_T^2}{r_0^2} = \frac{p^2}{p^2 + d_2^2}.$$

Symmetrical plane-plane resonator: with $d_1 = d_2 = d$, (3.60)

$$D_a = 1/d, \qquad D_b = 2/d, \qquad D_c = 0, \qquad D_d = 1/d,$$

$$\frac{\theta^2}{r_0^2} = \frac{D}{2d}, \qquad \frac{w_T^2}{r_0^2} = \frac{2 - dD}{2}.$$

The lowest beam quality (Max $(w_T\theta)$) is found at half of the maximum refracting power for D_b, at $D_{a,d} = 1/d$,

$$\text{Max}(\theta w_T) = \frac{r_0^2}{2d}.$$

The necessary resonator length increases with the square of the rod radius r_0.

Resonator with a constant divergence: $d_2 = R_2$, [3.11] (3.61)

$$D_a < D_b, \qquad d_1 = \frac{d_2}{D_d d_2 - 1}, \qquad R_1 = d_1 d_2 \frac{D_a - D_b}{D_a d_2 - 1},$$

$$D_a = \frac{1}{d_2} + \frac{1}{d_1 - R_1}, \qquad D_a = 0, \qquad \text{with} \qquad R_1 = d_1 + d_2 = \frac{D_b d_2^2}{D_b d_2 - 1},$$

$$D_b = \frac{1}{d_1} + \frac{1}{d_2}, \qquad \theta = \frac{r_0}{d_2}, \qquad \mathrm{Max}(w) = \frac{1}{2}(D_b - D_a) d_2 r_0 .$$

Resonator with a constant beam radius $d_2 = 0$, (3.62)

$$D_c < D_d, \qquad d_1 = \frac{R_2}{1 + D_d d_2}, \qquad R_1 = R_2 d_1 \frac{D_c - D_d}{1 + D_c R_2},$$

$$D_c = \frac{-1}{R_2} + \frac{1}{d_1 - R_1}, \qquad D_c = 0, \qquad \text{with } R_1 = R_2^2 \frac{-D_d}{1 + D_d R_2},$$

$$D_d = \frac{1}{d_1} - \frac{1}{R_2}, \qquad \mathrm{Max}(\theta) = \frac{1}{2}(D_d - D_c) r_0, \quad w = r_0.$$

R_2 is a free parameter and is only restricted by the mode volume, by the overall length of the resonator and by the availability of mirrors with the appropriate radii.

Dynamically Compensated Resonators

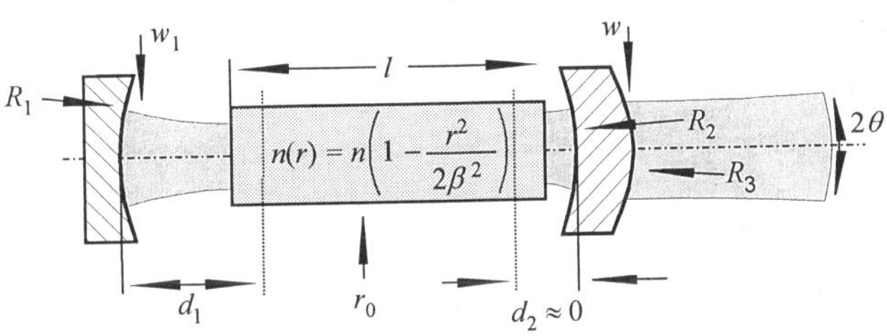

Fig. 3.16. Setup of compensated resonators

Under suitable conditions, insensitive resonators (Fig. 3.16) with $d_2 = 0$ and $w = r_0$ can be compensated dynamically, such that constant divergence and constant radius are achieved. The desired divergence θ is given. The divergence can be kept constant by:

1. *Shifting the plane HR mirror* ($R_1 = \infty$)

The divergence θ is left constant, if the distance d_1 of the HR mirror, dependent on the refracting power of the rod, is changed as follows (Fig. 3.17)

$$d_1(D) = \frac{r_0^2 D}{\theta^2 + r_0^2 D^2} .$$

The maximum shift d_{max} is

$$d_{max} = \frac{r_0}{2\theta}, \qquad \text{with } D_m = \frac{\theta}{r_0} . \tag{3.63}$$

With large values of D the shift becomes smaller again.

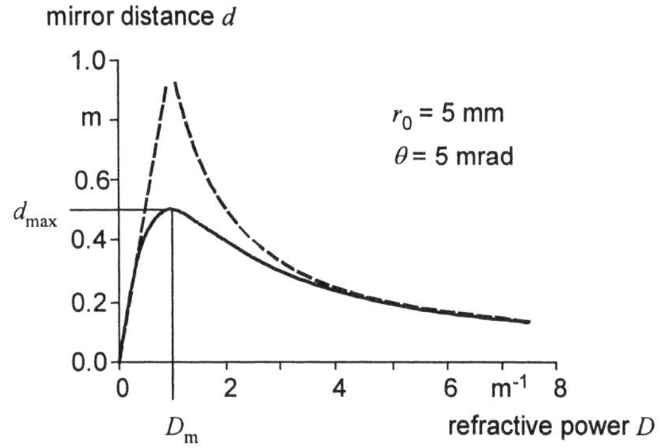

Fig. 3.17. Position of the HR mirror for constant divergence

2. *Variable mirror curvature* R_1 *with* d_1 = const.

The operating range is limited by the critical refracting power

$$D_c = 0, \qquad D_d = \frac{1}{d_{max}} = \frac{2\theta}{r_0} \tag{3.64}$$

and the dependence of the mirror curvature is

$$R_1 = \frac{d_1^2\theta^2 + r_0^2(1 - Dd_1)^2}{d_1\theta^2 - r_0^2 D(1 - Dd_1)} . \tag{3.65}$$

3. *Variable mirror curvature* R_2

The operating range depends on the interval ΔD

$$\Delta D = D_d - D_c = \frac{1}{d_1} - \frac{1}{d_1 - R_1}, \tag{3.66}$$

which is independent of R_2. The divergence determines via $D_d = 2\theta / r_0$ the stable range of operation ΔD at constant R_2. The divergence remains constant, if R_2 is changed as follows

$$\frac{1}{R_2} = \frac{1}{d_1} - \frac{\Delta D}{2} - D. \tag{3.67}$$

The parameter R_1 is given by

$$\frac{1}{R_1} = \frac{1}{d_1} - \frac{1}{\Delta D d_1^2}. \tag{3.68}$$

In order to avoid too small curvatures for R_2, d_1 is fixed by

$$\frac{1}{d_1} = \frac{\Delta D}{2} + \frac{D_{\max}}{2} \tag{3.69}$$

which gives $\dfrac{1}{R_2} = \dfrac{D_{\max}}{2} - D$.

To have constant divergence, the curvature $1/R_2$ of the output coupler has to be changed linearly with the refracting power of the rod, (Fig. 3.18).

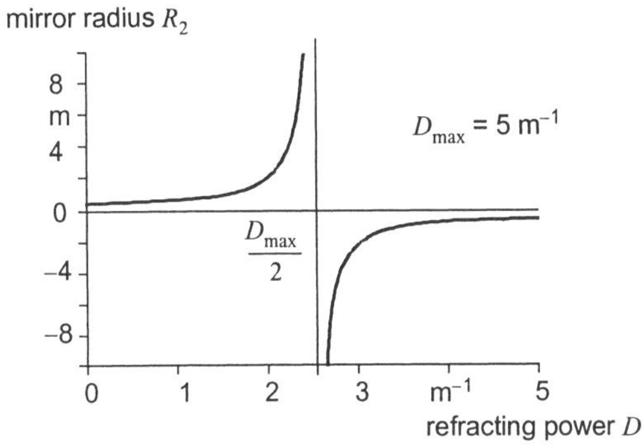

Fig. 3.18. Curvature R_2 versus refracting power D for constant divergence

By successively exchanging the output mirror or changing the mirror curvature the divergence changes can be kept small. Figure 3.19 shows the calculated divergence for an insensitive resonator with a beam radius independent of the pump power. Varying the mirror curvature of the output coupler with the pump power,

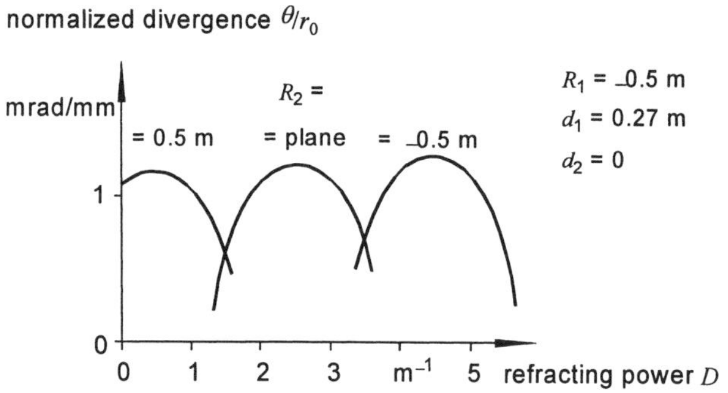

Fig. 3.19. Low variations of the divergence for different mirror radii R_2

approximately at maximum divergence, the beam divergence remains mostly constant.

3.5 Multicavity Resonators

Solid-state lasers with a single rod are limited in their output power and beam quality. YAG rods only exist up to a length of 200 mm with a maximal output power per rod of about 500 W. By using several pump cavities in series inside or outside the resonator the power can be increased to the kW-range.

Two External Mirrors and Two Rods

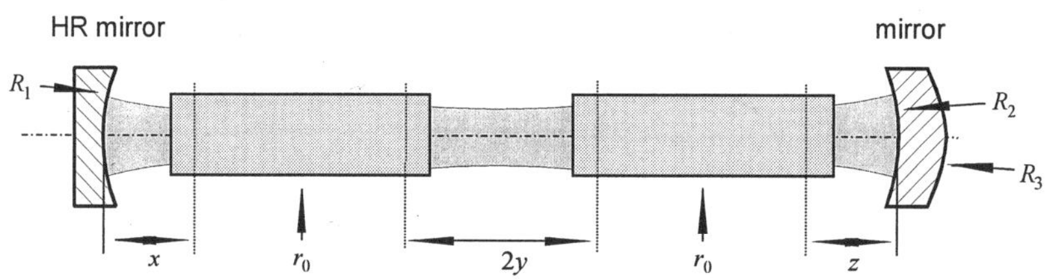

Fig. 3.20. Resonator with two cavities

Resonator matrix:

$$\begin{vmatrix} a & b \\ c & d \end{vmatrix} = \begin{vmatrix} 1 & 0 \\ -1/R_2 & 1 \end{vmatrix} \begin{vmatrix} 1 & z \\ 0 & 1 \end{vmatrix} \begin{vmatrix} 1 & 0 \\ -D & 1 \end{vmatrix} \begin{vmatrix} 1 & 2y \\ 0 & 1 \end{vmatrix} \begin{vmatrix} 1 & 0 \\ -D & 1 \end{vmatrix} \begin{vmatrix} 1 & x \\ 0 & 1 \end{vmatrix} \begin{vmatrix} 1 & 0 \\ -1/R_1 & 1 \end{vmatrix} \tag{3.70}$$

Mirror matrix for transmitted beam

$$M_{23} = \begin{vmatrix} 1 & 0 \\ \dfrac{n-1}{R_3} - \dfrac{n}{R_2} & 1 \end{vmatrix}, \qquad \text{sign for } R_2 \text{ and } R_3 \text{ as lens radii.} \quad (3.71)$$

In the case of almost identically rods and the plane mirrors very close to the rods $R_1 = R_2 = \infty$ and $x = z = 0$, the following simplifications result:

Matrix elements: $a = 1 - 2yD,$ $b = 2y,$ (3.72)

$c = 2D(yD - 1),$ $d = 1 - 2yD.$

Critical refracting power: $D_a = D_d = g\ 1/(2y),$ $D_c = 0$ or $D_c = 1/y.$ (3.73)

Resonator parameter:
$$p^2 = -\frac{ab}{cd} = -\frac{y}{D(yD-1)}. \quad (3.74)$$

Multimode divergence:
$$\frac{\theta^2}{r_0^2} = \frac{1}{p^2} = -\frac{D(1-yD)}{y}. \quad (3.75)$$

Multimode radius: $w = r_0.$ (3.76)

Beam quality:
$$\text{Max}(\theta w) = \frac{r_0^2}{2y}. \quad (3.77)$$

Since the same power is distributed between two rods, the refracting power of each single rod is only half the value. The range of stable operation ΔD with the same beam quality is double that of a single rod. In the middle of the stable range, however, is an unique unstable point, $D = 1/(2y)$. In general, it can be proved [3.9] that in the case of n rods the range of operation with the same beam quality increases by a factor n.

Multiple Cavities with Flat Mirrors

The optical matrix of a resonator with several cavities (Fig. 3.21) is of the type

$$M^n = \left(\begin{vmatrix} 1 & y \\ 0 & 1 \end{vmatrix} \begin{vmatrix} 1 & 0 \\ -D & 1 \end{vmatrix} \begin{vmatrix} 1 & y \\ 0 & 1 \end{vmatrix} \right)^n. \quad (3.78)$$

With the help of the theorem of Sylvester [3.17] this matrix can be evaluated

$$M^n = \begin{vmatrix} a & b \\ c & d \end{vmatrix}^n = \frac{1}{\sin\varphi}\begin{vmatrix} a\sin(n\varphi)-\sin(n-1)\varphi & b\sin(n\varphi) \\ c\sin(n\varphi) & d\sin(n\varphi)-\sin(n-1)\varphi \end{vmatrix} \tag{3.79}$$

and $\cos\varphi = (a+d)/2$.

For matrices with $\det(M) = 1$ the standardized critical refracting power can be calculated. Some values are given in Table 3.2.

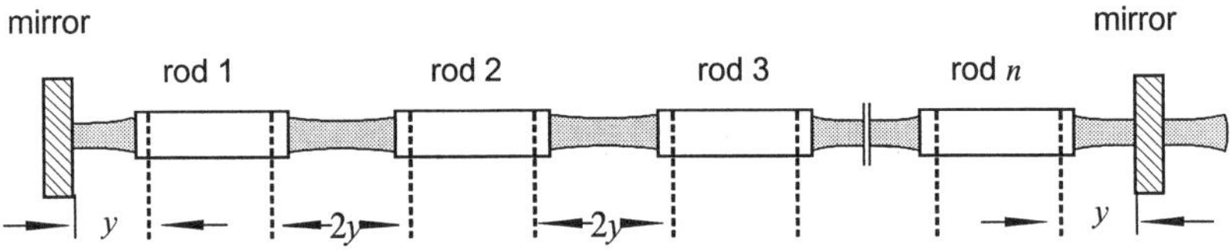

Fig. 3.21. Resonator with several pumping cavities

Table 3.2 Normalized critical refracting powers for symmetrical setups

Rods	$yD_a = yD_d =$	$yD_b =$	$yD_c =$
1	1	2	0
2	$1 \pm 0.5\sqrt{2}$	1, 2	0, 1
2	0.29, 1.71	1, 2	0, 1
3	1, $1 \pm 0.5\sqrt{3}$	1/2, 3/2, 2	0, 1/2, 3/2
3	0.13, 1, 1,87	0.5, 1.5, 3	0, 0.5, 1.5
4	$1 \pm 0.5\sqrt{2 \pm \sqrt{2}}$	1, $1 \pm 0.5\sqrt{2}$, 2	1, $1 \pm 0.5\sqrt{2}$, 2
4	0.08, 0.62, 1.38, 1.92	0.29, 1, 1.71, 2	0.29, 1, 1.71, 2
5	1, $1 \pm \sqrt{2(5 \pm \sqrt{5})/4}$	2, $(5 \pm \sqrt{5})/4$, $(3 \pm \sqrt{5})/4$	0, $(5 \pm \sqrt{5})/4$, $(3 \pm \sqrt{5})/4$

For unequal distances between the cavities or differences in the refracting powers of the rods the output power is reduced and instabilities may occur. Due to the symmetrical construction ($a = d$) and with (3.81), the resonator parameter for all resonators following this scheme is given by

$$p^2 = -\frac{b}{c} = \frac{2y - y^2 D}{D}. \tag{3.80}$$

Fig. 3.22. Laboratory model of a multicavity laser with 2 kW CW power, (HAAS-Laser)

A laboratory model of a laser with four cavities, each able to deliver more than 500 W average power, is shown in Fig. 3.22.

For distances which are large compared to the rod length the beam quality can be estimated to be

$$\frac{\theta^2 w^2}{r_0^4} = \frac{D(2 - yD)}{4y} \tag{3.81}$$

with the maximum being

$$\mathrm{Max}(\theta w) = \frac{r_0^2}{2y}. \tag{3.82}$$

3.6 Alignment Sensitivity of Stable Resonators

The relevant laser parameters, such as

- power or the energy,
- beam position relative to the geometric axis and
- mode structure

are dependent on

- the tilt of the resonator mirrors relative to each other,
- the tilt of the laser medium relative to the resonator,
- the displacement of the laser medium relative to the resonator axis,
- the position and diameter variations of additional apertures.

These changes of the resonator and beam parameters will be considered in turn and then, as far as possible, discussed for laser rods. More general considerations about misalignment stability may be found in [3.3].

Tilting the Rod

The reduction of the mode volume due to tilting the rod relative to the optical axis, can be estimated geometrically with the help of Fig. 3.23.

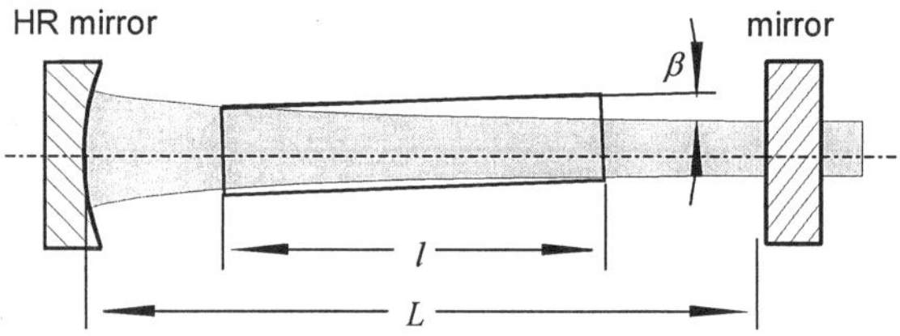

Fig. 3.23. Geometric reduction of the mode volume by tilting the rod

The effective rod diameter is reduced at small angles β to

$$2r_{\text{eff}} = 2r_0 - l\beta/n \tag{3.83}$$

and the relative mode volume decreases nearly linearly with the tilt angle

$$\frac{V_{m,\text{eff}}}{V_m} = 1 - \frac{l\beta}{nr_0}. \tag{3.84}$$

Equation (3.84) is only valid, if the aperture diameter is considerably larger than the corresponding TEM-00-mode radius.

Figure 3.24 shows the result of a such a measurement for a laser-glass rod and a YAG rod. The sensitivity of the resonator with respect to position displacements of the laser rod is considerably smaller than with respect to mirror misalignments. If the laser rod cannot be adjusted in its position independently of the mirrors, optimal adjustment can be reached by adjusting both mirrors synchronously in one direction.

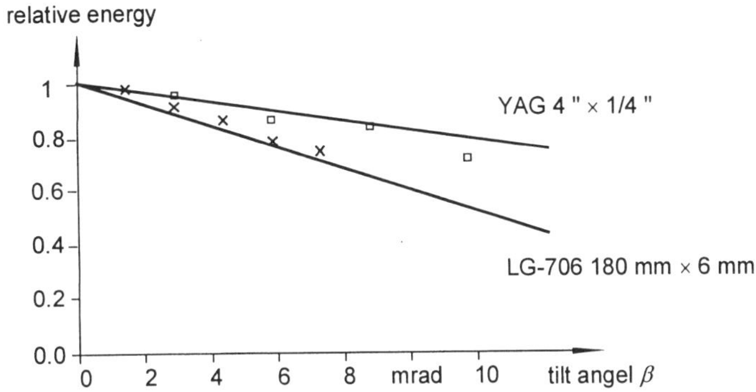

Fig. 3.24. Influence of the rod tilt on the pulse energy

Tilt of the Resonator Mirror

The resonator reacts in a very sensitive way to mirror misalignments. A mirror tilt by an angle α_i (Fig. 3.25) causes the reflection centers to shift with respect to the optical by the same angle α_i. The new optical axis then has an angle β relative to the geometrical axis

$$\beta = \frac{\alpha_1 R_1 + \alpha_2 R_2}{R_1 + R_2 - L}, \tag{3.85}$$

$R_i > 0$ for concave mirrors, $R_i < 0$ for convex mirrors.

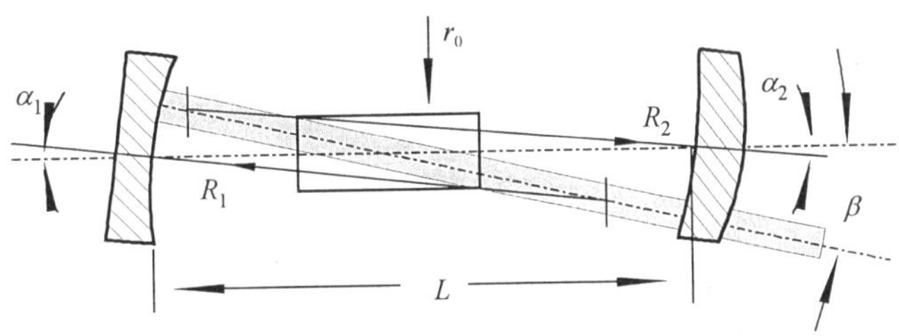

Fig. 3.25. Influence of the mirror misalignment

The shifting increases with the corresponding mirror curvature or the closer the curvature centers become ($L = R_1 + R_2$).

Two effects can reduce the utilizable laser power according to the above geometric consideration. The mode volume is reduced and the beam can be truncated by the mirror edges.

The effective mirror curvature is changed by the thermal lensing of the laser rod and thereby the sensitivity with respect to misalignments is reduced. For a plane-plane resonator in the pulsed operation the angle has been measured at which the energy decreases by about 10%. In Fig. 3.26 this angle is shown as a function of the average pump power. A linear relation between misalignment sensitivity and refracting power for a plane-plane resonator can be seen. The misalignment sensitivity due to angle misalignment within the whole stability diagram, g-diagram, is given in [3.12].

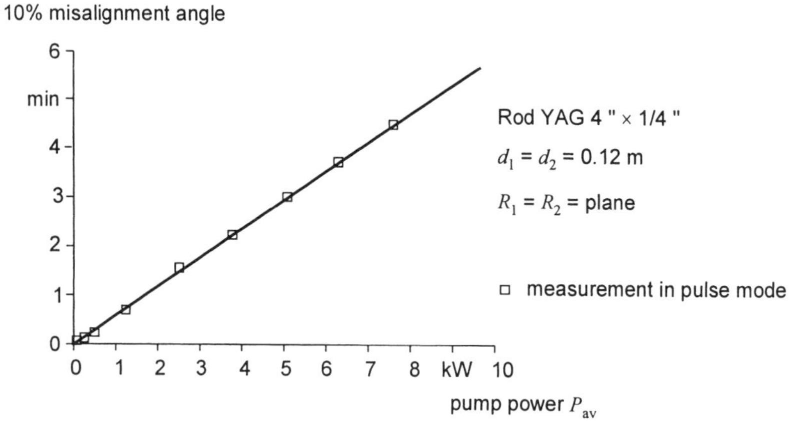

Fig. 3.26. Misalignment angle resulting in 10% energy drop as function of the pump power

Transverse rod translation in a plane-plane resonator only shifts the beam together with the rod. In resonators with curved mirrors the mode volume is reduced, and the rotational symmetry is disturbed.

An asymmetrical thermal load of the laser rod, e.g. in a single lamp cavity or by two different excitation lamps in a double-ellipse cavity, can result in a mechanical or optical bending of the laser rod. The beam migrates, and the mode volume and the beam quality changes.

Apertures in the Resonator

For cutting and marking applications the beam quality may not be sufficient in many cases. The mode order can be reduced by one or more apertures in the resonator, as given by (2.18).

With a high amplification (strong excitation, large effective cross section) particularly in the pulsed operation the laser starts in a resonator built by the HR mirror and the opposite rod end face. Therefore laser action in the unused material volume must be suppressed by apertures. The aperture must be centered relative to the optical axis. This becomes important especially in cutting applications because, on the one hand a very good beam quality is necessary, and on the other a high peak power is used. The aperture position in the z direction inside the resonator influences the transverse intensity distribution inside as well as outside the resonator [3.16].

3.7 Unstable Resonators

The condition for a beam in the geometric approximation, to reproduce itself after a complete resonator round-trip, leads to the eigenvalue equation [3.13]

$$M_f w = wA + \alpha B, \tag{3.86}$$

w = beam radius, Sect. 3.2,

$$M_f \theta = wC + \alpha D,$$

α = angle relative to the direction of propagation,

$$\begin{vmatrix} A & B \\ C & D \end{vmatrix} = \text{resonator round-trip matrix.}$$

Since the new beam has the radius $M_f w$ and the angle $M_f \alpha$ one defines M_f as the beam magnification. After a round-trip the geometric curvature $R = w/\alpha$ just has again its initial value. With the ABCD law (2.23)

$$\frac{1}{R} = \frac{C + D/R}{A + B/R} \tag{3.87}$$

and $AD - BC = 1$ follows

$$\frac{1}{R} = \frac{D - A}{2B} \pm \frac{1}{2B}\sqrt{(A + D)^2 - 4}\,. \tag{3.88}$$

A real solution of (3.90) exists, if

$$|A + D| \geq 2\,. \tag{3.89}$$

This type of resonator is called unstable. $A + D \geq 2$ is defined as the positive branch and $A + D \leq -2$ as the negative branch. Division of (3.87) by w yields the magnification

$$M_f = A + B/R \tag{3.90}$$

and with (3.89) follows

$$M_{f\pm} = \frac{A + D}{2} \pm \frac{1}{2}\sqrt{(A + D)^2 - 4} \quad \textit{magnification of the unstable resonator}. \tag{3.91}$$

The plus sign applies to the diverging wave, the minus sign to the converging wave. In [3.14] one finds various geometries for unstable resonators. In the following only the most common types will be examined further.

Confocal Unstable Resonator

The condition for a plane wave in the reference plane is, with (3.89)

$$AD = 1 \text{ (or } B = \infty). \tag{3.92}$$

The matrix for a round-trip of the diverging wave in the reference plane of the output mirror is given by

$$\begin{vmatrix} 1 & L \\ 0 & 1 \end{vmatrix} \begin{vmatrix} 1 & 0 \\ -2/R_1 & 1 \end{vmatrix} \begin{vmatrix} 1 & L \\ 0 & 1 \end{vmatrix} \begin{vmatrix} 1 & 0 \\ -2/R_2 & 1 \end{vmatrix} \tag{3.93}$$

with the matrix elements

$$A = 1 - 2L/R_1 - 4L/R_2 + 4L^2/(R_1R_2), \quad B = 2L - 2L^2/R_2, \tag{3.94}$$

$$C = -2/R_1 - 2/R_2 + 4L/(R_1R_2), \quad D = 1 - 2L/R_1.$$

The condition of (3.93) for a plane wave delivers

$$L = \frac{R_1 + R_2}{2}, \qquad \textit{confocal resonator.} \tag{3.95}$$

$L = R_1$ satisfies (3.93), but results in $B = 0$ and is no solution. The magnification M_f has two equivalent solutions in the confocal case

$$M_{f+} = -R_1/R_2 \qquad \text{and} \qquad M_{f-} = -R_2/R_1. \tag{3.96}$$

The negative branch (Fig. 3.27)

$-2 \geq A + D = -R_1/R_2 - R_2/R_1$ is fulfilled for $R_1 > 0$, $R_2 > 0$ and for $2L = R_1 - R_2$.

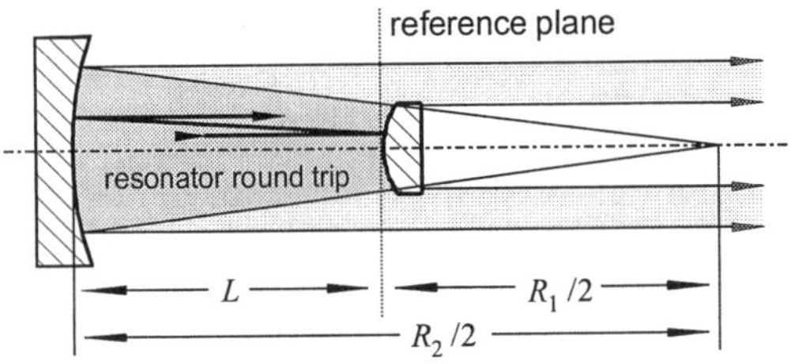

Fig. 3.27. Unstable confocal resonator, negative branch

The positive branch (Fig. 3.28) requires $2 \leq A + D = -R_1/R_2 - R_2/R_1$ and is fulfilled for $R_1 > 0$, $R_2 < 0$ and for $2L = R_1 + R_2$. Assuming a homogeneous intensity profile, the ratio of the output power to the total power is given by the ratio of the cross sections and reads, in the geometric approach:

$$T = \frac{M_f^2 w^2 - w^2}{M_f^2 w^2} = 1 - \frac{1}{M_f^2}, \text{ “}\mathit{transmission}\text{” } \mathit{of\ the\ right\ hand\ mirror.} \quad (3.97)$$

Therefore the equivalent reflection coefficient is

$$R = 1 - T = 1/M_f^2, \quad (3.98)$$

which in the case of a confocal resonator, (3.97) is

$$R = R_i^2/R_j^2. \quad (3.99)$$

It is only dependent on the mirror curvatures R_i, and not on the mirror diameter.

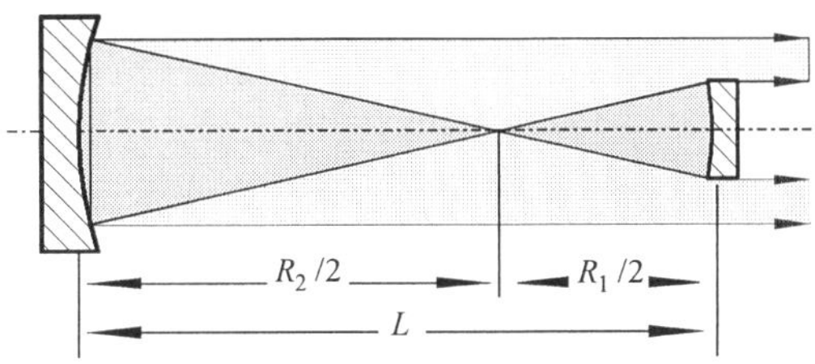

Fig. 3.28. Unstable confocal resonator, positive branch

Output Couplers for Unstable Resonators

Various constructions for the output mirror are sketched in Fig. 3.29, [3.15].

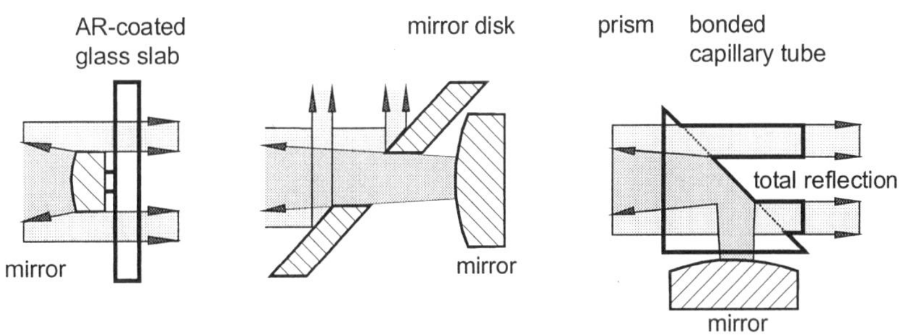

Fig. 3.29. Output couplers for unstable resonators

4 Thermal Effects

During optical pumping the laser material is heated [4.1] and must be cooled. The temperature gradient in the laser material initially induces a degradation of the beam quality and finally a fracture of the crystal. The heating therefore determines the beam quality and the maximal laser output power. The exact knowledge of the various heating mechanisms is therefore important and will be discussed in some detail.

4.1 Heat Efficiency

The main heating mechanisms are:

- The quantum defect, i.e. the energy difference between laser and pump photon, $\eta_{QD} = h\nu/h\nu_p$.

  For a loss-free oscillator and at high pump level, each pump photon $h\nu_p$ produces a laser photon $h\nu$ of lower energy. The energy difference in most cases is converted into heat. Therefore the minimal heating power P_H due to the quantum defect is related to the pump power by

  $$P_H = P(1 - \eta_{QD}) . \quad (4.1)$$

- Absorption of the laser radiation due to color centers or impurities in the crystal.

  Permanent color centers are generated by intense short-wavelength light in the crystal. They absorb in a broad range up to the laser wavelength. Measurements show that the YAG absorption coefficient α at 1 μm changes from 0.5×10^{-3} cm^{-1} to about 4×10^{-3} cm^{-1}. The additional heating power P_H due to the color center absorption for a typical rod of 15 cm length and a reflection coefficient of $R = 0.5$ is approximately

  $$P_H = P(1 - e^{-\alpha l})/(1 - R) = 0.12.$$

- Nonradiative transitions from the upper laser energy level. This is described by the quantum fluorescence efficiency η_Q.

- The direct absorption of the light outside and inside the pump range by impurities and color centers.
- Excitation from the pump level to higher levels by the pump light or by laser photons.
- Direct photon transitions from the pump levels.
- Excited state absorption, ESA. Excitation from the upper laser level by pump radiation or the laser radiation itself.

Taking all these effects into account, the heating power is of the same order as the excitation power by the lamps.

Example 4.1 Excitation of Nd:YAG

The main excitation bands are around 550 nm and 750 nm. Assuming that the two main pump bands contribute equally to the excitation then, with (4.1), one has

$P_H/P = (0.5\times1000/550) + (0.5\times1000/750) - 1 = 0.57.$

In this case the heating power is at least half of the laser power.

The heating power P_H in the laser material is nearly proportional to the pump power and therefore also proportional to the averaged electrical power

$$P_H = \eta_H (P_p - P_{p\,th}), \tag{4.2}$$

η_H = heat efficiency,

taking into account the threshold P_{pth} of the excitation source (arc lamp or laser diodes). However, for special materials, such as alexandrite, empirical results from [4.2] for the pulsed operation suggest the following relation

$$P_H = \eta_H F f(E_p)^{1.5}, \tag{4.3}$$

E_p = pump energy/W s,
f = pulse repetition rate.

In pulsed operation the pump light energy converted into heat is about 5% of the electrical pump power. Results for some examples are listed in Table 4.1.

The specific heat coefficient κ [4.1]

$$\kappa = \eta_H / \eta_{excit} \tag{4.4}$$

describes the amount of the stored energy that remains as heat in the laser material relative to the amount of energy that is stored in the upper laser level.

In general, the smaller the specific heat coefficient, the smaller is the thermal load and the influence by the thermal lens, for the same output power or energy.

Table 4.1 Specific heat coefficient and heat efficiency

Material	Concentration N_0 / %	Doping	Heating efficiency η_H / %	Heating coefficient κ	Ref.
glass Q-88	3	Nd	3.5	2.35	[4.3]
glass Q-98	3	Nd	4.6	2.85	[4.3]
glass LG-750	3	Nd	365	2.25	[4.3]
glass LG-760	3	Nd	2.3	1.5	[4.3]
glass LG-760	6	Nd	3.9	1.6	[4.3]
glass LHG-5	6	Nd	4.8	3.0	[4.3]
glass LHG-8	3	Nd	3.8	2.4	[4.3]
glass LSG-91H	2.7	Nd		2.45	[4.4]
alexandrite	0.13	Cr	7.5		[4.5]
YAG	1	Nd	5.05	3.25	[4.3]
YAG	1.4	Nd		3.4	[4.6]
GSGG	1.6	Nd/Cr		2.85	[4.6]

η_H refers to the electrical power

The temperature and the temperature distribution influence the laser performance.

- The temperature difference in the material between the center and the surface of the rod determines the maximum allowed load of the laser material and thereby the maximum available average output power.
- Generally the temperature gradient in the material leads to a lens effect during operation and must be taken into account.
  The absolute average temperature in the laser material determines the population of the laser energy levels. The efficiency, e.g. for Nd-YAG, is reduced with higher temperatures by the thermal distribution of the lower laser energy level, but increases for alexandrite up to an optimal temperature (see Fig. 4.3).

4.2 Temperature Profile

Steady-State Operation

Let the laser material with volume V be pumped with a power P_{in}. The amount P_H is heating the laser material. The surface O is held at a constant temperature T_0. A temperature profile establishes itself in the material and a temperature step occurs at the surface (Fig. 4.1).

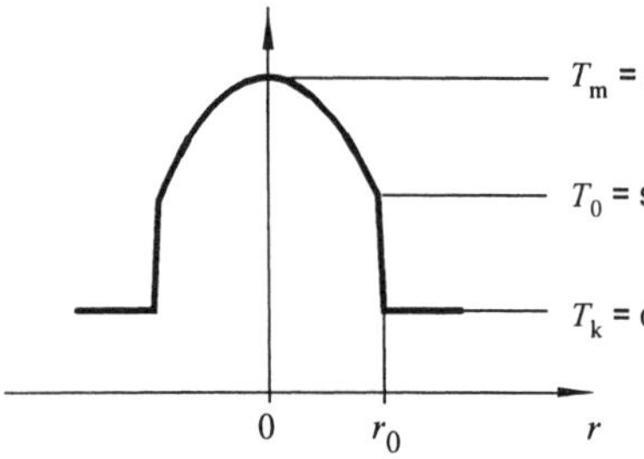

Fig. 4.1. Temperature distribution in the laser medium

The Poisson equation describes the temperature profile in the range $T_m \geq T \geq T_0$

$$\frac{dT}{dt} = \frac{k}{\rho c} \Delta T + \frac{P_H}{V \rho c}, \tag{4.5}$$

k = heat conductivity,
ρ = specific weight,
c = specific heat,
P_H = heating power,
V = material volume.

The stationary case for cylindrical symmetry reads [4.15]

$$0 = \frac{d^2 T}{dr^2} + \frac{1}{r}\frac{dT}{dr} + \frac{P_H}{kV}. \tag{4.6}$$

At the surface $r = r_0$ Newton's cooling law applies

$$\left.\frac{dT}{dr}\right|_{r=r_0} = \frac{\lambda}{k}(T_k - T_0), \tag{4.7}$$

r_0 = rod radius.

The heat transfer coefficient λ is defined by

$$P_H = \lambda O (T_0 - T_k), \tag{4.8}$$

O = cooling surface,
T_k = coolant temperature.

and can be regulated in certain limits by coolant, flow tube diameter and coolant velocity. The temperature difference $(T_m - T_0)$ inside the material determines the material load as well as the lens effect. The average temperature T_{av} controls the population of the lower laser level. Depending on the pump light distribution in the crystal and the cooling, a temperature profile is established in the laser material.

Table 4.2 Temperature profiles for rod and slab

rod

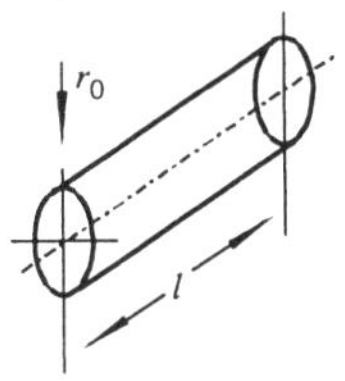

cooling surface $O = 2\pi r_0 l$, volume $V = \pi r_0^2 l$

$$\frac{T_m}{2} + T_0 = \frac{3}{2}T_k + \frac{P_H}{4\pi l}\left(\frac{3}{2\lambda r_0} + \frac{1}{2k}\right)$$

temperature profile	center temperature T_m $(r = 0)$	average temperature
$T(r) - T_0 = \frac{P_H}{4kV}(r_0^2 - r^2)$	$T_m - T_0 = \frac{P_H}{4kV}r_0^2 = \frac{P_H}{4\pi kl}$	$T_{av} = \frac{P_H}{2\pi l}\left(\frac{1}{4k} + \frac{1}{\lambda r_0}\right) + T_k$
$T_0 - T_k = \frac{P_H}{2\pi\lambda r_0 l}$	$T_m - T_k = \frac{P_H}{2\pi r_0 l}\left(\frac{r_0}{2k} + \frac{1}{\lambda}\right)$	

slab

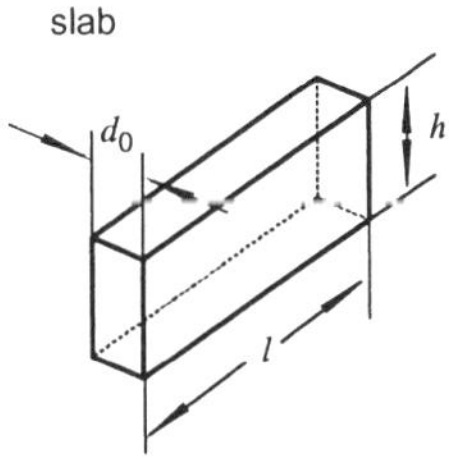

cooling surface $O = 2hl$, volume $V = d_0 hl$

temperature profile	center temperature T_m	average temperature
$T(x) - T_0 = \frac{P_H}{8kV}(d_0^2 - 4x^2)$	$T_m - T_0 = \frac{P_H}{8kV}d_0^2 = \frac{P_H d_0}{8khl}$	$T_{av} = \frac{P_H}{2hl}\left(\frac{d_0}{3k} + \frac{1}{\lambda}\right) + T_k$
$T_0 - T_k = \frac{P_H}{2\lambda hl}$	$T_m - T_k = \frac{P_H}{hl}\left(\frac{d_0}{8k} + \frac{1}{2\lambda}\right)$	

In the stationary case and for a homogeneous heating in rod and slab systems, a parabolic temperature profile arises. These temperature profiles are given in Table 4.2, neglecting the endface cooling. In both cases one has

$$T_m - T_o = \frac{1}{4}\frac{d_0}{O}\frac{1}{k}P_H \tag{4.9}$$

d_0 = material thickness in cooling direction,
T_m = center temperature,
T_0 = surface temperature,

$$T_0 - T_k = \frac{1}{\lambda k} P_H . \tag{4.10}$$

The temperature difference between center and cooling surface is proportional to the heating power P_H, to the total material thickness d_0 in direction of the temperature gradient and inversely proportional to the heat conductivity k and the cooling surface O.

The system-specific efficiency η_H determines the material load and the lens effect, the heat-transfer coefficient λ with η_H the average temperature and thus the laser efficiency. Both quantities have to be determined experimentally [4.5].

Example 4.2 Nd:YAG rod

Pulsed operation, pulse period small compared to the thermal relaxation time. With $P_{in} = 2000$ W, $\eta_H = 5\%$, $l = 5$ cm, $r_0 = 0.2$ cm, $k = 0.1$ W/(cm °C), $\lambda = 0.3$ W/(cm^2 °C) it follows: $P_H = 100$ W, $T_m - T_0 \approx 16$°C, $T_0 - T_k \approx 53$°C.

The linear relation between rod temperature and pump power in the cw-mode can be seen in Fig. 4.2. The lamps have a power offset of about 0.5 kW before they emit sufficient light to heat the laser rod noticeably.

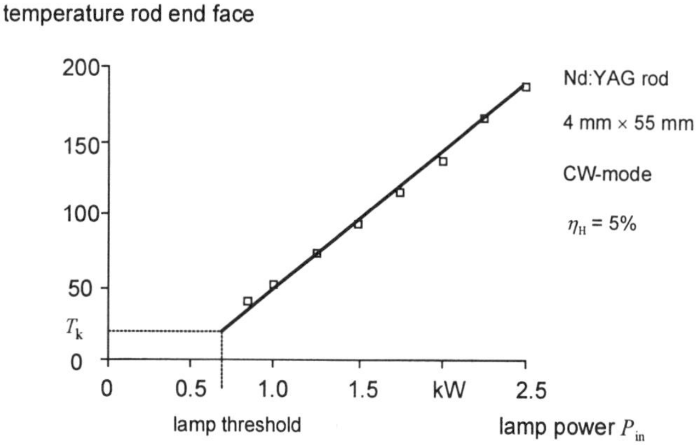

Fig. 4.2. Temperature at the rod surface versus electrical pump power (cw-mode)

4.3 Transient Thermal Effects

The temporal behavior of the temperature in the pulsed operation is mainly determined by two time constants. The first time constant τ characterizes the time in which the temperature relaxation in the laser material takes place. It is only dependent on the material parameters (heat conductivity) and the geometry. The second time constant τ_1 determines the time for the temperature equalization with the coolant. It is determined by the heat transfer coefficient λ.

Temperature Relaxation in the Material

Ensuring a well defined cooling by a high flow rate through the flow tubes, the intrinsic temperature relaxation time can be determined experimentally. The time-dependent temperature can be measured either by direct or indirect methods, by the absorption of He-Ne-laser light in alexandrite [4.5], by the birefringence [4.7] relaxation or the transient thermal lens. This time constant is given in [4.8] as

$$\tau = \frac{c\rho}{k} r_0^2 . \tag{4.11}$$

In Table 4.3 are listed some values for a rod radius of 3 mm and various materials.

Table 4.3 Time constant of temperature relaxation in a rod with 6 mm diameter

Material	$c\rho/k$ s/mm^2	Theory τ/s	Measurement τ/s	Ref.
Nd:glass (silicate)	2.8	25.2		
Nd:glass EV-4	2.4	21.6	5.25	[4.7]
Nd:YAG	0.196	1.76	1.6	[author]
ruby	0.071	0.64		
alexandrite	0.135	1.21		
Nd:GGG	0.412	3.7		

Temperature Relaxation with Coolant

The transient temperature distribution in the laser medium and the time evolution, immediately after the pump pulse [4.5], can be described by

$$T(r,t) - T_\mathrm{k} = (T_0 - T_\mathrm{k}) \sum_1^\infty A_n J_0(\mu_n r / r_0) \mathrm{e}^{-t/\tau_n} , \tag{4.12}$$

$$\tau_n = \frac{c\rho}{k\mu_n^2} r_0^2 \qquad \text{and} \tag{4.13}$$

$$\frac{\mu_n J_1(\mu_n)}{J_0(\mu_n)} = \frac{\lambda r_0}{k} \tag{4.14}$$

with $J_{0,1}$ being the Bessel functions of 0-th and 1-st order respectively. For a parabolic temperature distribution the expansion coefficients are given in [4.9] as

$$A_n = \frac{8J_1(\mu_n)}{\mu_n^3\left[J_0^2(\mu_n)+J_1^2(\mu_n)\right]}.$$

Since the expansion coefficients become very small with increasing n, only $n = 1$ is considered

$$\tau_1 = \frac{c\rho}{k\mu_1^2} r_0^2 \tag{4.15}$$

and

$$\frac{\mu_1 J_1(\mu_1)}{J_0(\mu_0)} = \frac{\lambda r_0}{k}. \tag{4.16}$$

From measurements of τ_1 and with (4.15) μ_1 can be determined and with Table 4.4 (according to [4.9]) the heat-transfer coefficient λ. The relation between the coolant flow rate and the thermal exchange constant is discussed in Sect. 5.3.

Table 4.4 Determination of the heat-transfer constant, according to [4.9]

λ_p/k	0	0.01	0.04	0.1	0.4	1.0	4.0	10	40	100	∞
μ_1	0	0.14	0.28	0.44	0.85	1.25	1.91	2.17	2.35	2.38	2.40

4.4 Thermal Load

Due to the temperature difference $(T_m - T_0)$ a surface tension σ is induced:

$$\sigma = F\frac{\alpha E}{1-\nu}\left(T_m - T_0\right). \tag{4.17}$$

F = geometry factor, F(rod) = 1/2, F(slab) = 1/3 [4.12]
α = expansion coefficient
E = modulus of elasticity (Young's modulus)
ν = Poisson number.

For safety reasons the tension should be smaller than the fracture tension by a factor of at least 5 [4.10]

$$\sigma \le \frac{1}{5}\sigma_{max} \tag{4.18}$$

The maximum temperature difference between the center and the surface of a rod may therefore not be larger than

$$\left(T_{\mathrm{m}}-T_0\right) \leq \frac{(1-\nu)}{5} \frac{\sigma_{\max}}{F \alpha E} . \tag{4.19}$$

For homogeneous heating it follows, with (4.9)

$$\frac{\sigma_{\max}}{5}=\frac{F}{4} \frac{\alpha E}{(1-\nu)} \frac{d_0}{k O} P_{\mathrm{H}}, \tag{4.20}$$

O = cooling surface,
d_0 = material thickness in direction of temperature gradient.

The thermal shock parameter R_{T} is defined as

$$R_{\mathrm{T}}=\frac{k(1-\nu)}{\alpha E} \sigma_{\max} . \tag{4.21}$$

The maximal heating power $P_{\mathrm{H,max}}$ in the laser medium can be expressed as

$$P_{\mathrm{H,max}}=\frac{4}{5 F} \frac{O}{d_0} R_{\mathrm{T}}, \tag{4.22}$$

and the corresponding acceptable pump power $P_{\mathrm{p,max}}$ as

$$P_{\mathrm{p,max}}=\frac{1}{\eta_{\mathrm{H}}} \frac{4}{5 F} \frac{O}{d_0} R_{\mathrm{T}} . \tag{4.23}$$

Material properties and surface processing methods determine the thermal shock parameter $R_{\mathrm{T}} = \sigma_{max}\, k(1-\nu)/\alpha E = \sigma_{\max} M$. The material parameter R_T is dependent only on material properties. The fracture tension $\sigma_{\max}$ depends also strongly on the surface finish.

The maximal average laser power $P_{\max}$ for pulsed operation, limited by the safe thermal load, can then be estimated, with (4.2), (4.3) and with (1.152) or (1.149), as

$$P_{\max}=\frac{4}{5 F} \frac{O}{d_0} \frac{\eta_{\mathrm{extr}}}{\eta_{\mathrm{H}}} R_{\mathrm{T}} . \tag{4.24}$$

for YAG: $\eta_{\mathrm{extr}} \approx 0.5$, [4.11],
$F(\mathrm{rod}) = 1/2$, $F(\mathrm{slab}) = 1/3$, [4.12],
$\eta_{\mathrm{H}} = \kappa \eta_{\mathrm{excit}}$,

If measurements of the slope-efficiency of a laser system are available, the maximal attainable laser power $P_{\max}$ can also be estimated as

$$P_{\max}=\frac{4}{5 F} \frac{O}{d_0} \frac{\eta_{\mathrm{slope}}}{\eta_{\mathrm{H}}} R_{\mathrm{T}}, \qquad \text{with } P_{\max} \approx \eta_{\mathrm{slope}} P_{\mathrm{p}} . \tag{4.25}$$

High average laser power without fracture can only be obtained by a large cooling surface O or by a short cooling distance d_0. However, the statistical risk increases with the dimension of the surfaces. For a large surface a fracture induced by surface defects occurs already at a smaller load. For a large surface the acceptable load should be lowered. The maximal attainable laser power therefore does not increase linearly with the cooling surface. The allowed load by flash lamp pumping is considerably lower than the comparable pure thermal load, according to [4.10]. This is taken into account by the empirical safety factor of 5.

The slope efficiency and the heat efficiency are correlated. An improvement in the pump setup will increase both unless a spectral selection does not improve the ratio. If the pumping direction is parallel to the direction of cooling (in the direction of d_0), one has a minimal d_0 because of the necessary absorption depth. High laser powers can then only be obtained with a large surface O or with other materials (R_T).

Table 4.5 Laser power and performance of laser rods

		Silicate-glass	Phosphate-glass	YAG	Alex.	GGG/GSGG	YLF	Ruby	BEL
α	10^{-6}/°C	10	12	8.2	6.8	9.5	13	5.8	8
E	GPa	90	50	317	445	210	76	352	100
ν		0.24	0.26	0.25	0.25	0.28	0.33	0.25	0.25
k	W/m °C	1.35	0.7	13	23	7	6	42	5
σ_{max}	MPa	170	90	280	280	120	50	550	100
η_H	%	5	5	5	7	5	5	5	5
η_{slope}	%	3	5	4.5	2.5	7	3	0.8	2.75
R_T	kW/m	0.19	0.08	1.05	1.6	0.30	0.20	8.49	0.47
T_m-T_0	°C	162	126	91	79	49	38	229	106
P_{in}/l	W/cm	195	78	1056	1605	305	205	8531	471
P_{max}/l	W/cm	5.8	3.9	47.5	40.1	21.3	6.1	68.3	13.0
l_{max}	cm	50	50	20	15	20	10	15	15
P_{max}	W	292	195	950	602	427	61	1024	194

ref. in Chaps. 12 and 13.

For rods, (4.25) reads

$$\frac{P_{max}}{l} = \frac{8\pi}{5}\frac{\eta_{slope}}{\eta_H} R_T \qquad \text{with} \qquad P_{max} \approx \eta_{slope} P_p. \tag{4.26}$$

The permitted laser power can only be increased by longer rods. An overview of the attainable laser power P_{max} is given in Table 4.5 based on rod lengths l_{max} that are commercially available at the moment.

The maximal attainable laser power P_{max}/l can be increased by surface treatment. [4.13] reports an increase of the fracture limit by almost an order of magnitude due to polishing and a special etching process for glass, GSGG and YAG.

4.5 Influence on the Level Population

The average laser medium temperature influences the term population. Depending on the operation mode there is an optimal or maximal operating temperature. The laser efficiency can increase initially with increasing temperature for alexandrite up to approximately 130°C in the pulsed operation (Fig. 4.3) – or decrease as for Nd:YAG and Nd:GGG (Fig. 4.4).

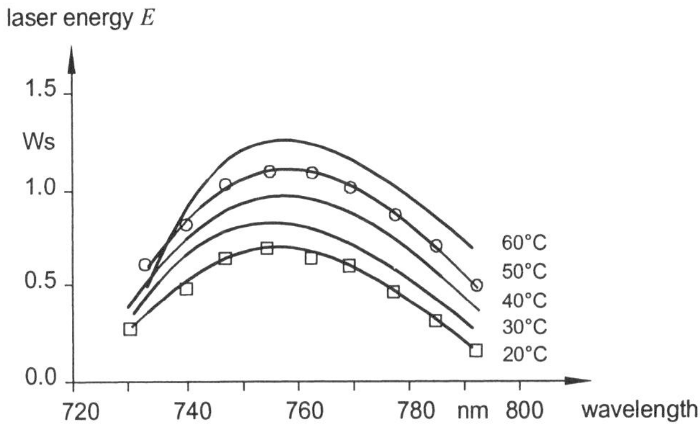

Fig. 4.3. Efficiency versus temperature for alexandrite at different temperatures, [4.14]

In laser glass the amplification is also reduced with increasing temperature due to the increasing population of the lower laser energy level. This reduction depends on the average temperature in the medium and on the ion concentration. The population of the lower level in YAG and the resulting change in efficiency is discussed in Sect. 1.5 (Fig. 1.8).

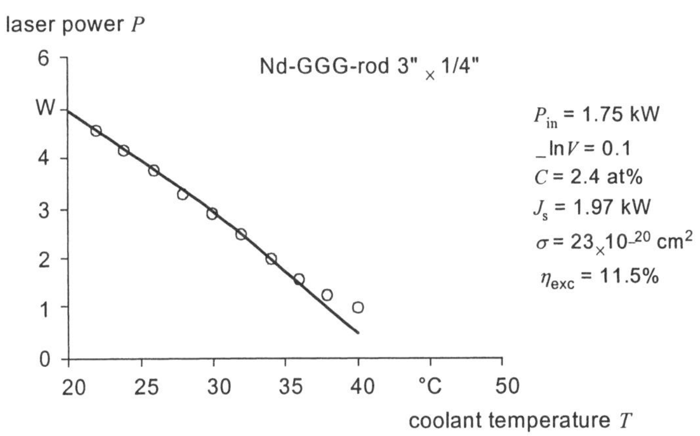

Fig. 4.4. Temperature dependence of the laser output power of GGG

4.6 Thermal Lensing

The temperature difference ($T_m - T_0$) between center and surface of the laser medium causes a lens effect. Three effects contribute to the refracting power:

- the endface deformation by longitudinal expansion (α),
- the radial temperature gradient of the refractive index (dn/dT) and
- the photoelastic effect ($C_{r,\phi}$).

α, dn/dT and $C_{r,\phi}$ are the characterising quantities.

Surface Deformation of a Rod

If a rod can expand freely the endface deformation can be estimated as $\Delta l = \alpha l(T_m - T_0)$. According to Koechner [4.8] the restricted expansion value is considerably smaller. For a parabolic temperature profile l has roughly to be replaced by the rod radius r_0

$$\Delta l = \alpha r_0 (T_m - T_0) = \frac{\alpha P_H}{4\pi k} \frac{r_0}{l} . \qquad \text{(Fig. 4.5)} \tag{4.27}$$

The curvature radius R is approximately

$$R = \frac{r_0^2 + \Delta l^2 / 4}{\Delta l} \approx \frac{r_0^2}{\Delta l} \tag{4.28}$$

and the curvature $1/R$ becomes

$$\frac{1}{R} = \frac{\alpha r_0 P_H}{4kV} \tag{4.29}$$

$V = \pi r_o^2 l$ = rod volume.

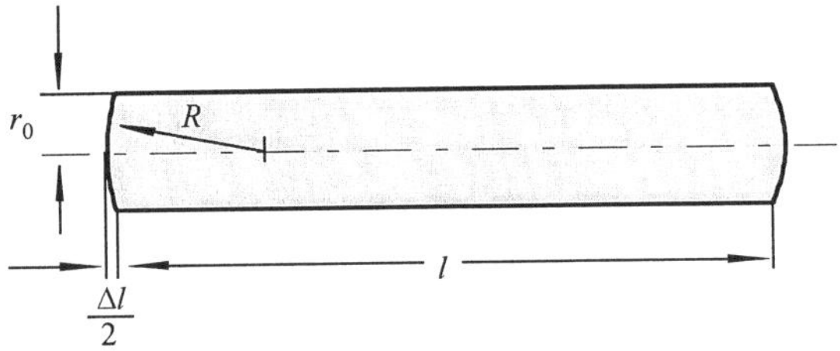

Fig. 4.5. Endface radius caused by thermal expansion

The rod thus becomes a thick lens. The actual longitudinal deformation is very small and can be neglected as long as $l >> r_0$.

Refractive Index Gradient

Due to the temperature distribution, the refractive index of the active medium becomes space-dependent. Neglecting the z dependence, one has:

$$n(r) - n_{\mathrm{m}} = \frac{\mathrm{d}n}{\mathrm{d}T}\left(T(r) - T_{\mathrm{m}}\right), \tag{4.30}$$

n_{m} = refractive index in the center,
$\mathrm{d}n/\mathrm{d}T$ = temperature coefficient of the refractive index.

If a homogeneous heat distribution is assumed, a parabolic refractive index profile arises in a rod:

$$n(r) = n - \frac{\mathrm{d}n}{\mathrm{d}T}\frac{P_{\mathrm{H}}}{4kV}r^2 . \tag{4.31}$$

Photoelastic Effect

Mechanical stress in tangential ϕ and in radial r directions also induces a change in the refractive index. Since the stress is quadratically dependent on the radial coordinate r for homogeneous heating, the photoelastic part of the refractive index acquires the same dependence [4.8],

$$n_r = \frac{n^3 \alpha P_{\mathrm{H}} C_r}{2kV} r^2 , \tag{4.32}$$

$$n_\phi = \frac{n^3 \alpha P_{\mathrm{H}} C_\phi}{2kV} r^2 . \tag{4.33}$$

Differences in n_r and n_ϕ cause birefringence. The tension coefficients C_r and C_ϕ are calculated for YAG with the photoelastic coefficients p_{ij} as follows [4.14]:

$$C_r = \frac{(17\nu - 7)p_{11} + (31\nu - 17)p_{12} + 8(\nu + 1)p_{44}}{48(\nu - 1)} , \tag{4.34}$$

$$C_\phi = \frac{(10\nu - 6)p_{11} + 2(11\nu - 5)p_{12}}{48(\nu - 1)} . \tag{4.35}$$

The optical constants for various laser materials are listed in Table 4.6. The corresponding references are given in Chaps. 10 and 11.

Table 4.6 Poisson number ν and photoelastic coefficients

	YAG	Glass	GGG	GSGG
ν	0.2500	0.2500	0.28	0.28
p_{11}	−0.0290	0.1190	−0.086	−0.012
p_{12}	0.0091	0.2610	−0.027	0.019
p_{44}	−0.0615	−0.0191	−0.078	−0.0665
C_r	0.0172	0.0814	0.011	0.0235
C_ϕ	−0.0025	0.0663	−0.0164	0.0015

In isotropic media $p_{44} = (p_{11} - p_{12})/2$

Refractive Index Change

The two essential changes in the refractive index add up to

$$n_{r,\phi}(r) = n - \left(\frac{\mathrm{d}n}{\mathrm{d}T}\frac{P_\mathrm{H}}{4kV} + \frac{n^3\alpha P_\mathrm{H} C_{r,\phi}}{2kV}\right) r^2 = n\left(1 - \frac{r^2}{2\beta_{r,\phi}^2}\right), \tag{4.36}$$

$\beta_{r,\phi}$ is the characteristic focal length of the thermal lens. The material and geometrical constants can be combined as

$$\frac{1}{\beta_{r,\phi}^2} = \frac{P_\mathrm{H}}{kV}\left(\frac{1}{2n}\frac{\mathrm{d}n}{\mathrm{d}T} + n^2\alpha C_{r,\phi}\right) = \frac{P_\mathrm{H} M_{r,\phi}}{V} \tag{4.37}$$

and the material parameter $M_{r,\phi}$ is given by

$$M_{r,\phi} = \frac{1}{k}\left(\frac{1}{2n}\frac{\mathrm{d}n}{\mathrm{d}T} + n^2\alpha C_{r,\phi}\right). \tag{4.38}$$

The term due to the thermal refractive index change normally dominates and to a good approximation one has

$$\frac{1}{\beta^2} = \frac{1}{2nk}\frac{\mathrm{d}n}{\mathrm{d}T}\frac{P_\mathrm{H}}{V}. \tag{4.39}$$

Example 4.3 Nd-YAG rod

The material constants $\mathrm{d}n/\mathrm{d}T = 7.3\times10^{-6}/°\mathrm{C}$, $n = 1.83$, $\alpha = 6\times10^{-6}/°\mathrm{C}$, $C_r = 0.017$, and $C_\phi = -0.0025$ results in

$\frac{1}{2n}\frac{\mathrm{d}n}{\mathrm{d}T} = \frac{2\times10^{-6}}{°\mathrm{C}}$ and $n^2\alpha C_r \approx 0.34\times10^{-6}/°\mathrm{C}$, $n^2\alpha C_\phi \approx -0.05\times10^{-6}/°\mathrm{C}$

With $P_\mathrm{H} = 500$ W, $r_0 = 5$ mm and $l = 150$ mm follows $\beta = 0.37$ m.

4.7 Lens Effects in the Resonator

The thermal lens of a laser rod influences the beam quality of the resonator. Some special cases will be considered in more detail.

Rod with external mirrors

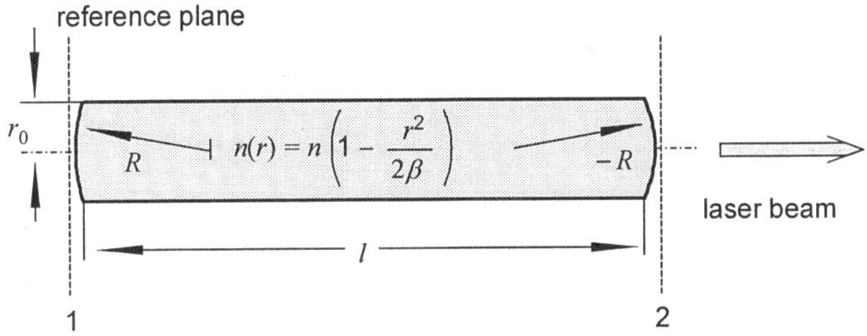

Fig. 4.6. Resonator with external plane mirrors, reference planes outside the rod, and a curvature R caused by thermal distortion

The equivalent matrix for a transition (Fig. 4.6) from 1 to 2 consists of a transition 1 with the curvature R, the propagation in the rod and the transition 2 with the curvature $-R$:

$$M_{12} = \begin{vmatrix} a & c \\ b & d \end{vmatrix} = \begin{vmatrix} 1 & 0 \\ \frac{n-1}{-R} & n \end{vmatrix} \begin{vmatrix} \cos(l/\beta) & \beta\sin(l/\beta) \\ \frac{-1}{\beta}\sin(l/\beta) & \cos(l/\beta) \end{vmatrix} \begin{vmatrix} 1 & 0 \\ \frac{1-n}{nR} & \frac{1}{n} \end{vmatrix} .$$

The matrix then elements are:

$$a = d = \cos(l/\beta) + \frac{1-n}{nR}\beta\sin(l/\beta), \qquad b = \frac{\beta}{n}\sin(l/\beta),$$

$$c = \frac{2(1-n)}{R}\cos(l/\beta) + \frac{(1-n)^2}{nR^2} - \frac{n}{\beta}\sin(l/\beta),$$

and for $l << \beta$ in good approximation:

$$a = d = 1 - \frac{n-1}{nR}l, \qquad b = \frac{l}{n}, \qquad c = -\frac{2(n-1)}{R} - \frac{nl}{\beta^2} + \frac{(1-n)^2}{nR^2}.$$

Rod with Fixed Mirrors

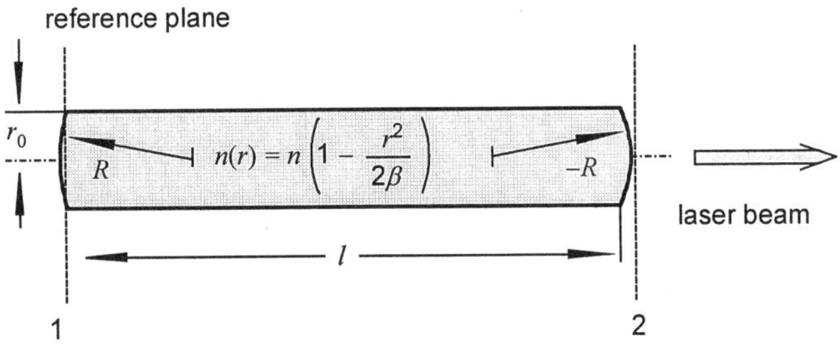

Fig. 4.7. Rod resonator with one fixed mirror (coated endface) and one plane external mirror

The equivalent matrix of the rod resonator in Fig. 4.7 is similar to the previous one, except for the first transition.

$$M_{12} = \begin{vmatrix} a & c \\ b & d \end{vmatrix} = \begin{vmatrix} 1 & 0 \\ \frac{n-1}{-R} & n \end{vmatrix} \begin{vmatrix} \cos(l/\beta) & \beta\sin(l/\beta) \\ \frac{-1}{\beta}\sin(l/\beta) & \cos(l/\beta) \end{vmatrix} \begin{vmatrix} 1 & 0 \\ \frac{-1}{R} & 1 \end{vmatrix}.$$

The matrix elements are:

$$a = \cos(l/\beta) + \frac{\beta}{R}\sin(l/\beta), \qquad b = \beta\sin(l/\beta),$$

$$c = \frac{2(1-n)}{R}\cos(l/\beta) - \left(\frac{n}{\beta} - \frac{n-1}{R}\right)\sin(l/\beta),$$

$$d = n\cos(l/\beta) - \frac{(n-1)\beta}{R}\sin(l/\beta),$$

and for $l \ll \beta$ in good approximation:

$$a = 1 - l/R, \qquad b = l,$$

$$c = -\frac{2(n-1)}{R} - \frac{nl}{\beta^2} + \frac{(n-1)}{R^2}l, \qquad d = n - \frac{(n-1)}{R}l.$$

Determination of the Refracting Power

In both cases, with the approximation $l \ll \beta$ the last term in the matrix element c can be neglected and for rods with flat endfaces the refracting power $D = -c$ follows with (4.36) as

$$D_{r,\phi} \approx \frac{nl}{\beta_{r,\phi}^2} + \frac{2(n-1)}{R}. \tag{4.40}$$

The refracting power due to the curved endfaces is, in general, small compared to the other two effects and (4.40) simplifies to

$$D_{r,\phi} \approx \frac{nl}{\beta_{r,\phi}^2} = \frac{P_\mathrm{H}}{\pi r_0^2} M_{r,\phi} \tag{4.41}$$

with

$$M_{r,\phi} = \frac{1}{2k}\left(\frac{\mathrm{d}n}{\mathrm{d}T} + 2n^3 \alpha C_{r,\phi}\right). \tag{4.42}$$

The refracting power is proportional to the heating power P_H and inversely proportional to the cross section of the rod. In some laser glasses the thermal refractive index change compensates the photoelastic effect, i.e. for these glasses the first order approximation is

$$\frac{\mathrm{d}n}{\mathrm{d}T} = -2n^3 \alpha C_{r,\phi} \qquad \text{and} \qquad D_{r,\phi} \approx 0.$$

Table 4.7 Material parameter for the refracting power and thermal lens coefficient

		LG-680 SCHOTT	LG-706 SCHOTT	LG-750 SCHOTT	LG-760 SCHOTT	YAG	GGG
k	W/m°C	1.35	0.69	0.52	0.60	13.0	6.43
$\mathrm{d}n/\mathrm{d}T$	10^{-6}/°C	1.46	−4.55	−6.5	−8.19	5.66	17.4
n		1.56	1.504	1.516	1.508	1.82	1.943
α	10^{-6}/°C	9.3	10.9	11.4	12.5	6.9	5.67
C_r		0.067	0.085	0.081	0.074	0.017	0.011
C_ϕ		0.051	0.073	0.074	0.064	−0.0025	−0.164
$2n^3\alpha C_r$	10^{-6}/°C	4.73	6.30	6.43	6.38	1.41	0.92
$2n_3\alpha C_\phi$	10^{-6}/°C	3.60	5.41	5.88	5.49	−0.21	−13.64
$\alpha(n-1)$	10^{-6}/°C	0.17	0.18	0.20	0.21	0.19	0.18
M_r	10^{-6}m/W	2.36	1.40	0.13	−1.33	0.35	1.44
M_ϕ	10^{-6}m/W	1.94	0.76	−0.41	−2.08	0.27	0.31

The material parameters for the refracting power are given in Table 4.7. The thermal behavior can be deduced from the material parameter $M_{r,\phi}$. For a 3 mm rod radius and a pump powers of 1 kW the corresponding refracting powers D_r and D_ϕ are given in Table 4.8.

Under the assumption of homogeneous heating condition (4.41) is valid for many laser materials in rod form with conventional stable resonators in thermal equilibrium. For the beam parameter product $w\theta$ of rods in the multimode operation it was shown in (3.57) that for a conventional stable resonators with external mirrors one has

$$\mathrm{Max}(w\theta) \geq r_0^2(D_i - D_j)/4 \tag{4.43}$$

With (4.41) it follows that

$$\mathrm{Max}(w\theta) \geq \frac{M_{r,\phi}(P_{\mathrm{H}i} - P_{\mathrm{H}j})}{4\pi}. \tag{4.44}$$

To a first approximation the beam quality is not influenced by the choice of the rod geometry. As pumping power P_{in} and laser power P are proportional to each other a compromise between beam parameter product ($w\theta$) and operation range $\Delta P = (P_i - P_j)$ has to be accepted. For an operation range of the laser from 0 to P_{max} the beam quality is limited by the ratio of excitation efficiency to heat efficiency,

$$\mathrm{Max}(w\theta) \geq \frac{2}{5\sqrt{\pi}} \frac{\eta_{\mathrm{slope}}}{\eta_{\mathrm{H}}} MR_{\mathrm{T}}. \tag{4.45}$$

Table 4.8 Thermal lensing coefficient of various materials

		LG-680	LG-706	LG-750	LG-760	YAG	GGG
η_{H}	%	5	5	5	5	5	5
P_{in}	kW	1	1	1	1	1	1
r_0	mm	3	3	3	3	3	3
$D_r(P_{\mathrm{in}},r)$	m^{-1}	4.17	2.48	0.22	−2.36	0.62	2.54
$D_\phi(P_{\mathrm{in}},r)$	m^{-1}	3.43	1.34	−0.72	−3.67	0.48	0.54

The heat efficiency η_{H} and the refracting powers D_r and D_ϕ are dependent on the cavity and cxcitation.

Transient Lensing Effect

Up to now only the average refracting power of the thermal lens in the stationary mode or in the pulsed operation at a sufficiently high repetition rate ($f >> 1/\tau$) has been considered. At low pulse repetition rate the temporal behavior of the refracting power during a single pulse is of interest.

The transient behavior of the thermal lens is described in [4.16] for example. During the pulse the lens takes on both negative and positive focal lengths (Fig. 4.8). In addition, the lens is dependent on the pumping light distribution in the laser medium.

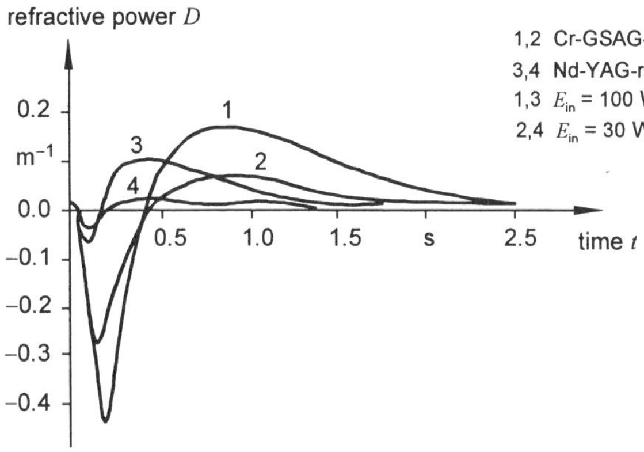

Fig. 4.8. Transient refracting power [4.16]

4.8 Experimental Determination of the Refracting Power

For resonator calculations the average refracting power, dependent on the operating and material parameters, must be known.

An experimental setup, as in Fig. 4.9, allows a simple determination of the average refracting power D_d for YAG or glass rods.

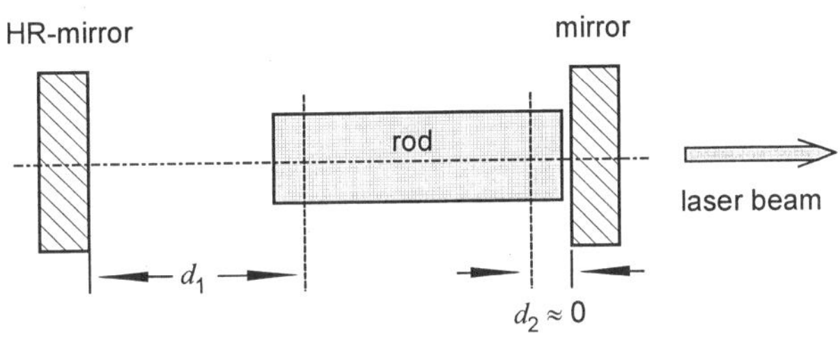

Fig. 4.9. Resonator to determine the critical refracting power

For a plane-plane resonator the critical refracting power is $D_d = 1/d_1$. The average laser power as a function of the pumping power with the distance d_1 as parameter is measured.

When the critical pumping power is achieved, the laser power decreases since the resonator becomes unstable. For different distances d_1 the resonator becomes unstable at different pump powers P_{crit}. From the gradient of $1/d_1$ as a function of P_{crit} the thermal lens coefficient can be extracted

$$D_d = \eta_H \frac{M_{r,\phi} P_{crit}}{\pi_0^2} . \quad (4.46)$$

One defines $\frac{D_\mathrm{d}}{P_\mathrm{crit}} = \frac{M_{r,\phi}\eta_\mathrm{H}}{\pi_0^2}$ as the *lens coefficient.* (4.47)

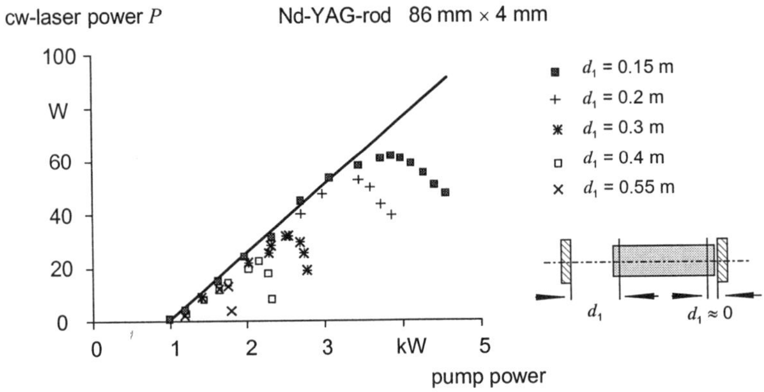

Fig. 4.10. Determination of the average refracting power for Nd:YAG rods

In Fig. 4.10 the laser power has been measured as a function of the pumping power for different reflecting distances d_1. Deviations from the straight line indicate the beginning of the unstable region for the r or ϕ component of the refracting power.

Another method [4.2] (Fig. 4.11) is the determination of the focal length of a He-Ne test laser or Moiré reflectometry [4.17]. These methods allow the investigation of the transient behavior.

Table 4.9 gives an overview of the refracting powers determined experimentally by various methods.

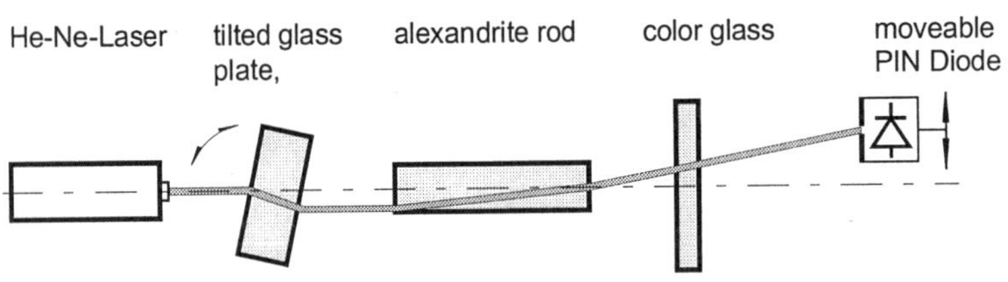

Fig. 4.11. Experimental setup to determine the refracting power [4.2]

The previous results indicate the power-dependent average value for the refracting power. According to (4.41) the refracting power for the two components r and ϕ can differ. These differences have been estimated in [4.20] for YAG and GSGG to be

$$D_r/D_\phi = 1.13 \text{ for Nd:YAG}$$

and

$$D_r/D_\phi = 1.22 \text{ for Cr:Nd:GSGG.}$$

The relative difference in the refracting powers of GSGG and YAG are:

$D_{GSGG}/D_{YAG} \approx 10$ [4.21]

and

$D_{GSGG}/D_{YAG} \approx 5.9$ [4.20].

The ratio of the refracting power components in r and ϕ directions also depends on the pump light distribution. As in many crystals the heat conductivity is temperature dependent, the refracting power too will depend on the crystal temperature, which is due to the cooling-water temperature and the water flow rate. Values of the refracting power per kW electrical pump power D/P_{elect} are compiled in Table 4.9.

Table 4.9 Experimentally measured refracting powers for various materials

Material	Rod $2r_0 \times l$ / (mm×mm)	Slope η_{slope}/%	Refract. power D/P_{elect} / m^{-1}/kW	Ref.
cw mode				
YAG	4 × 85	2.7	0.73	author
YAG	5 × 80	3.6	0.45	[4.18]
YAG	1/4 " × 4 "	2.8	0.38	author
pulse mode				
APG-1	6 × 180	2.5	5	author
YAG	1/4 " × 4 "	3.24	0.5	author
YAG	3/8 " × 6 "	4.0	0.4	author
GGG	1/4 " × 3 "	4.5	2.16	author
YAP	1/4 " × 3 "	2.7	0.43	author
YAG	10 × 76	-	0.17	[4.19]
YLF	10 × 76	-	0.02	[4.19]
YLF	10 × 76	-	0.01	[4.19]
GSAG	5 × 56	0.38	1.5	[4.16]
alexandrite	1/4 " × 4 "	0.5	1.1×10^{-5} [a]	author
alexandrite	8 × 100		1.8×10^{-5} [b]	[4.2]

[a] m^{-1}/Hz $(W\ s)^{3/2}$ for $E = 328$ W s, $f = 10.5$ Hz.

[b] m^{-1}/Hz $(W\ s)^{3/2}$, average value for both crystals axes a and b. The c axis lies in the rod axis

4.9 Thermally Invariant Resonators

The lens effect on the laser beam during operation can be avoided, in principle, if the temperature gradient is parallel to the beam propagation. Solutions of this kind are "disk amplifiers" or "active mirrors" [4.22] (Fig. 4.12).

A disadvantage of these solutions is that the laser beam has to traverse the cooling medium. This leads to disturbances and absorption by water or gas.

Combining several active mirrors leads to the "multislab laser" [4.24], i.e. a "slab" [4.23] with multiple internal reflections (Fig. 4.13).

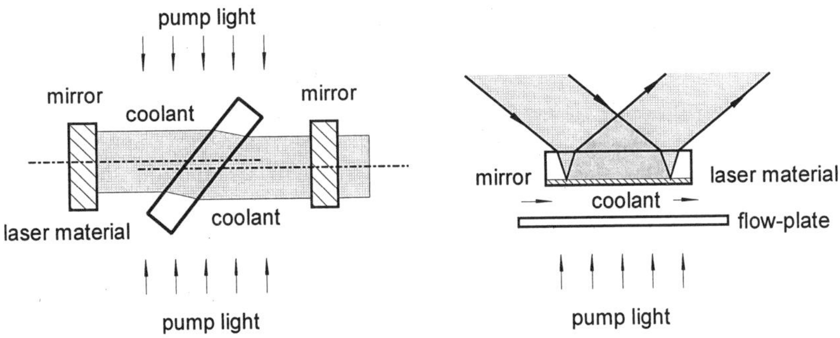

Fig. 4.12. Thermally invariant constructions

The slab is thermally isolated in the y direction, and therefore the temperature remains constant in the y direction. The external cooling is in the x direction, creating a temperature gradient. Due to the zigzag path in the z direction these influences average out to a first order and the laser beam is not disturbed by thermal effects, except for those that are caused by the temperature distribution at the slab tips. Since the slab is available on the market as a technical solution it will be discussed in some detail in the following chapter.

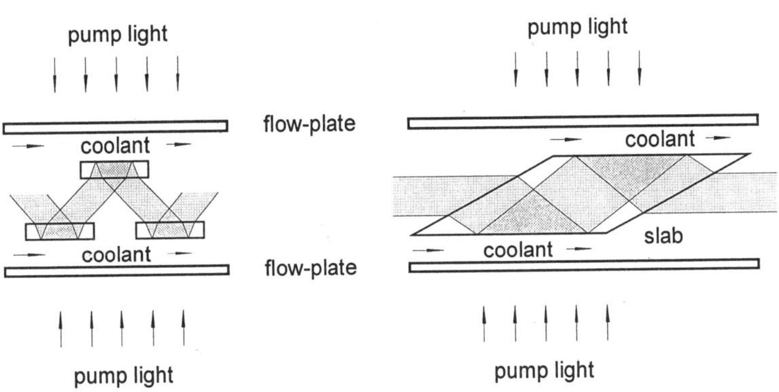

Fig. 4.13. Disk resonator and slab

Transparent cooling media with good heat conductivity are suggested in [4.25], where sapphire plates are alternated with the active laser material (Fig. 4.14). The heat flow occurs mainly in the beam direction.

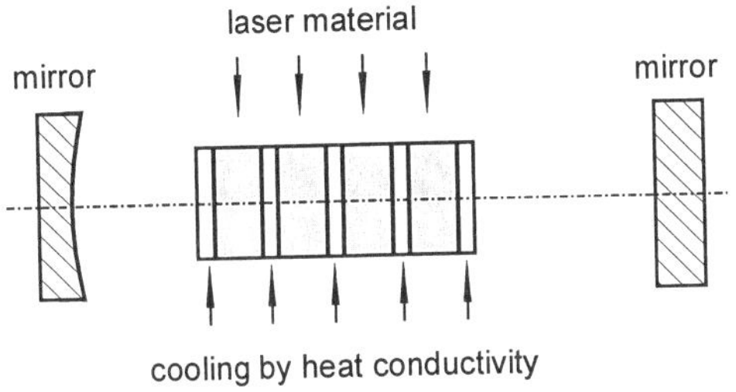

Fig. 4.14. Contact-cooled laser

Part II

Technical Realization

5 Beam Source

The principle setup of a pulsed solid state laser for material processing is shown in the following block diagram (Fig. 5.1). The setups for cw, pulsed or Q-switched operations are similar.

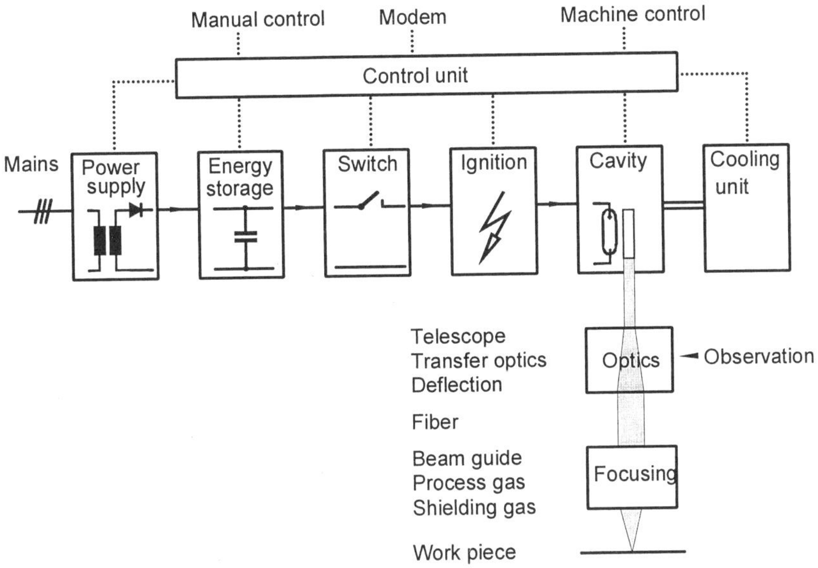

Fig. 5.1. Setup of a pulsed solid-state laser

After voltage transformation the energy from the mains is at first stored in a capacitor bank. In this way the high peak power can be obtained in pulsed operation. The modulator, a fast semiconductor switch, regulates the laser beam intensity and pulse-train duration by modulation of the flash lamp current. The flash lamp converts the electrical energy into pump light for the laser medium. The total laser efficiency is about 10%. The heat produced in the beam source must therefore be dissipated by a cooling unit.

The laser beam is guided by suitable optics or beam control systems to the processing position and there, generally, focused on to the workpiece surface.

In industrial use the laser is considered as a machine tool. Its handling should be without any difficulties and disturbance during production. This results in the following requirements for the laser beam source:

- *Compact and small geometry:*
  It should be possible to integrate the laser into the processing machine.
- *Orientation independence:*
  The beam source is moved during the processing in many applications.
- *Constant and reproducibly adjustable beam parameters and also programmable parameters:*
- *Service- and maintenance-friendly:*
  The necessary lamp change must be simple to carry out and should not lead to any long machine service period.
- *Construction as a component:*
  This permits to a large extent the reusability as a tool in the case of machine conversions.
- *High process efficiency.*

All of these requirements are not independent from each other and must be taken into account during beam source engineering, according to their weight and feasibility.

Various requirements on the laser medium, for example large cooling surface, high efficiency, best beam quality, etc., have caused very different special technical solutions. To reach high laser power an annular laser rod is suggested in [5.1]. High beam quality can be achieved with slab lasers or by combining active mirrors [5.2]. The limited lifetime of lamps leads to solutions with electrodeless lamps [5.3] at high average power and makes excitation with laser diodes or even with light-emitting diodes (LED) at low powers meaningful.

The specific problems will be discussed further by looking at the most frequently used setups with arc or flash lamps and a laser medium in the rod geometry.

5.1 Pump Cavity for Rods

In the cavity the pump light of the excitation source is transferred to the laser medium and more or less absorbed. To obtain a high efficiency, the light of the source has to reach the laser medium nearly loss-free. The geometric losses as well as reflection losses should be kept low, and the pump light should be adapted to the desired intensity distribution. Also, the pump light spectrum should coincide with the excitation spectrum and simultaneously minimize the quantum defect.

For pumping laser rods of high or medium output powers a double ellipse cavity with two excitation lamps is generally used, as outlined in Fig. 5.2. In contrast to a single ellipse the efficiency is reduced, but the load per lamp is lower and therefore two lamps have a significantly longer lifetime.

The reflector surface is gold-plated in most cases. Silver, however, has a better reflection coefficient but oxidizes in industrial atmospheres and its protection is expensive. Special ceramics with glazing and in particular coated glasses are also used. The latter solutions offer the advantage that in the case of a lamp break with non-insulated electrodes a short circuit via the metallic reflector is prevented.

All optical parts which are exposed to the light, in particular the flow-tubes should be antireflective coated for the dominating pump bands, and reflective for the other parts of the lamp spectrum.

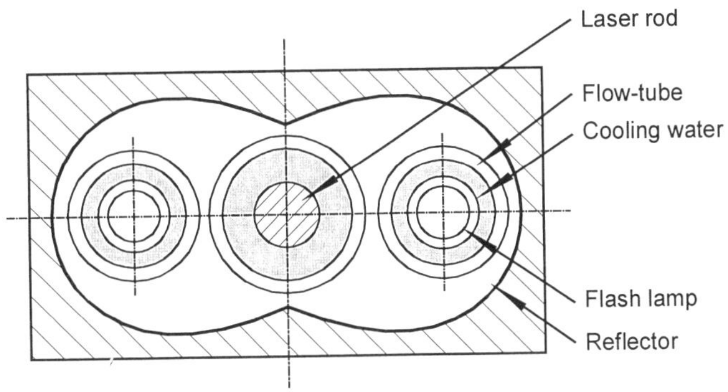

Fig. 5.2. Sectional view of a pump cavity with a double ellipse, [5.4]

The pump light should have a short path through the cooling water. This is in contradiction to the necessity to cool the reflector surface. The reflector then has to be cooled at the backside and thereby can become bent if strongly heated.

At small average power the quartz flow-tube is doped with cerium or titanium to convert useless spectral parts into useful pump light. For systems with high average power good cooling must be ensured. Here, the UV absorption in the quartz tube has to be avoided because it would heat up the quartz wall even more. To prevent the generation of color centers by UV-light, this light must be filtered outside the lamp, e.g. by the flow-tubes.

Special application problems normally set some boundary conditions such as laser power and operating period. The desired operating period determines the lamp size and the laser power the rod length. Thereby some conditions concerning the dimensioning of the pump reflector follow [5.5].

Pump Reflector

- Rod and lamp should be assembled as close as possible to each other.
- The reflector form should have the shortest distance possible to rod and lamp.
- The reflector form must be designed such that the desired illumination of the rod is achieved.
- The maximum efficiency can be reached with a single ellipse, the efficiency decreases somewhat for multiple ellipses [5.6].
- The rod should be chosen with a larger diameter than the lamp.
- For fixed diameter longer rods are more efficient [5.7].

Various reflector forms are possible. At small average power single ellipses are used because they offer the best efficiency and the smallest expenditure. The inhomogeneous illumination of the rod leads to a one-sided load and the beam may shift with increasing load (banana effect). With double ellipses the load is distributed over two lamps, the operating time is greater and a more homogeneous illumination of the rod can be achieved. Brass forms electrolytically gold-plated or special glazed ceramic reflectors are the most common ones. Ray-tracing programs can be helpful to optimize the system [5.8].

Rod Holders

Figure 5.3 shows a typical construction of a rod holder, according to [5.9].

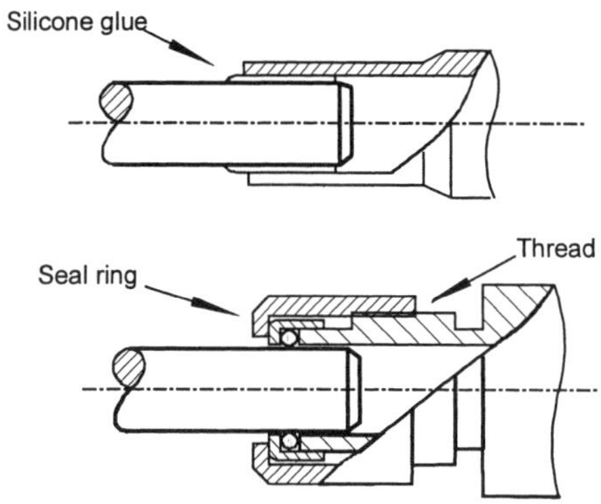

Fig. 5.3. Rod-holder construction, [5.9]

The requirements on the active medium mount in the case of the rod are:

- Sealing of the rod endfaces against the cooling water pressure.
- "Soft" guiding of the rod so that no edge damage appears while handling the rod.
- Resistance of the of seal material against pump and laser radiation. If this is not possible, the gasket must be protected against the radiation.

Rod Dimension

Basically the desired average laser power determines the rod length. For mechanical reasons the ratio of diameter to length is restricted. As a certain thickness is necessary for efficient absorption, the attainable average laser power will depend on the rod volume. In Fig. 5.4 the obtained laser power is plotted versus the rod volume.

The diameter depends on the desired time constants and the doping. For a given pump light absorption the rod can be thinner the higher the doping.

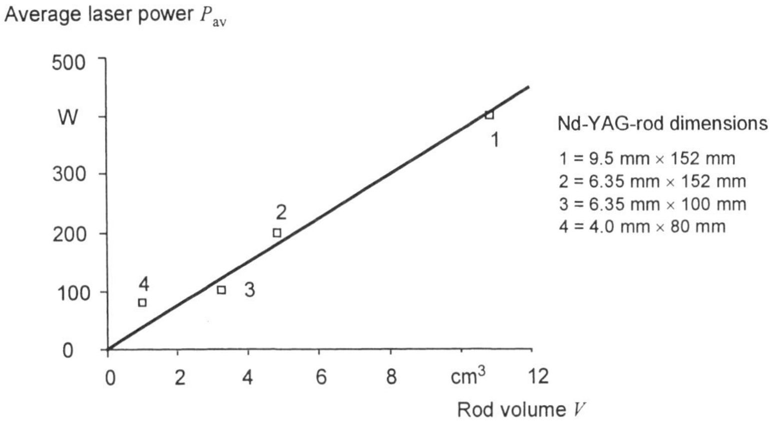

Fig. 5.4. Average laser power as a function of the rod volume

Pumping Light Distribution

The extraction efficiency of a pump light cavity is determined by the adaptation of the inversion density to the laser light distribution in the resonator. The desired inversion profile can be achieved by constructive design (geometry), by the pump light distribution, by the choice of the material thickness or by the doping concentration.

For a given radiation profile $J(r)$, e.g. Gaussian mode, in the resonator, for the case of cw or pulsed operation with long pulses ($T >> \tau$), the optimized pump light profile $J_p(r)$ is given by (1.47)

$$J_p(r) = \frac{1}{\eta_Q} \frac{\lambda}{\lambda_P} \frac{\ln(V\sqrt{R})}{\ln(\sqrt{R})} \left(J(r) - \ln(\sqrt{R}) J_s \right), \tag{5.1}$$

λ = laser wavelength
λ_P = pump wavelength
J_s = saturation intensity.

The threshold has been assumed as constant over the cross section. The pump light must have a constant pedestal J_s and correspond to the desired radiation profile $J(r)$. In this case the threshold inversion N_{th} is established everywhere in the active medium for stationary operation. The z dependence was neglected.

The optimal doping concentration N_0 is subject to several boundaries conditions:

- on the one hand, it should be high to absorb the pump light completly,
- on the other hand, the desired pump light profile is not only influenced by the cavity but also by the doping.

- in addition, the lifetime of the upper laser energy level is reduced with increasing concentration (concentration quenching) [5.10] and this has to be taken into account by J_s in (5.1),

$$\tau = \frac{\tau_0}{1+(C_0 / C_{0.5})^2}, \tag{5.2}$$

τ_0 = lifetime at $N_0 = 0$,
$C_{0.5}$ = concentration, at which the lifetime is half τ_0.

The optimal doping is dependent on the geometry (rod, slab), the dimensions of the active medium and the effective cross section for the pump light absorption. Also, the mode of operation and the mode structure (monomode or multimode) can play a role. For example, the quenching can be neglected if the pulse length is short with respect to the lifetime τ. These relations are very complex and analytical formulae do not yet exist.

For Nd-YAG the following rules can be employed to reach 00-mode:

- use lamps as thin as possible and image them well onto the rod,
- choose rod radius small and
- choose a high doping concentration.

Absorption Coefficient

Table 5.1 Pump light absorption by laser LG-680 glass, [5.11]

Wavelength λ/nm	Absorption cross section $\sigma_p/10^{-20}$ cm^2	Absorption $\alpha = N_0\sigma_p$/cm^{-1}	Penetration depth $1/\alpha$ /mm
514	0.19	0.537	18.6
534	0.401.125	8.8	
572	2.43	6.84	1.5
586	1.91	5.4	1.8
741	0.65	1.83	5.5
810	0.64	1.80	5.5

Laser glass LG-680 with $C = 3$ wt% Nd, $\alpha = FC\sigma_\alpha$, $F = 0.94\times10^{20}$/wt%/cm^3

The absorption of the pump light for different wavelengths is given in Tables 5.1 and 5.2 for the silicate glass LG-680.

In addition to the average absorption of the pump band in Table 5.2 the relative amount of the pump light for a Kr arc lamp in this band is indicated. For every pump band the effective excitation by an arc lamp can be determined in this way. By cutting off the short wavelengths of the pump light the ratio of laser efficiency to thermal load can be optimized.

Table 5.2 Pump light absorption by Nd:laser glass, [5.12]

Pump range area $\Delta\lambda$/nm	Kr lamp efficiency $\eta(\lambda)$ %	Absorption cross section σ_p 10^{-20} cm^2	Spectral excitation $\eta(\lambda)\sigma_\alpha$ % 10^{-20} cm^2	Absorption $\alpha = N_0\sigma_p$ cm^{-1}	Penetration depth $1/\alpha$ mm^{-1}
530 ±20	3.3	0.39	1.3	0.73	13.7
580 ±20	4.8	1.17	5.6	2.19	4.6
750 ±20	11	0.72	8	1.35	7.4
805 ±20	19	1.44	27.4	2.69	3.7
880 ±20	11.3	0.74	8.4	1.39	7.2

Silicate-laser glass with $C = 4$ wt% Nd, $\alpha = FC\sigma_\alpha$, $F = 0.94\times10^{20}$/wt%/cm^3

Table 5.3 Pump light absorption in Nd:YAG, [5.13]

Pump wavelength λ/nm	Absorption cross section $\sigma_p/10^{-20}$cm^2	Absorption coeff. $\alpha = N_0\sigma_p$/cm^{-1}	Penetration depth $1/\alpha$ /mm
358	1.79	2.72	3.7
530	1.66	2.52	4.0
580	2.17	3.29	3.0
680	0.27	0.41	24.4
750	3.30	5.01	2.0
790	3.50	5.31	1.9
880	0.71	1.08	9.3

YAG with $C = 1.1$ at% Nd, $\alpha = F_1C\sigma_\alpha$, $F_1 = 1.38\times10^{20}$/wt%/cm^3

Table 5.4 Pump light absorption in Nd:YAG, [5.12]

Pump range area $\Delta\lambda$ nm	Kr lamp efficiency $\eta(\lambda)$ %	Absorption cross section σ_p 10^{-20} cm^2	Spectral. excitation $\eta(\lambda)\sigma_\alpha$ % 10^{-20} cm^2	Absorption $\alpha = N_0\sigma_p$ cm^{-1}	Penetration depth $1/\alpha$ mm^{-1}
530 ±20	3.3	0.09	0.3	0.17	58.8
580 ±20	4.8	0.24	1.1	0.44	22.7
750 ±20	11	0.65	7.2	1.22	8.2
805 ±20	19	0.69	13	1.28	7.8
880 ±20	11.3	0.07	0.8	0.13	76.9

Nd:YAG with C = 0.98 wt% Nd, $\alpha = F_1F_2C\sigma_\alpha$, $F_1 = 1.38\times10^{20}$/at%/cm^3, $F_2 = 1.38$ at% / wt%

For several lines in the main pump bands the penetration depth as well as the absorption is given for Nd:YAG in Tables 5.3 and 5.4.

In [5.14] the effective pump cross section is $\sigma_p = 0.3\times10^{-20}$ cm^2 for a flash lamp operating in the range from 330 nm to 900 nm. If it is pumped with a narrow line, for example with laser diodes, then it is mostly the absorption maximum which is of interest. These are indicated in Table 5.5 for some materials, [5.14].

Table 5.5 Peak absorption of the pump light in Nd:YAG, [5.14]

Crystal	Pump wavelength λ nm	Range $\Delta\lambda$ nm	Absorption cross section σ_p 10^{-20} cm^2	Penetration depth [c] $1/\alpha$ mm^{-1}
YAG	808	1	6.1	1.61
GSGG	808	1	6.4	1.56
YLF	792	2	1.9	5.26
YVO_4	808	8	6.9[a]	1.45
BEL	811	6	11.6[b]	0.86
LOS	817	1.5	6.5	1.54
ED-2	810	15	2.2	4.55

[a] in π direction, [b] in y direction, [c] $N_0 = 1\times10^{20}$ ions/cm^3

Excitation

The most important excitation sources for material processing lasers are, at present discharge lamps, with all their disadvantages as limited lifetime or a nonoptimized adapted spectrum that additionally heats the laser medium. They will be considered separately in the following section.

Two developments stand out in avoiding these disadvantages: first the electrodeless discharge lamps excited by high frequency, and secondly laser diode arrays.

The high frequency lamp works with continuous discharging – pulsed enhancement around a factor two seems possible – and is predestined for high average power. In Fig. 5.5 the concentric construction of a laser with HF discharge lamp is shown. A laser output power of almost 300 W has been reached so far.

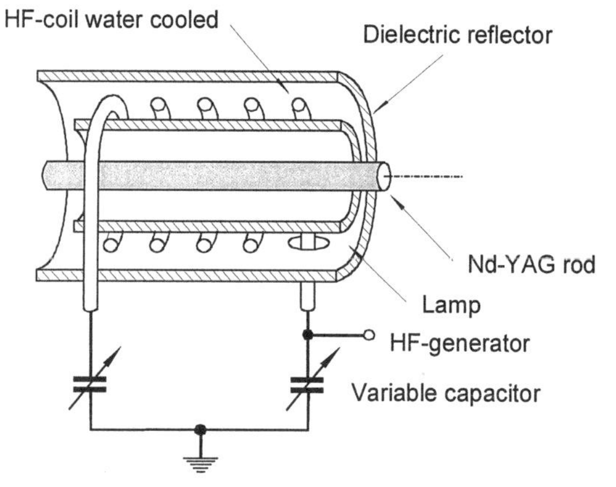

Fig. 5.5. Laser arrangement with high frequency excitation of an electrodeless arc lamp, [5.3]

The laser diodes operate in a very narrow frequency range well absorbed by the solid laser material. In addition, the pump frequency is close to the solid-state laser frequency. Thus the thermal load is significantly lower than for lamps. Lasers with low average power but with very high beam quality are possible. The excitation with diode lasers is considered in detail in Chap. 7.

5.2 Pump Cavity for Slabs

Thermally invariant constructions which show no influence of the lens effect on the beam quality can be achieved if the optical path is the same for every single ray, for example if the temperature gradient is parallel to the beam propagation.

A beam propagating in a zigzag through the laser material has for all rays the same averaged optical path, despite a small shift between the individual rays. The thermal lens effect is compensated by this path. A one-dimensional solution for the zigzag slope is a slab [5.15]. Figure 5.6 shows the principle construction and Fig. 5.7 the beam path.

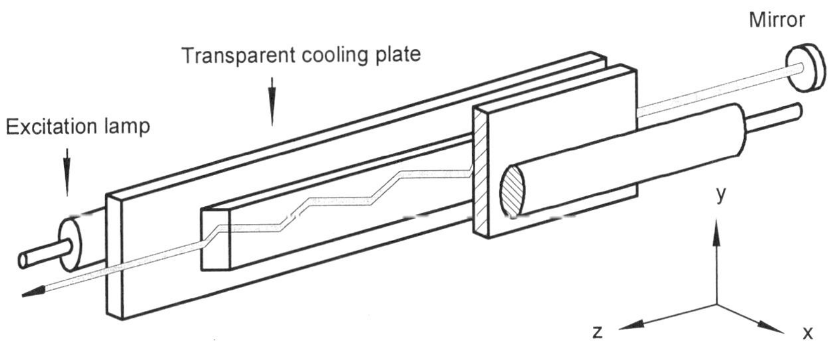

Fig. 5.6. Schematic construction of a slab laser, [5.16]

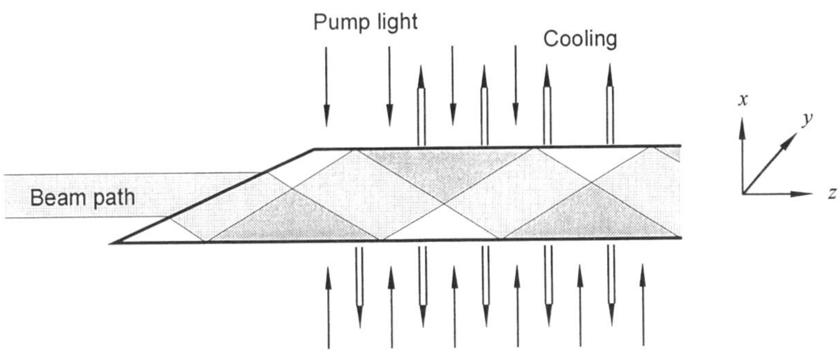

Fig. 5.7. Beam propagation in a slab, [5.17]

The ideal thermal insulation at the small faces causes the isotherms to run parallel to the y axis. There is a temperature gradient only in the x direction that causes a thermal lens.

Up to now, these slab conceptions with excitation lamps are only occasionally converted into technical solutions, which are available on the market. The reason may be the difficulty to find a suitable way to cool the laser material. Moreover, the beam quality of slab lasers requires a very homogeneous pump light distribution and homogeneous cooling. In future, the slab design, however, may become more important if laser diodes are available as excitation sources. The relevant data of slab-arranged laser constructions are given in Table 5.6.

The slab delivers a rectangular or square beam. However, for cutting and welding the width must be independent of the processing direction. Therefore, in most cases the slab is useful only in conjunction with a fiber.

Table 5.6 Performance of slab lasers

Material	Dimensions $t{\times}h{\times}l$ mm^3	Power P_{av} W	Efficiency η_{slope} %	η_{total} %	Beam quality $(w\theta)_x$ mm×mr	$(w\theta)_y$ mm×mr	Ref.
Nd:GGG	6×10×100	45	1.7	1.05	6×1	10×3	[5.18]
Nd:GGG	10×18×167	145	2.85	2.4			[5.18]
Nd:GGG	7×35×115	26	1.2				[5.19]
Nd:GGG	5×70×180	500					[5.20]
Nd:glass	6×20×300	55					[5.21]
Nd:YAG	6×15×100	250	2.4		6×2	15×2	[5.21]
Nd:YAG	6×15×100	300	3.3				[5.22]
Nd:YAG	5.6×18.4×154	500	2.4	2.0			[5.23]
Nd:YAG	6×25×162	250	3.3				[5.24]
Nd:YAG	6×25×210	1000	8.5	6.0	5		[5.25]

Slab Geometry

The necessary geometric boundary conditions of a slab are considered in the following. The geometry must satisfy the optical and the cooling conditions so that a maximum laser power with an excellent beam quality can be obtained.

From Fig. 5.8 one can derive the following:

- for the angles in the triangle A-B-C,

$$\gamma + (90° + \beta) + (90° - \theta) = 180°, \qquad \beta + \gamma = \theta. \tag{5.3}$$

- when the beam is incident parallel to the slab axis one has

$$\gamma + \alpha = 90°, \tag{5.4}$$

– from the refraction at the entrance,

$$n \sin\beta = \sin\alpha, \tag{5.5}$$

– follows, if the beam enters under the Brewster angle,

$$n = \tan\alpha_B. \tag{5.6}$$

– And, if the volume is fully used,

$$\gamma = 90° - \theta. \tag{5.7}$$

For a slab with a fully used volume it follows, with (5.3) – (5.7),

$$\alpha = \beta, \qquad \beta = \theta - \gamma = 2\alpha - 90°, \qquad -n\cos(2\alpha) = \sin\alpha. \tag{5.8}$$

An approximate solution to (5.8) is

$$\alpha = \arctan\left(\frac{\sqrt{2}\sqrt{8n^2+1}+1}{2\sqrt{\sqrt{8n^2+1}+4n^2-1}}\right). \tag{5.9}$$

If the angle of incidence is chosen as the Brewster angle, in order to keep the losses small,

$$-\cos(2\alpha_B) = \cos\alpha_B. \tag{5.10}$$

The angles become

$$\alpha_B = \theta = 60°, \ \gamma = 30°, \tag{5.11}$$

and the refractive index

$$n = \sqrt{3} \approx 1.73\,. \tag{5.12}$$

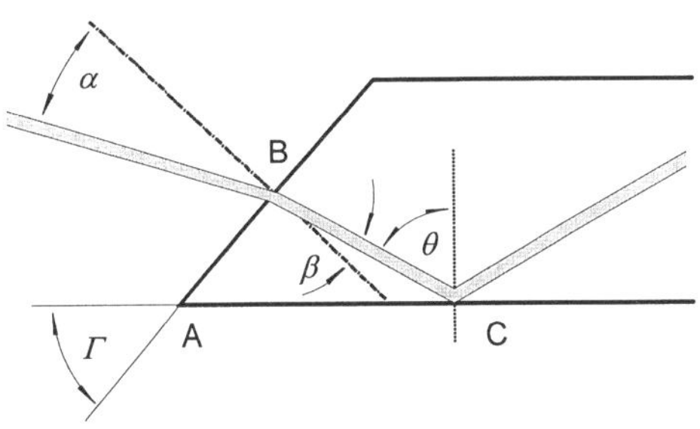

Fig. 5.8. Slab with inclined ends

There is only one definite refractive index for which the incident beam under the Brewster angle α_B is parallel to the longitudinal axis of the slab and at the same time the full volume is used.

Symmetrical Brewster Slab

In the following, a slab with Brewster ends and symmetrical, parallel beam path will be considered further. This slab geometry is the most common one at present. With (5.3) and (5.4) it follows that

$$\theta = \beta - \alpha + 90°, \tag{5.13}$$

and using (5.6) and (5.7) results in:

$$\tan\theta = \frac{\sin(\beta - \alpha + 90°)}{\cos(\beta - \alpha + 90°)} = \frac{2n}{n^2 - 1}, \tag{5.14}$$

$$\tan\gamma = \tan(90° - \alpha) = 1/n, \tag{5.15}$$

$$\tan^2\beta = \frac{\sin^2\beta}{1 - \sin^2\beta} = \frac{1}{n^2}. \tag{5.16}$$

The angles then are

$$\alpha = \arctan(n), \qquad \beta = \arctan(1/n), \qquad \gamma = \beta,\ \theta = 2\beta. \tag{5.17}$$

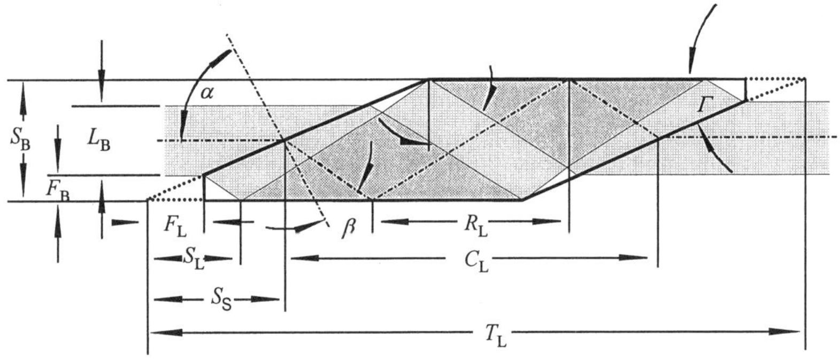

Fig. 5.9. Geometry of the symmetrical slab with Brewster ends

From Fig. 5.9 the dimensions of the symmetrical slab with Brewster ends can be derived:

$$R_L = \frac{2n}{n^2 - 1} S_B, \qquad R_L = \text{reflection length},\ S_B = \text{slab width}, \tag{5.18}$$

$$F_B = \frac{n^3 - 3}{n^2 + 1} S_B \,, \qquad F_B = \text{bevel width}, \tag{5.19}$$

$$F_L = nF_B \,, \qquad F_L = \text{bevel length}, \tag{5.20}$$

$$S_S = \frac{n}{2} S_B \,, \qquad S_S = \text{slab tip length}, \tag{5.21}$$

$$C_L = 2NnR_L \,, \qquad C_L = \text{center length at } 2n\text{-th reflection}, \tag{5.22}$$

$$T_L = C_L + 2S_S \,, \qquad T_L = \text{total length from tip to tip}, \tag{5.23}$$

$$O_L = \frac{nC_L}{\sin\theta} \,, \qquad O_L = \text{optical path length}, \tag{5.24}$$

$$L_B = S_B - 2F_B \,, \qquad L_B = \text{laser width outside the slab}, \tag{5.25}$$

$$L_F = \frac{L_B}{\sin\theta} S_H \,, \qquad S_H = \text{slab height}, \tag{5.26}$$

L_F = cross-sectional area of the laser in the slab.

Temperature Distribution

Analogously to the rod, a parabolic temperature profile occurs in the slab in the cooling direction. The temperature gradient and the slab thickness determine the maximal permitted pump power and thus also the attainable laser output power.

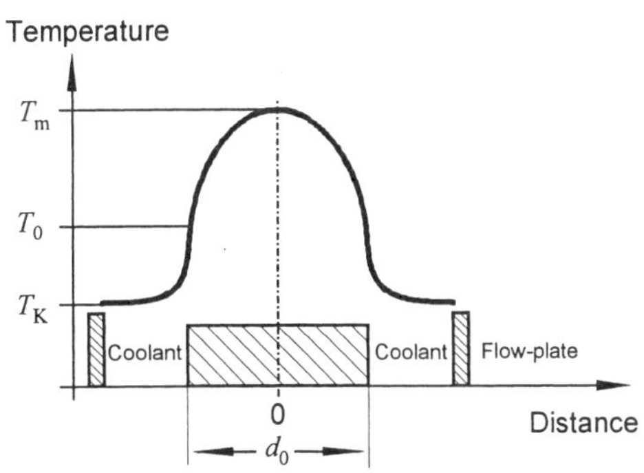

Fig. 5.10. Principle temperature distribution in the slab

Figure 5.10 shows the principle temperature distribution in a slab.

$$T_m - T_0 = \frac{d_0^2}{8kV} P_H \,, \tag{5.27}$$

T_m = temperature in slab center,
T_0 = temperature on slab surface,
V = slab volume,
d_0 = slab thickness,
k = thermal conductivity.

The maximal permitted pump power P_p is, according to (4.23)

$$P_p = \frac{12}{5} \frac{O}{\eta_H d_0} R_T , \tag{5.28}$$

$O = 2hl$ = cooling surface,
h = slab height,
l = slab length,
R_T = 1.04 W/mm (YAG),
η_H = thermal efficiency.

The slab thickness should be small on the one hand to allow high pump power, on the other hand large for high absorption. A thickness between 5 and 7 mm is common. For a stable double-folded resonator a h/d_0 ratio of 2, for an unstable one a ratio of 3, is favorable.

Example 5.1 Laser power of a slab

A slab with height h = 15 mm, thickness d_0 = 5 mm and a thermal efficiency η_H = 5% has a maximal pump power per length of

$$\frac{P_p}{l} = \frac{12}{5} \frac{2 \times 15 \times 1.04}{5 \times 0.05} \frac{\text{W}}{\text{mm}} \approx 300 \frac{\text{W}}{\text{mm}}$$

and with a laser efficiency of 3% a laser output power of approximately P/l = 9 W/mm can be achieved.

This implies a maximum pump power of 40 kW and a laser output power of maximum 1350 W for a 15 cm long slab (without Brewster ends).

5.3 Cooling

Since the efficiency of solid state lasers is in the per cent range, almost all the supplied power must be extracted from the pump cavity as heat by the coolant. Also, a considerable amount of heat arises in the power supplies, particularly at higher average powers. At low average laser output powers (around 10 W) a compressor cooling unit can be used without additional water connection. The power supply is air-cooled by fans. At high average powers (approximately more than 300 W) beam source and power supply need to be water-cooled, variants with additional compressor and the use of available return cooling units are also common.

Cooling Power

In solid-state lasers, which are pumped by lamps, the main part of the heat arises next to the lamps, smaller amounts at the laser medium and at the reflector.

$$P_{\mathrm{H}} = \frac{\mathrm{d}V}{\mathrm{d}t} \rho c \Delta T \,, \tag{5.29}$$

$\mathrm{d}V/\mathrm{d}t$ = coolant flow,
ρ = specific weight,
c = specific heat,
ΔT = temperature difference.

Example 5.2 Heating power in a rod laser cavity

To determine the different amounts of heat in a flash lamp pumped Nd:YAG laser the water flow and temperature rise of the cooling water for the laser rod and the lamps including reflector have been measured.

Flash lamps:	$\mathrm{d}V/\mathrm{d}t = 230$ l/h,	$\Delta T = 8°\mathrm{C}$,	$P_{\mathrm{H}} = 2139$ W (84%).
Laser rod:	$\mathrm{d}V/\mathrm{d}t = 120$ l/h,	$\Delta T = 3°\mathrm{C}$,	$P_{\mathrm{H}} = 418$ W (16%).

Since the laser threshold depends on temperature, one chooses the coolant temperature at the entrance into the cavity similar to that of the laser medium. The dew point, however, should always be below the cooling temperatures. Temperatures above 50°C should be avoided at the open cooling circuit to prevent condensation at cooler parts. Turbulent flow is necessary to achieve a high heat-transfer coefficient and avoid formation of steam bubbles at the lamp surface. Laser materials such as Nd:YAG and Nd:glass in pulse mode show a slight dependence of the laser energy on the cooling temperature in the range of 5 to 50°C. The cooling water at the exit may have a temperature of approximately 40°C and the temperature difference between cooling water in and out lies around 20°C.

For the usual cooling of a rod with flow-tubes (Fig. 5.11) the necessary relations are represented in the following. The temperature T_0 at the medium (laser rod, lamp) follows with (4.10) as

$$T_0 = T_{\mathrm{K}} + \frac{P_{\mathrm{H}}}{\lambda O} \,, \tag{5.30}$$

λ = heat-transfer coefficient,
O = cooling surface,
P_{H} = heat power in the medium,
T_{K} = temperature of the coolant.

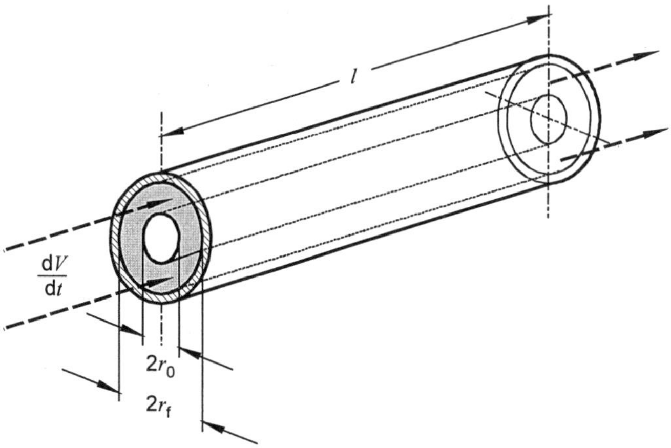

Fig. 5.11. Flow-tube cooling

By choice of the flow speed and cooling section the heat-transfer coefficient and therefore the average temperature is influenced in the laser medium, not, however, the temperature distribution in the medium.

The flow rate $\mathrm{d}V/\mathrm{d}t$ is

$$\frac{\mathrm{d}V}{\mathrm{d}t} = \pi(r_\mathrm{f}^2 - r_0^2)u\,, \tag{5.31}$$

r_f = flow-tube radius,
r_0 = rod radius,
u = flow speed.

The hydraulic diameter D is defined as the ratio of flow section to wetted circumference and is, for flow-tubes

$$D = \frac{4\pi(r_\mathrm{f}^2 - r_0^2)}{2\pi(r_\mathrm{f} + r_0)}\,. \tag{5.32}$$

The heat-transfer coefficient λ is determined for laminar flow, $900 < N_\mathrm{Re} < 2000$, as [5.26]

$$\lambda = 1.02 N_\mathrm{Re}^{0.45} N_\mathrm{Pr}^{0.33} N_\mathrm{Gr}^{0.05} \frac{k}{D}\left(\frac{D}{l}\right)^{0.4}\left(\frac{r_\mathrm{f}}{r_0}\right)^{0.8} \tag{5.33}$$

and for turbulent flow, $12\,000 < N_\mathrm{Re} < 22\,000$, as

$$\lambda = 0.02 N_\mathrm{Re}^{0.8} N_\mathrm{Pr}^{0.33} \frac{k}{D}\left(\frac{r_\mathrm{f}}{r_0}\right)^{0.53} \tag{5.34}$$

with

$N_{Pr} = c\mu/k$ = Prandtl number,
$N_{Re} = \rho u D/\mu$ = Reynolds number,
$N_{Gr} = \Gamma\rho^2 g D^3(T_0 - T_K)/\mu^2$ = Grashof number,
Γ = cubic expansion coefficient,
$g = 9.81\,\mathrm{m/s^2}$,
μ = viscosity.

The measurements in Fig. 5.12 show a typical result for the heat-transfer coefficient. The coefficient is determined from the temperature decay time. According to (5.34) it is dependent in a nonlinear way on the water flow between flow-tube and rod.

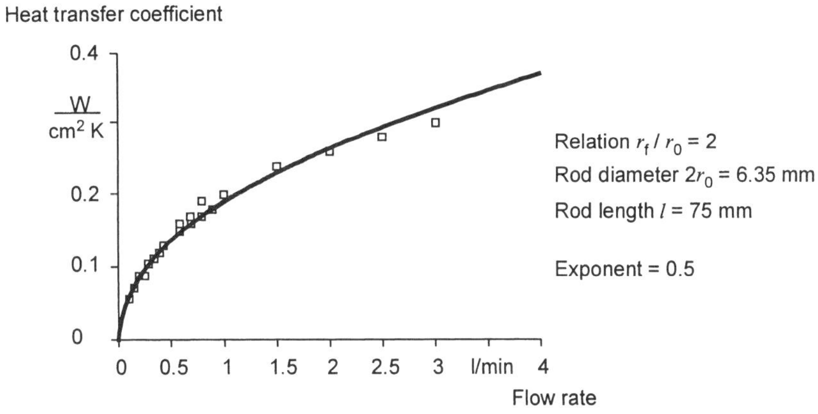

Fig. 5.12. Heat-transfer coefficient as function of the flow rate

Example 5.3 Cooling of a Nd:YAG rod with a flow-tube

With $dV/dt = 500$ l/h $\approx 140\,\mathrm{cm^3/s}$, $r_0 = 0.5$ cm, $r_f = 0.75$ cm follows $u \approx 140$ cm/s, $N_{Re} \approx 7000$ and $\lambda \approx 0.67\,\mathrm{W/(cm^2\,°C)}$.

For the most important cooling media, water and air, the characteristic parameters are given in Table 5.7.

Table 5.7 Characteristic cooling parameters for water and air

	c W s/(g °C)	μ g/cm s	k W/(cm °C)	ρ g/cm^3	Γ 10^{-3}/°C
Water	4.186	10^{-2}	6×10^{-3}	1	0.280
Air	1	17×10^{-4}	2.4×10^{-4}	12.9×10^{-3}	3.66

	N_{Pr}	N_{Re}	N_{Gr}
Water	7	$100\dfrac{uD}{\mathrm{cm^2/s}}$	$2.74\times10^6\,\dfrac{D^3(T_0-T_K)}{\mathrm{cm^3\,°C}}$
Air	7.08	$7.6\dfrac{uD}{\mathrm{cm^2/s}}$	$0.2\times10^6\,\dfrac{D^3(T_0-T_K)}{\mathrm{cm^3\,°C}}$

A Reynolds number of 20 000 should be used for rod cooling. Lamp manufacturers recommend a necessary flow rate, as given by (5.29), and a Reynolds number of at least 10 000.

Cooling for Laser Diodes

Diode lasers must be cooled well to attain a high lifetime. The assembly of the power diodes is usually done on a flat cooling surface that is cooled by water on the back side. The heat-transfer resistance of the heat source, in this case the junction layer, to the cooling medium can be reduced by a suitable microstructuring so that besides copper silicon can also be used. This implies an essential simplification for the integration of the laser diodes on the cooling body and also for the cooler production.

The temperature rise of the junction layer during the operation of the diodes is determined by the heat power P_H and the heat-transfer resistance R

$$\Delta T = RP_H. \tag{5.35}$$

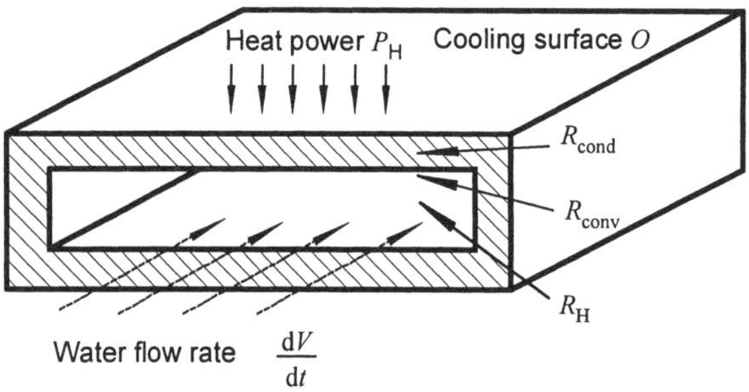

Fig. 5.13. Heat-transfer resistance of a cooler

The heat resistance consists of three parts [5.28] $R = R_{cond} + R_{conv} + R_H$ (Fig. 5.13):

- the amount due to heat conduction,

$$R_{cond} = \frac{d}{k_M WL}, \tag{5.36}$$

d = distance between heating and cooling,
$W \times L$ = heat exchange area,
k_M = heat conduction of the cooler material,

- the amount due to the heating of the cooling medium,

$$R_H = \frac{1}{\rho c_p} \frac{dt}{dV}, \tag{5.37}$$

ρ = specific weight,
c_p = specific heat,
dV/dt = coolant flow rate,

– and the amount due to the transition from the cooling surface to the cooling medium

$$R_{conv} = \frac{1}{\lambda O}, \tag{5.38}$$

λ = heat-transfer coefficient,
O = heat transition surface.

If we construct a very thin cooler (100 μm is possible), then R_{cond} is 0.0067°C/W with silicon k = 1.48 W/(°C cm) and an area of approximately 1 cm^2. With a flow rate of 10 cm^3/s and water cooling, ρc_p = 4.187 W s/(°C cm^3) a resistance of R_H = 0.024°C/W can be achieved.

The largest contribution to the thermal resistance is given by the transition to the cooling medium. This can be improved by structuring the surface with cooling channels, see Fig. 5.14. The heat-transfer coefficient λ, the transitional surface O and the coolant flow dV/dt depend on the number and the dimensions of these cooling channels.

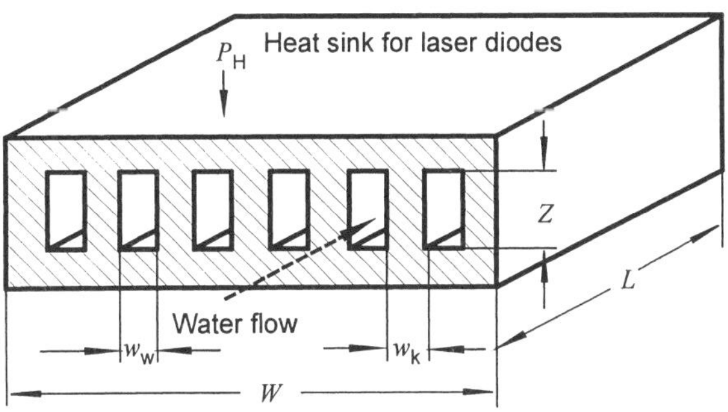

Fig. 5.14. Microchannel cooler

An optimization for laminar flow yields the ratio of the channel width to channel separation as [5.28]:

$$w_W = w_K = w, \tag{5.39}$$

w = channel width,

the flow rate

$$\frac{dV}{dt} = \frac{w^3 \alpha}{24\mu} \frac{W}{L} \rho, \tag{5.40}$$

with $\alpha = \sqrt{\dfrac{k_M}{k_K N_{Nu}}}$,

$\alpha = z/w$ = aspect ratio,
k_K = heat conductivity of the coolant (water),
N_{Nu} = Nusselt number,
μ = viscosity of the coolant, $\mu(H_2O) = 10^{-2}$ g/cm s,
p = pressure,

and the two thermal resistances

$$R_H = \frac{1}{\rho c_p} \frac{24\mu}{w^3 \alpha} \frac{L}{W} \frac{1}{p}, \tag{5.41}$$

$$R_{conv} = \frac{1}{0.76} \frac{2}{k_K N_{Nu}} \frac{1}{LW} \frac{w}{\alpha}. \tag{5.42}$$

Both components are dependent on the channel width. The sum of the two has a minimum (Fig. 5.15) at a channel width of [5.28]

$$w = 2.29 \left[\frac{\mu k_K L^2 N_{Nu}}{\rho c_p p} \right]^{1/4}. \tag{5.43}$$

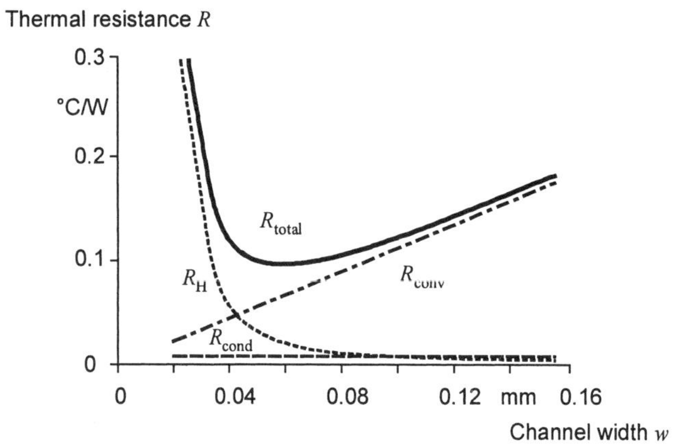

Fig. 5.15. Thermal resistance dependence on the channel width

5.4 Cooling Unit

Beam Source Cooling

From the above considerations and measurements the following rules for the design of a cooling unit have to be obeyed.

- Above a water temperature of approximately 40°C the cooling unit must be switched on. To avoid condensation from the air it should not be cooled below the surrounding temperature. At an excess temperature of 50°C the laser should be switched off by an automatic control and the failure of the cooling unit has to be signaled.

- To guarantee a continuous turbulent flow, the flow and the pumps should be oversized, particularly at low power. The ratio of the water flow rates for lamps and laser medium is approximately 4:1. Both flow rates must be monitored.

- Glass particles of broken lamps must be retained by a sieve. They should not cause downtimes.

Figure 5.16 shows the principle construction of a cooling unit with a heat exchanger against waters.

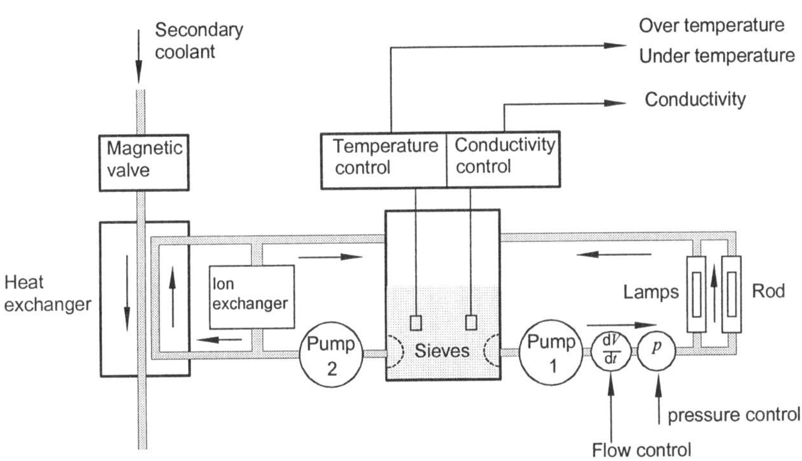

Fig. 5.16. Construction of a cooling unit, water against water

Given a desired average temperature in the laser medium, the heat-transfer coefficient λ, the setup (r_f, r_o), the necessary flow speed u and the power balance determine the minimal flow rate necessary according to (5.29) and (5.31). The necessary pump pressure p follows with the flow resistance R

$$p = R\left(\frac{\mathrm{d}V}{\mathrm{d}t}\right)^2 , \tag{5.44}$$

R = flow resistance.

The flow resistance is characteristic for beam source, pump and cooling unit. The flow resistance can be kept small by the choice of the cross section, avoiding sharp edges and diversions in the cooling water current. Typical values are in the range of 0.5 to 2×10^{-6} bar/(l/h)^2.

Dew Point

The temperature of optical components, such as laser mirrors or lenses, which require cooling, should not fall below the dew point temperature. Figure 5.17 shows the dew point temperature at atmospheric humidity and air temperature. For example the cooling water temperature must be above 15°C at an air temperature of 20°C and a humidity of 50%.

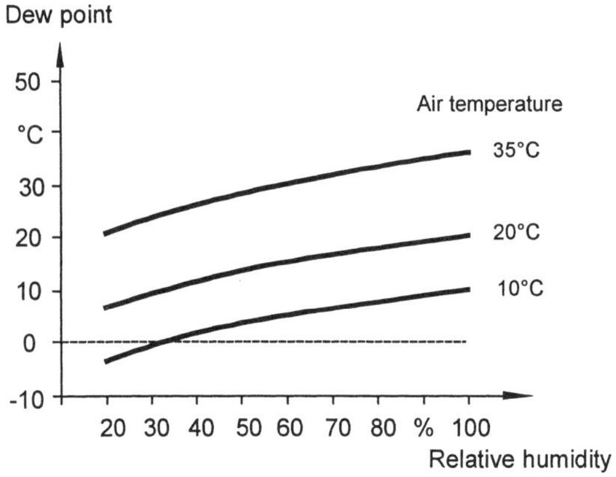

Fig. 5.17. Condensation diagram [5.27]

Power Supply Cooling

The efficiency of the power supply, of the electronic control and of the power switch is around 90%. Depending on laser power, this implies a heat power of some 100 W up to and greater than 1 kW for the power supplies. At low power level the heat may be extracted with ventilators to the air, at high power levels an internal cooling unit and external cooling against water are common.

Since a laser equipment for material processing is considered as a material processing machine, it must fulfill the corresponding safety and noise regulations.

With an external cooling the system-unit cover can be kept closed and the requirements (e.g. IP 54) can be easily fulfilled.

If no closed-loop cooling device is available at the end, then the cooling unit has to be designed such that the water consumption of the secondary loop becomes minimal. At high pressure this can be realized by several heat exchangers in series. The additional investment costs amortize by saving cooling water. If a closed-loop cooling device already exists, the flow does not play an important role. Since the pressure is low in this case, the heat exchangers are set up in parallel.

6 Gas Discharge Lamps

Discharge lamps for pumping solid-state lasers are widely used and will be discussed in some detail. In order to be able to design the lamps for optimal performance in applications and in order to design the circuit for a specific lamp, one must know the lamp parameters and their functional relationships. The following sections define the lamp-specific parameters in relation to the operating parameters and – as far as they are known – show the general relation existing between them. In many cases it is only possible to work with an empirical description of the functional relations. The necessary parameters, including the geometric variables, are listed separately for flash lamps and arc lamps in Table 6.4. The parameters should be stated for the worst case scenario in order to guarantee in all situations reliable operation of the electrical circuits that are dimensioned on the basis of these parameters.

Figure 6.1 shows the typical construction of an arc lamp for cw-excitation. The construction of a flash lamp is almost identical. The electrical connection is made by the two tungsten rods, which are sealed in the quartz-glass. To compensate for the different temperature expansion between quartz and tungsten an additional transition-glass is necessary. The electrodes are water-cooled in the shrink-down zone, in which the quartz fits as tight as possible the tungsten rod. A detailed description of construction and operation is given for example in [6.1].

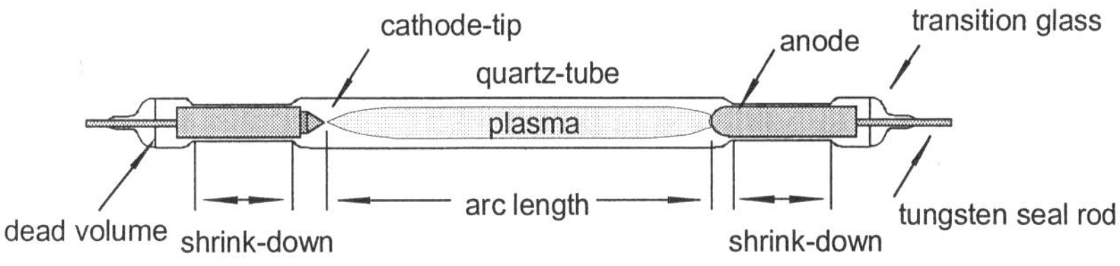

Fig. 6.1. Typical design of an arc lamp, [6.1]

The operation of the lamps can be divided into three phases, which are described explicitly in the following:

- *Ionization*
  By applying a high voltage to the lamp, sufficient free charge carriers are created. An arc can be observed and a current flows through the lamp connections.

- *Expansion of the plasma to form a stable simmer arc*
  After ionization, the lamp will be kept in operation with a low current in the mA- to A-region. The electrodes, especially the cathode, are kept at high temperature in order to guarantee a long lifetime.

- *Operating current*
  With increasing current the lamp changes from the simmer into the operating region. This occurs in pulsed or pure cw-mode. Due to the supplied energy the simmer arc heats up and expands until it nearly fills the inner diameter of the tube. The plasma arc is then stabilized by the inner tube wall.

6.1 Ionization for Flash and Arc Lamps

The ignition of a noble gas lamp follows in several steps. These steps and the corresponding trigger circuits are discussed in the following section.

Ignition

Ignition starts if the applied voltage is high enough to drive a current through anode and trigger wire. Light is first emitted at the anode but not yet at the cathode. The plasma moves along the wall towards the cathode [6.21]. Reaching it, a current can be detected at the cathode and an ionization channel between anode and cathode is built up. During the whole trigger phase, energy is absorbed from the trigger circuit, partially in order to create the plasma channel and partially to load the capacity between plasma channel and ground. After reaching the cathode the distance between anode and cathode becomes low-impedance and the ground capacity will be partially unloaded. This can be proved by the reversed current in the trigger electrode.

Instant of Ignition

The start of the steep rise of the cathode current can be defined as the instant of ignition. Simultaneously the current changes the polarity through the ignition aid and light can be detected directly at the cathode.

The ignition voltage is mainly determined by the dimensions, the gas pressure, the type of gas and the cathode material, as well as by any ignition aids used, such as anode rings and the surrounding.

Figure 6.2 shows the temporal behavior of current, voltage and light for both cases. The energy absorbed by the lamp during the trigger phase is approximately 1.5 mW s for the dynamic and 0.3 mW s in the static case. From the stored charge the intrinsic capacity can be estimated to be $C_T \approx 1\ \text{A} \times 0.5\ \mu\text{s} = 125\ \text{pF}$.

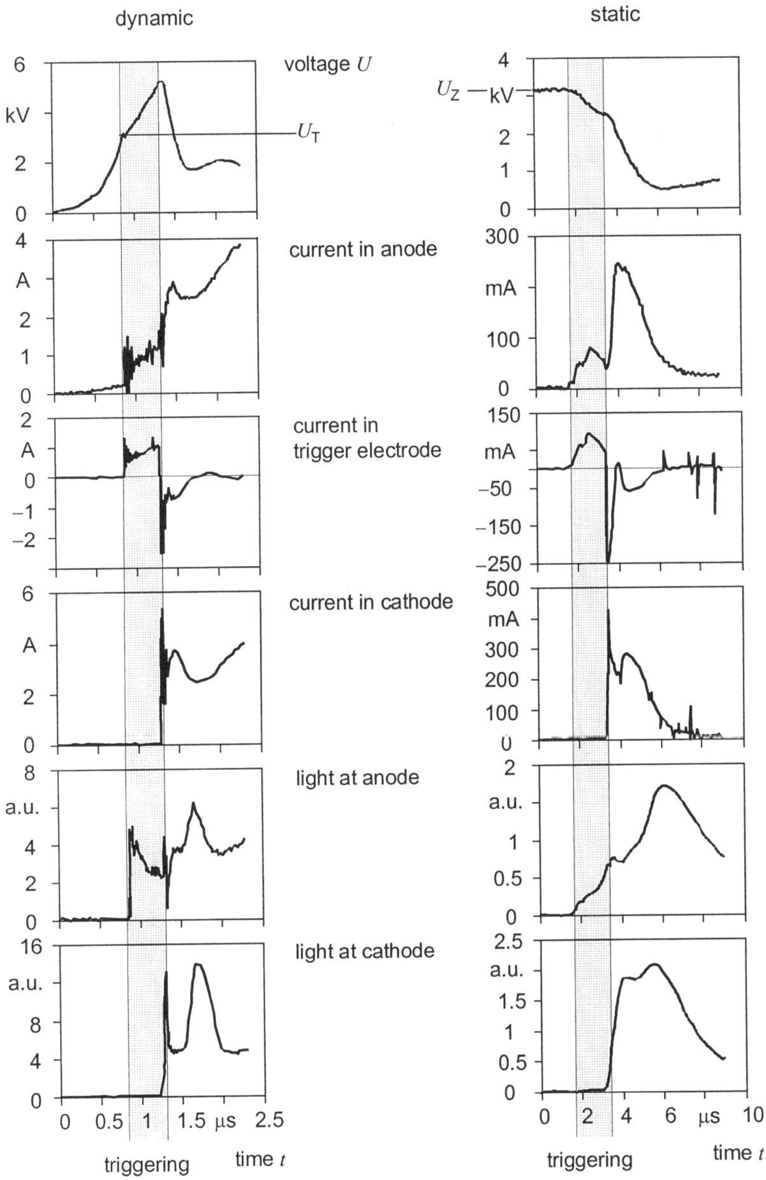

Fig. 6.2. Ignition voltage and ignition current versus time

Blocking Voltage U_{th}

This is the maximum dc voltage that can be permanently applied to the lamp without causing self-ionization to take place and igniting the lamp.

Ignition Voltage U_Z

This is the voltage that is required in order to allow a trigger current to flow through the cold and dark cathode. This voltage is dependent on the surroundings of the lamp, e.g. water or air, or on a possible ignition aid.

In a measurement setup the trigger voltage can be determined by variation of the applied voltage. In the following, two principle trigger circuits, as shown in Fig. 6.3, will be discussed.

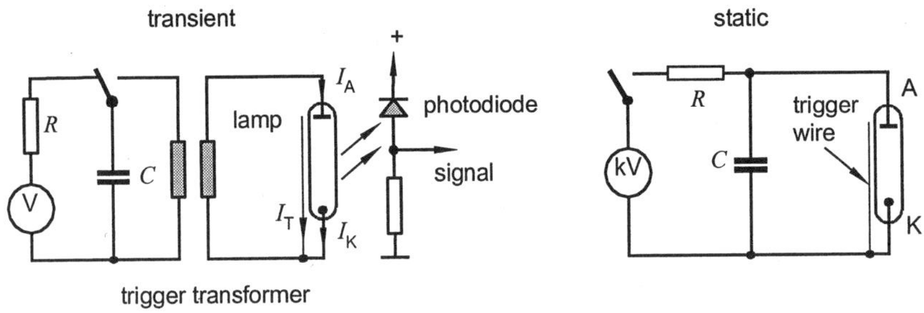

Fig. 6.3. Circuits to determine the static and dynamic trigger voltage

The empirical relation between the static ignition voltage U_Z the lamp length and the cold fill pressure for the circuit shown in Fig. 6.3 is given by (6.1). This relation is in agreement with the experience of manufacturers and is also stated in [6.2]:

$$U_{Z=} \propto l\sqrt{p_0}\,, \qquad \textit{static trigger voltage,} \tag{6.1}$$

l = arc length,
p_0 = cold fill pressure.

The trigger voltage is sensitive to noble gas impurities and to N_2 or O_2 additives, according to [6.2]. The trigger voltage rises with the concentration of these molecular gases. For approximately 5% N_2 the trigger voltage increases by around $\sqrt{2}$.

$$U_{Z=} = K_Z F_g \sqrt{1 + F_M C_M}\, l\sqrt{p_0}\,, \tag{6.2}$$

K_Z = trigger constant,
F_g = factor for type of gas,
F_M = factor for molecular gas = 20 for N_2,
C_M = concentration of the molecular gas.

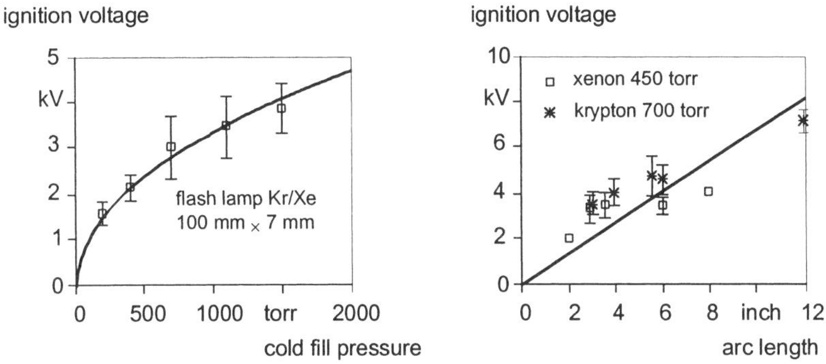

Fig. 6.4. Ignition voltage dependence on lamp parameters

In order to take account of the variance of sample values and other statistical deviations, the ignition voltage U_Z should be higher by a factor of 1.5 when designing the ignition circuit. Figure 6.4 shows the dependence of the static ignition voltage on the pressure and also how it changes as a function of the arc length.

If a fast rising voltage dU/dt is applied to the lamp for ignition purposes, the ignition voltage is shifted towards higher values. An empirical estimate of the necessary ignition voltage U_Z is given by (6.3).

$$U_Z - U_{Z=} + \sqrt[3]{K\frac{\mathrm{d}U}{\mathrm{d}t}}, \qquad \textit{dynamic ignition voltage,} \qquad (6.3)$$

K = dynamic constant.

Figure 6.5 compares the measured values with the curve obtained with (6.2). It shows the mean value and the related standard deviation of the ignition voltage. For a given ignition circuit, dU/dt is fixed and the necessary ignition voltage U_Z can be determined with the help of (6.3).

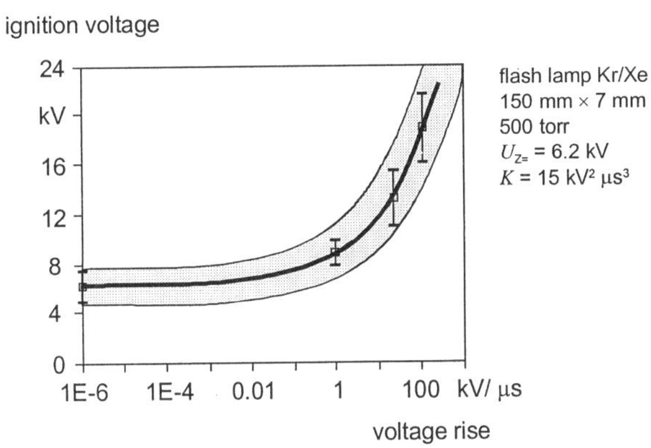

Fig. 6.5. Ignition voltage rise versus voltage slope

Increase of the Ignition Voltage for Lamps in Series

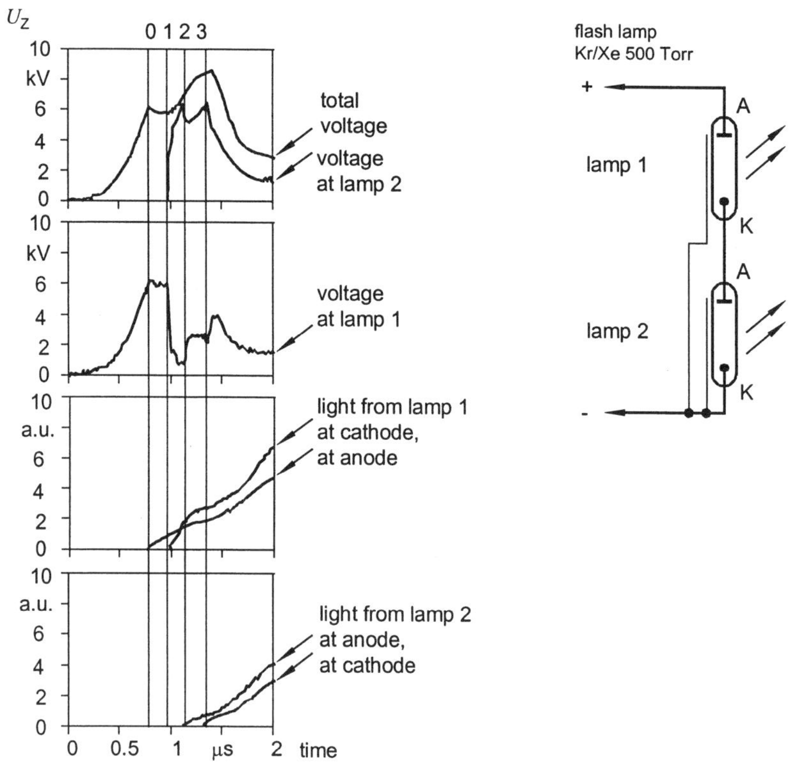

Fig. 6.6. Ignition with two lamps in series

If two identical lamps are connected in series both with a trigger wire they ignite sequentially (light may be detected at the cathode). The voltage at lamp 1 drops to a small value during the time interval 1 to 2. At point 2 the second lamp starts triggering and finishes at point 3. Figure 6.6 shows the time correlation for the triggering of two flash lamps.

The necessary ignition voltage contains three parts: the normal ignition voltage U_Z for one lamp, the voltage at the first ignited lamp and the increased ignition voltage of the second lamp due to the fast rising voltage after the first lamp was ignited. The first lamp plays the role of a fast switch.

With N identical lamps connected in series, the voltage increases in accordance with the empirical results shown in Fig. 6.7.

$$U_Z(N) = U_Z + \Delta U_Z(N-1), \tag{6.4}$$

U_Z = ignition voltage for one lamp,
ΔU_Z = voltage drop.

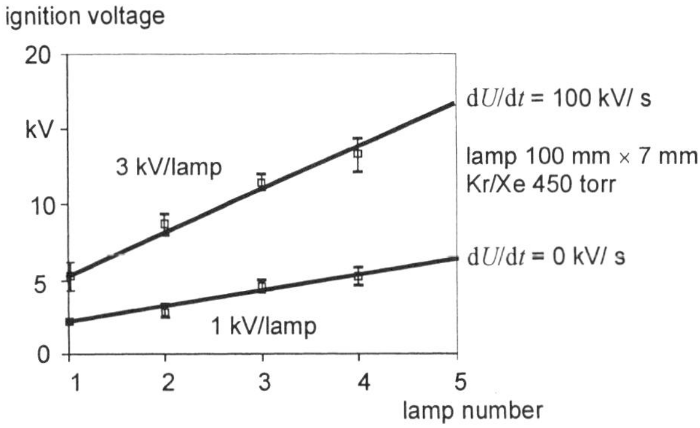

Fig. 6.7. Increasing ignition voltage for lamps connected in series

6.2 Expansion of the Plasma Arc (Booster Circuit)

After the first ionization the plasma arc will expand and in order to stabilize in diameter it has to be supplied with energy from an external source. Depending on the actual diameter the plasma has a corresponding impedance. An additional current is possible only for a sufficiently high voltage from a booster source. By means of a so-called booster circuit (Fig. 6.8), the lamp is punched into the corresponding current range.

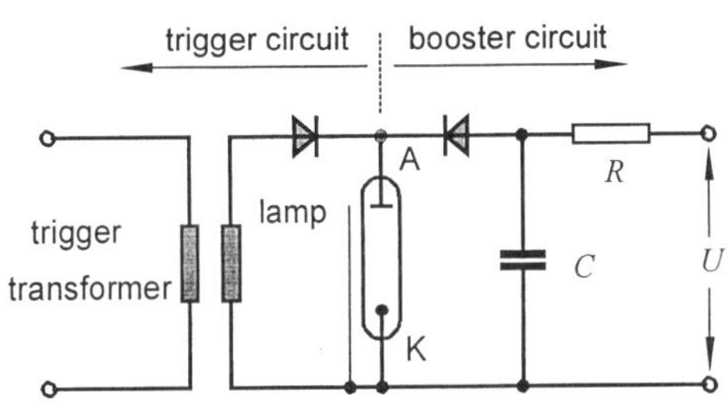

Fig. 6.8. Trigger and booster circuit

Two conditions have to be fulfilled to enable the discharge of capacitor C:

- The capacitor voltage U must be greater than a critical, lamp-specific voltage U_B (booster voltage),

- and, depending on the capacitor voltage $U > U_B$, a certain energy $E = \int U(t) \times I(t) \mathrm{d}t$ must be supplied to the lamp (Fig. 6.9).

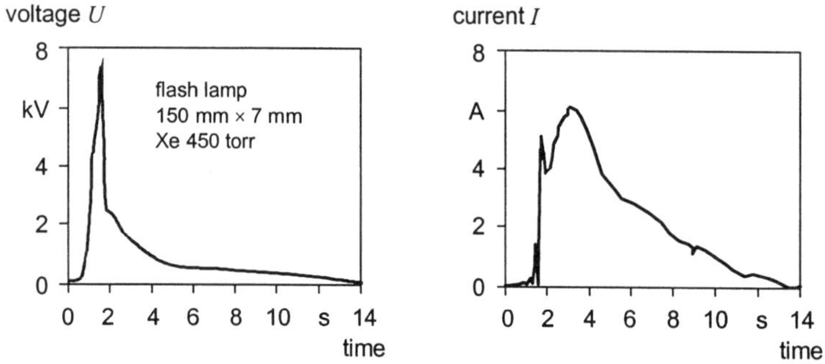

Fig. 6.9. Trigger voltage and trigger current of a flash lamp to determine the trigger energy

Equation (6.5) provides an empirical description of this relation and Fig. 6.10 shows a measurement of the necessary capacitor voltage U as a function of the ignition charge.

$$(U - U_B)^2 E_Z = K_B, \tag{6.5}$$

U_B = booster voltage,

K_B = booster constant,

$$E = \int UI\mathrm{d}t\,.$$

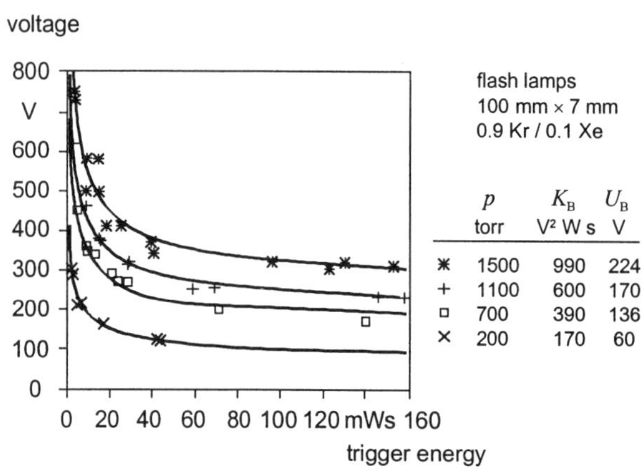

Fig. 6.10. Booster condition

The booster parameters are approximately linearly dependent on arc length and pressure.

After a flash lamp has been triggered, the temporal behavior of voltage and current at the booster capacitor can be seen in Fig. 6.11.

A residual voltage U_{min} is left in the capacitor after flashing the lamp. This voltage is the lowest voltage in the static UI characteristic of the lamp and the voltage may not fall below this value during operation. Particularly after a pulse one has to take care that the applied voltage is always higher than U_{min} during the transition from the high current range into the simmer range.

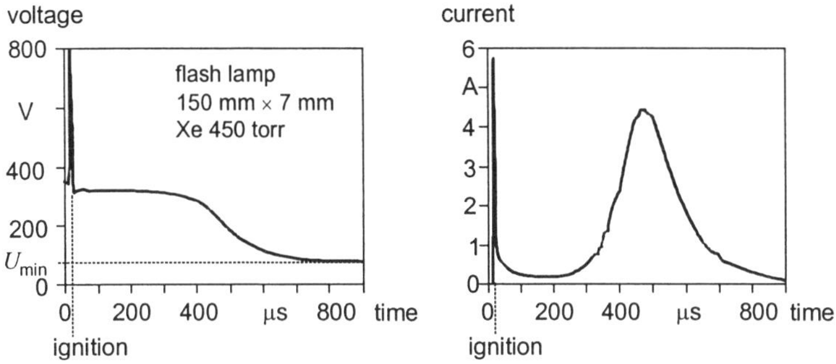

Fig. 6.11. Booster voltage, current and minimal voltage of the lamp characteristic, circuit according to Fig. 6.8

6.3 Voltage Current Characteristic

The static U I curve of a lamp is characterized by three principle regions (Fig. 6.12) which can be described adequately, as

- the simmer range from some mA to some A,
- the cw range from some A up to several 10 A,
- and a high current range up to some 100 A.

The solid line in Fig. 6.12 represents the resulting characteristic.

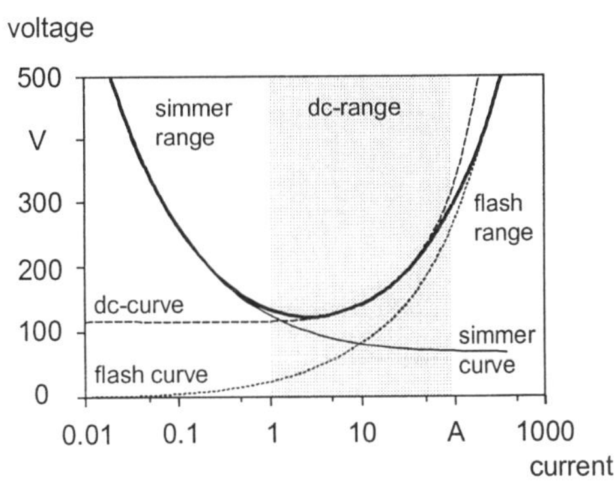

Fig. 6.12. Characteristic regimes of a lamp

Static Simmer Mode (50 mA to 5 A)

After the ionization the flash lamp should be on standby with a constant simmer current I. The simmer voltage depends on the electrode position at which the arc begins and may change if the arc jumps during operation.

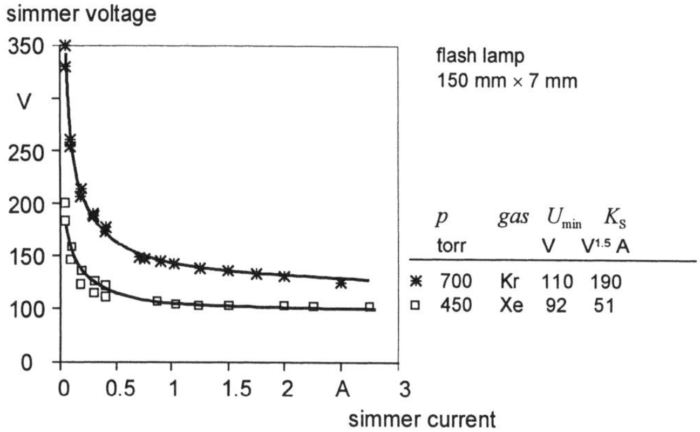

Fig. 6.13. Typical static simmer characteristic

The simmer characteristic shows a negative differential resistance (Fig. 6.13), so that the use of a constant current source is recommended.

For the static case the following dependence of the simmer characteristic was found empirically (see also [6.3]).

$$(U - U_{\mathrm{min}})^{\alpha} = \frac{K_{\mathrm{S}}}{I}, \tag{6.6}$$

U = voltage /V, $\alpha \approx 1.5$,
I = current/A,
K_{S} = simmer constant.

U_{min} is dependent on the used noble gas pressure ($\sim p^{0.4}$) and is for xenon a little lower than for krypton.

CW Operation (5 A to 40 A)

The characteristic is almost linear in this range and can be described by [6.4]:

$$U = U_0 + rI, \tag{6.7}$$

$r = r(d, l)$ differential impedance,
$U_0 = U_0(d, l)$ base voltage.

Figure 6.14 shows a typical characteristic.

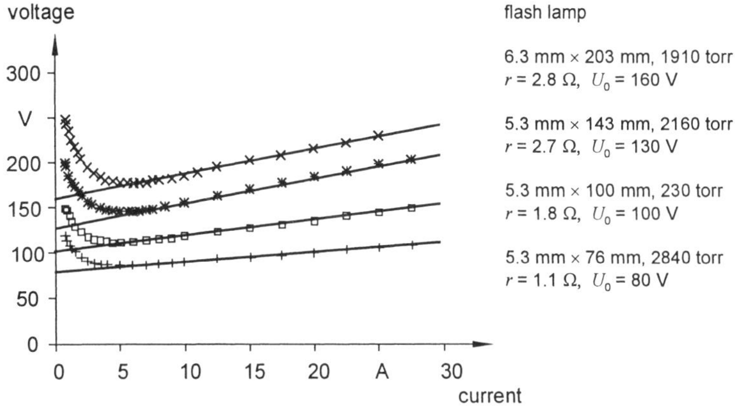

Fig. 6.14. Characteristic of two arc lamps in series

The pressure dependence of current and voltage for two lamps with various filling pressure is, according to [6.5],

$$\frac{I_1}{I_2} = \sqrt[4]{\frac{p_2}{p_1}} \text{ and } \frac{U_1}{U_2} = \sqrt[4]{\frac{p_1}{p_2}}, \tag{6.8}$$

p_1, p_2 = lamp pressure.

The recommended operational limits are dependent on the lamp diameter and listed in Table 6.1.

Table 6.1 Limitations for cw operation, for a cooling water velocity of at least 6 m/s for a lifetime of approximately 300 h [6.6]

Inner lamp diameter		4 mm	5 mm	6 mm
Current range	A	15 – 24	22 – 33	32 – 45
Voltage range/length	V/cm	14.2 – 19.8	12.6 – 16.5	11.8 – 16.0
Nominal power/length	W/cm	336	440	560
Maximum power/length	W/cm	430	520	650
Differential impedance	Ω/cm	0.26	0.192	0.14

High Current Range in the Pulsed Operation

The general formulation for the description of the lamp characteristic in the high current range (I = 20 to 100 A) is [6.7]:

$$U = Fp^{\alpha} \frac{l}{d} I^{\beta}, \tag{6.9}$$

F = factor, d = inner lamp diameter,

p = fill pressure /torr,
$\beta \approx 0.5$,
$\beta = 0.19 \ln(i) - 0.37$,
l = electrode distance.
$\alpha \approx 0.29$,
I = current /A
i = current density in A/cm^2 [6.8],

Lamp Parameter K_0

The lamp-specific constants are summarized in a parameter K_0 and (6.8) becomes with the approximation $\beta \approx 0.5$ [6.9]:

$$U = K_0 \sqrt{I}\,, \tag{6.10}$$

$$K_0 = F\left(\frac{p}{g}\right)^{\alpha} \frac{l}{d},$$

$F = 1.28$ V/√A,
p = fill pressure,
$\alpha = 0.25$,
g = 450 torr for Xe and 805 torr for Kr.

K_0 is normally defined with the exponent $\beta = 0.5$. With this definition K_0 must be measured under fixed operating conditions, for example in the single-pulse operation. The K_0 spread of new lamps around the mean average value is about ±5%.

At constant current the lamp voltage changes during the pulse. This change also depends on other parameters such as average electrical power and dead volumes. Equation (6.9) has to be taken as an approximation. In Fig. 6.15 at constant current the voltage can take on values that lie in the shaded range, dependent on the operation and lamp parameters. To keep the laser pulse power constant during a pulse, the decrease of the capacitor charge then must be compensated with increasing current. This has to be taken into account in the design of the power supply.

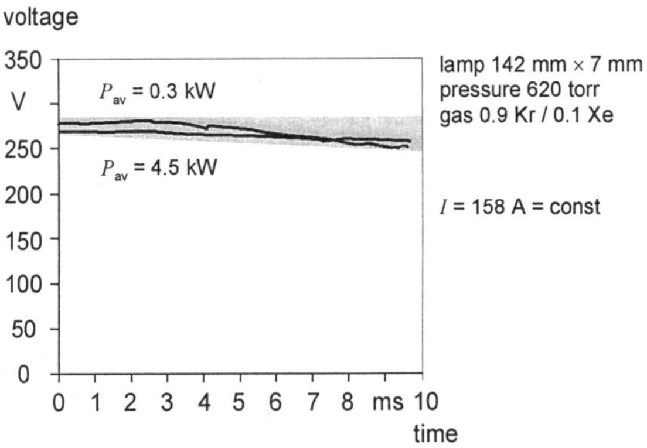

Fig. 6.15. Voltage range of a flash lamp at constant current dependent on the operation parameters

Figure 6.16 shows the measurement of a flash lamp characteristic. The measurement is fitted best with an exponent β slightly above 0.5.

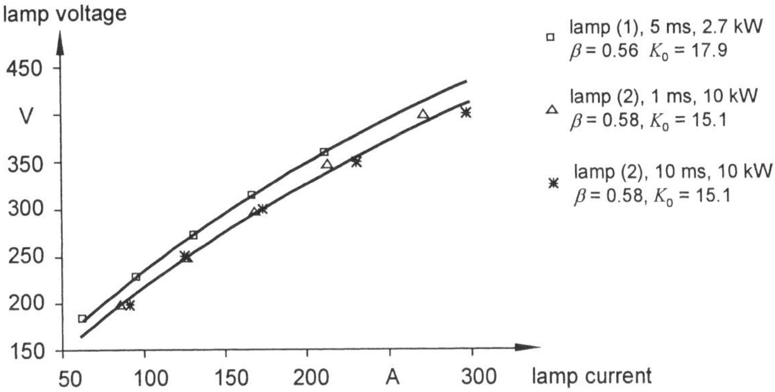

Fig. 6.16. Characteristic of two flash lamps in the high current range

6.4 On- and Off-Switching Characteristics

Turn-on Characteristics

The lamp shown in Fig. 6.17 is on standby operation due to the simmer current. The discharge current I from the capacitor C starts with a delay after switching. Starting from the simmer arc, the discharge channel expands to a new diameter [6.10].

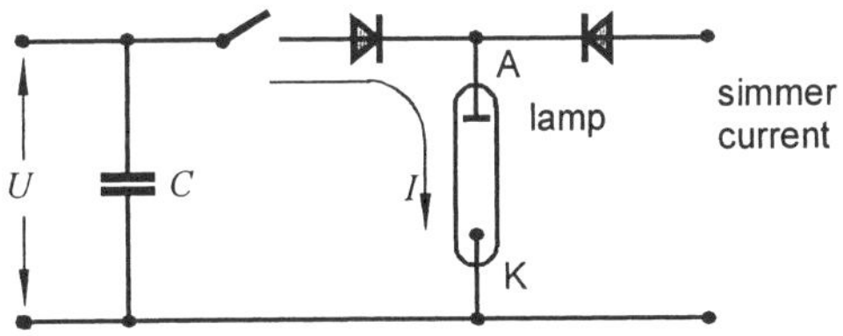

Fig. 6.17. Circuit to discharge energy capacitor in the pulsed operation

Two phases characterize the transient current. During the first phase the plasma arc with low temperature expands to the inner lamp diameter due to the supplied energy. The current rises exponentially. The plasma diameter is limited by the lamp wall in the second phase. The additional energy input basically increases the temperature and the pressure in the plasma until a stationary condition is reached.

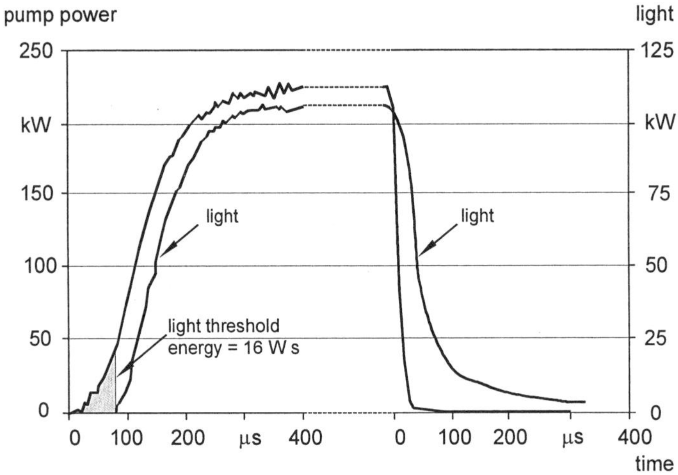

Fig. 6.18. Pump power and light of a flash lamp during switching on and off

Figure 6.18 shows a typical course of the pump power and of the light during on- and off-switching. After the switch is closed, the full constant voltage is applied. Therefore the pump power has the same temporal behavior as the current. The light is delayed after a certain amount of energy has reached the lamp. This delay limits the maximum repetition rate of the laser.

After switching off the pump power, the plasma continues to emit light for some time. The time constant for the decreasing light is about 60 µs in Fig. 6.18. This switching behavior of a lamp determines the minimal pulse duration for a given maximal optical power.

Delay and Threshold of the Pump Light

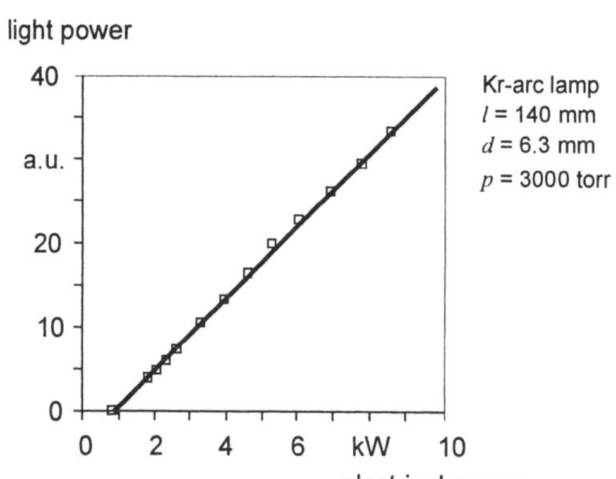

Fig. 6.19. Measurement of the light threshold with two arc lamps

As for the laser the excitation lamp has to absorb a certain amount of energy to heat the plasma before a noticeable amount of light is emitted [6.22]. This threshold energy E_{th} is proportional to the discharge volume and is almost independent of the operation parameters. The cooling of the lamp constantly extracts energy from the plasma. To compensate these losses, a minimal power for maintaining the plasma temperature has to be transferred to the lamp. The measurements in Fig. 6.19 show the light emission during cw operation, where the light intensity is drawn as a function of the pump power. A threshold of almost 1 kW pump power can be clearly seen.

The formulation for the description of the light energy respective light power is

$$E_p = \eta_{Lamp}(E_{in} - E_{pth}), \tag{6.11}$$

E_{pth} = energy threshold for the pump light, $\eta_{Lamp} \approx 0.5$

for the pulsed operation and

$$P_p = \eta_{Lamp}(P_{in} - P_{pth}),$$

P_{pth} = power threshold for the pumping light

in the cw operation.

The threshold E_{pth} and P_{pth} are dependent on the lamp parameters. They are proportional to the arc length with respect to the total volume. For a lamp diameter of 7 mm typical results are approximately 0.2 – 0.5 W s/cm^3 or about 35 W/cm.

In Fig. 6.18 the time to reach light threshold has been measured in pulsed operation. The threshold energy to the light is about 1 W s, the total energy for onset of lasing is approximately 2.5 W s. Both results are almost independent of the capacitor voltage and the average input power to the lamp. Considerable losses can appear in the fast-pulsed operation (kHz range) due to the threshold. For a description of the thermal lens the threshold of the excitation lamp must be taken into account (compare with measurements in Fig. 4.2).

The delay time until the maximum light power is reached can be estimated with the following considerations:

- The energy from the electrical circuit must heat the plasma to the operating temperature first. This energy is dependent on the plasma volume, the pressure and the plasma temperature.

$$E_{pth} = p_0 V_P \left(\frac{T_{th}}{T_0} - 1 \right), \tag{6.12}$$

p_0 = cold fill pressure, V_P = plasma volume
T_{th} = plasma temperature at light threshold, T_0 = room temperature.

- After reaching the light threshold of light emission the supplied power heats the plasma and is emitted as light.

$$E_{\text{Plasma}} + E_{\text{Light}} = E_{\text{Plasma}} + \eta_{\text{Lamp}} \int_0^\tau UI(t)\mathrm{d}t = \int_0^\tau UI(t)\mathrm{d}t \tag{6.13}$$

- Assuming a linear current increase up to the final value I_0 during the whole delay time and that half of the energy is emitted as light, an estimation for the delay time can be obtained:

$$\tau = \frac{4E_{\text{pth}}}{UI_0}, \tag{6.14}$$

$I(t) = I_0\, t/\tau$

typically $E_{\text{pth}} \approx 25$ W s and $\tau = 300$ µs.

To get very short pulses, a high voltage must be applied at the beginning of the pulse. The delay time of the light can be reduced slightly by using a high simmer current (Fig. 6.20).

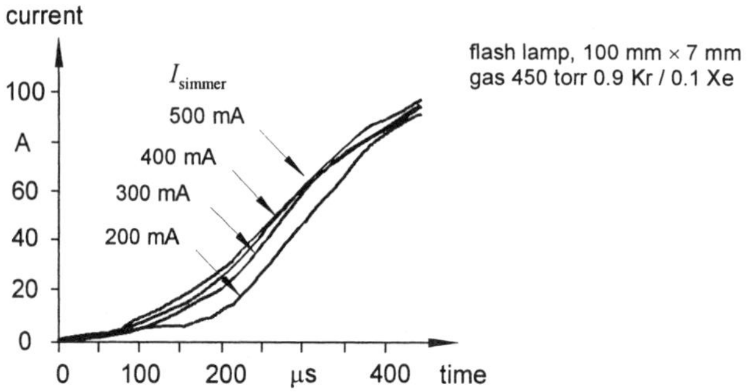

Fig. 6.20. Transition from the simmer regime to the high current regime at a constant voltage and various simmer currents

Description of the Current Progress

When switched on, the lamp behaves like a time-dependent inductance. This time dependence can be described as [6.11]:

$$L(t) = L_0 \mathrm{e}^{-t/\tau_{\text{d}}}, \tag{6.15}$$

L_0 = lamp inductance for $t = 0$,
τ_{d} = delay time.

The characteristic of the lamp, (6.10), has to be supplemented by the inductive part to

$$U = K_0\sqrt{I} + L(t)\frac{\mathrm{d}I}{\mathrm{d}t}. \tag{6.16}$$

The transient effect and the stationary behavior can be described approximately by (6.16). No inductive behavior of the lamp can be observed when turning off. Both parameters L_0 and τ, are, however, dependent on the simmer current [6.12] and on the applied voltage. An empirical determination for a 4" flash lamp, according to Fig. 6.21 yields

$$L_0 = K_\mathrm{L}U^{-\alpha}, \quad K_\mathrm{L} \approx 1.5\times10^8\ \mathrm{mH}\times\mathrm{V}^\alpha, \quad \alpha \approx 3.5,$$

$$\tau_\mathrm{d} = K_\tau U^{-\beta}, \quad K_\tau \approx 4.7\times10^3\ \mathrm{m}\times\mathrm{V}^\beta, \quad \beta \approx 2.0.$$

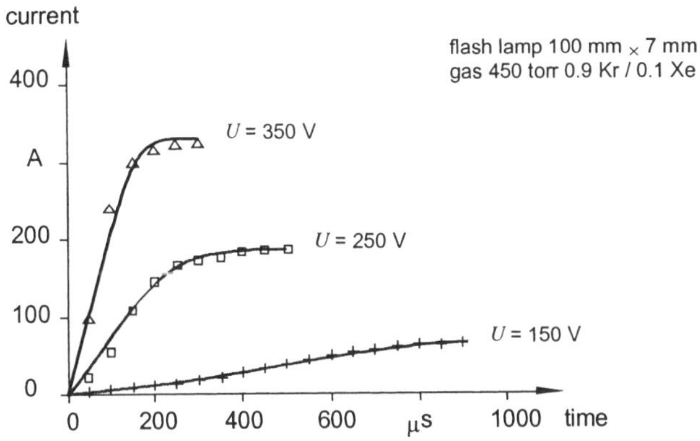

Fig. 6.21. Transition of the simmer current into the high current range at constant simmer current and various voltages of the charging capacitor

Recovering Time

After a high current the lamp needs a certain recovery time before a voltage can be applied again, otherwise the lamp would discharge immediately.

$$\tau_\mathrm{r} = \tau_\mathrm{r}(U, I, K_\mathrm{r}), \tag{6.17}$$

K_r = recovering constant.

The necessary period for recovery depends on lamp-specific parameters, but also on external parameters such as reverse voltage and pulse current.

6.5 Transient Progress

In the pulsed operation at high current the characteristic changes for the simmer mode. After a pulse the current decreases with a different characteristic than in the stationary operation. With a constant simmer current, dependent on the pulse duration and the applied voltage, the voltage shows a short overshooting ΔU [6.12]. This must be taken into account when designing the simmer power supply. Figure 6.22 shows the principal temporal behavior.

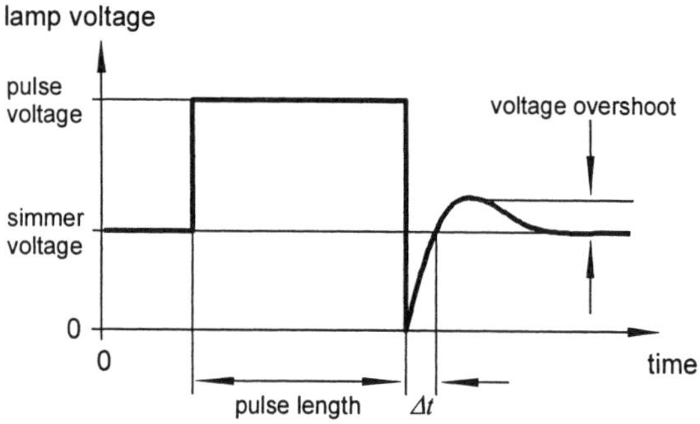

Fig. 6.22. Principal evolution of the lamp voltage after a high current pulse

The empirical relations for the delay time Δt and the voltage overshoot ΔU are

$$\Delta U = \frac{K_\mathrm{d}(U - U_\mathrm{d})T^\alpha}{I_\mathrm{S}}, \tag{6.18}$$

K_d = dynamic lamp constant, T = pulse duration,
U = operating voltage during the pulse, U_d = dynamic voltage,
$\alpha \approx 2$, exponential factor, I_S = simmer current,

$$\Delta t = \frac{K}{T^\beta}, \tag{6.19}$$

β = exponential factor, K = empirical constant.

For the dynamic case the simmer power supply must deliver the voltage $U + \Delta U$. This requires the additional power $\Delta P = \Delta U \times I_\mathrm{S}$ by the power supply or otherwise has to be provided by a separate capacitor.

Efficiency

Higher fill pressure improves the efficiency of a lamp. This also causes an increase of the trigger and operating voltages. In Fig. 6.23 the relative dependence of the 1 μm fluorescence in Nd:YAG on the Kr-fill pressure is shown, according to [6.18].

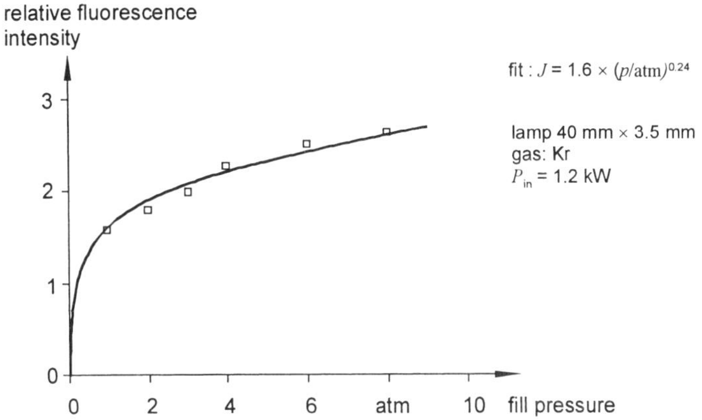

Fig. 6.23. Efficiency of an arc lamp for various cold fill pressures, [6.18]

6.6 Lifetime

The on-time of a solid state laser in industrial use is actually determined only by the lifetime of the excitation lamps. The lifetime is defined as the time in which the lamp is either destroyed mechanically or the light intensity falls below a certain percentage of the initial of value.

Both cases for the lamp failure can be controlled electronically. In the first case the simmer current is interrupted, in the second the laser radiation falls below an adjustable value and the necessary exchange of the lamps can be signaled.

Except for those lamps that fail relatively early, all lamps behave during operation similarly. In the course of time, particles sputtered from the cathode are deposited at the wall near the cathode. On the one hand this leads to a progressive wall darkening, on the other hand the quartz material becomes chemically attacked. In most cases the lamp remains operational, although the light has dropped by about 20%. If one continues to operate the lamp, microcracks arise starting from the cathode region until the lamp shatters more or less completely into fragments.

Cathode Ageing

Empirical attempts to determine the lifetime are given by (6.20) to (6.24). The necessary measurements, however, have been carried out only with a small number of lamps. They require a lot of effort and are very expensive. Results can be collected only as empirical observations of many users. The relations follow from physical considerations or empirical results.

In Fig. 6.24 the relative laser energy decrease per pulse dependent on the total pulse number of the flash lamps is shown. The light efficiency of the flash lamps decreases with time due to the chocolate-colored depositions on the inner wall of the lamp. Nevertheless, the lamps continue working.

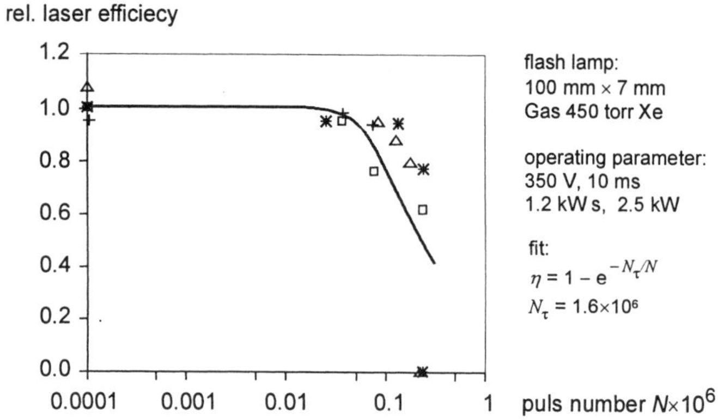

Fig. 6.24. Flash lamp lifetime, measurement results and empirical approximation

Lifetime in cw operation

The lifetime of the cathode is reduced with increasing temperature. The temperature is assumed to increase proportionally to the average power.

$$\tau \approx \frac{E_\tau}{P_{\mathrm{av}}}, \tag{6.20}$$

E_τ = lifetime constant, $\quad E_\tau \sim dl$,

P_{av} = average power.

Lifetime in pulsed operation, ms-range, 10 to 100 kW pulse power

In the pulsed operation the lifetime is limited by the depositions on the inner wall. The amount, which is emitted by the electrodes per pulse, depends essentially on the pulse energy. According to measurements at constant average pump power an empirical approximation for the attainable pulse count is

$$N = \left(\frac{E_\tau}{E_{\mathrm{p}}}\right)^{\beta}, \tag{6.21}$$

E_τ = lifetime constant, $\quad \beta = 2 - 2.5$ empirical constant.

The on-time of the lamps is the total number N of the pulses times the pulse distance $1/f$

$$\tau = \frac{N}{f} = \frac{1}{f}\left(\frac{E_\tau}{E_p}\right)^\beta, \tag{6.22}$$

$E_p = UIT$ = pump pulse energy, T = pulse duration.

Lifetime in the pulsed operation, μs range

In the μs-pulse operation a certain pulse energy may not be exceeded to achieve a fixed pulse number N. The pump energy at which the lamp will be just destroyed is called the explosion energy E_x, which is, according to [6.13]

$$\tau = \frac{N}{f} = \frac{1}{f}\left(\frac{E_\tau}{E_p}\right)^\beta, \tag{6.23}$$

K_x = explosion constant, E_x = explosion energy,
d = inner lamp diameter, l = arc length

and the total pulse number is given as

$$N = \left(\frac{E_\tau}{UIT}\right)^\beta, \tag{6.24}$$

$\beta = \beta(d)$ is 8 to 10 for $d \approx 8$ to 15 mm.

Cooling Temperature Influence

The cooling of the outer lamp surface also somehow determines the temperature inside the lamp. If the cooling is too low, the annealing temperature or a chemical process temperature of quartz might be reached. The UI characteristic of a lamp is also influenced by the cooling. Manufacturer recommendations for the maximum wall load are provided in Table 6.2, dependent on the cooling method.

Table 6.2 Maximal wall load of lamp materials, according to [6.13, 6.14, 6.15]

		Synthetic quartz	Doped quartz	Sapphire
Shock parameter R_T	W/cm	14.5		100
Convection cooling	W/cm^2	15		
Air cooling (forced air)	W/cm^2	30		
Water cooling	W/cm^2	300	160	2750

In stationary conditions the temperature difference between the inner wall side and coolant is

$$\Delta T = \frac{ks}{O} P_H, \tag{6.25}$$

P_H = supplied heat power, k = thermal conductivity.

The thermal conductance increases with the surface area O and is inversely proportional to the wall thickness s.

During operation, the electrodes, particularly the anode, become very hot. When designing the lamps one has to take into account that for cooling the connections need a sufficiently large cross section.

Critical Acoustic Frequencies

An acoustic resonance can appear in a flash lamp, if the pulse frequency f creates a stationary acoustic wave in the lamp. With short rectangular pulses even subharmonical pulse frequencies might activate these resonances. The arc is spatially changed by the resonances, and adopts the periodical structure of the stationary wave. In these frequency domains laser power and energy may be changed due to the changed pumping light conditions. The lowest frequencies for the single modes are, [6.16]:

$$f_l = n\frac{u}{2l}, \qquad \textit{longitudinal,} \tag{6.26}$$

u = sound velocity,
$n = 1, 2, \ldots$
l = distance between the electrodes

$$f_\phi = 1.84\frac{u}{2\pi r}, \qquad \textit{azimuthal,}$$

$2r$ = inner lamp diameter

$$f_r = 3.83\frac{u}{2\pi r}, \qquad \textit{radial.}$$

The resonant frequencies can superpose to $f_2^2 = f_l^2 + f_r^2$. The sound velocity u is given by

$$u = \left(\frac{c_\mathrm{p} p}{c_\mathrm{v} \rho}\right)^{0.5}, \tag{6.27}$$

c_i = specific heat at constant pressure or volume,
p = operation pressure,
ρ = gas density.

Using thermodynamic quantities (6.27) can be rewritten as

$$u = \left(\frac{c_\mathrm{p} RT}{c_\mathrm{v} M}\right)^{0.5}, \tag{6.28}$$

$c_\mathrm{p}/c_\mathrm{v} = 5/3$ for noble gases,
$R = 8.314$ W s/(mol K) = gas constant,

T = mean gas temperature in K,
M = averaged atomic or molecular weight, 131 for Xe and 84 for Kr.

For the longitudinal resonances essentially the temperature along the lamp axis, for the radial and azimuthal resonances the average temperature between wall and lamp center is decisive.
In [6.17] it is shown that the mean temperature, in the measured range, increases linearly with the pump energy E_p.

$$T = T_0 + FE_p/V, \qquad (6.29)$$

T_0 = environmental temperature (coolant),
$V = \pi r^2 l$ (volume),
F = 244 K/(W s/cm^3), factor, [6.17].

Typical results for the sound velocity and the temperature during the operation are given in [6.16] as u = 500 to 590 m/s and T = 2000 to 3000 K.

Example 6.1 Resonant frequencies of an arc lamp with 6 mm inner diameter and 150 mm arc length

For u = 500 m/s follows
$f_l = n \times 166$ Hz, n = 1, 2, 3. . ., f_ϕ = 4.9 kHz, f_r = 10.3 kHz.

6.7 Spectral Characteristics

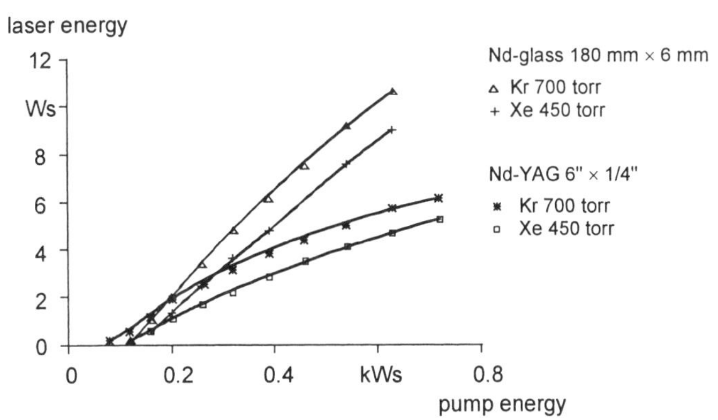

Fig. 6.25. Nd:YAG and Nd:glass laser efficiency dependent on the flash lamp gas

The spectral emission of the lamp must be matched as well as possible to the laser material used. Radiation outside the pump bands leads to unnecessary heat in the laser medium and should therefore be suppressed. Lamp radiation at the laser wavelength becomes amplified in the laser material and reduces the useful inversion.

In Fig. 6.25 the efficiencies of two lamps with Xe and Kr filling are compared. Although the overall light efficiency of Xe lamps is better, a higher laser energy with Kr lamps is achieved since Kr lamps are spectrally better matched to the main absorption bands of Nd.

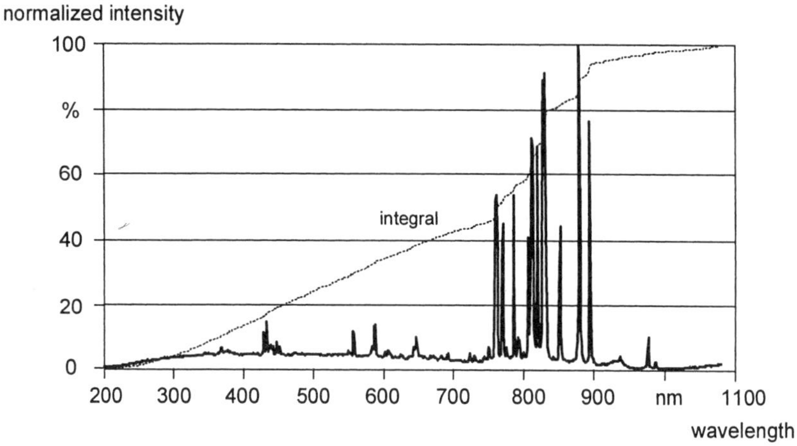

Fig. 6.26. Emission spectrum of a Kr arc lamp in cw operation at P_{in} = 4.5 kW with a fill pressure of 2290 torr and the dimensions l = 14 cm and d_i = 6.3 mm [6.23]

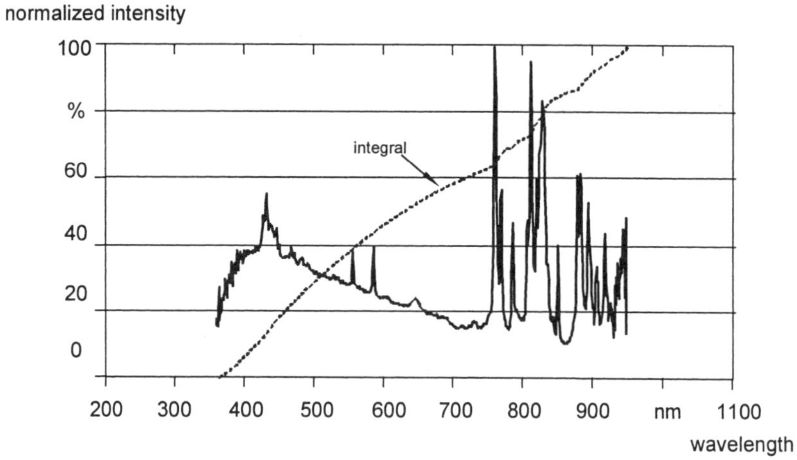

Fig. 6.27. Flash lamp spectrum at an average pump power of 3 kW and a pulse duration of 4 ms, arc length l = 100 mm, inner diameter d_i = 7 mm, 450 torr Kr with 10% Xe, K_0 = 16.3, [6.23]

Beside the spectral intensity in Figs. 6.26 and 6.27 the integrated optical power up to the respective wavelength is also shown. The optical power of a cw lamp up to 700 nm is approximately 42%, in the range 700 to 900 nm 52% and 6% above 900 nm. In a pulsed lamp 50% of the light power is emitted below 650 nm.

The better spectral overlap of Kr lamps with the excitation spectrum of Nd:YAG can be seen when comparing the emission spectra. In addition, for the Xe lamp the lamp emission band at 1 μm has to be suppressed in order to avoid the amplification by the laser medium.

Further spectral qualities such as transmission of the flow tubes and filter are provided with the corresponding curves in Chap. 12. Therefore, in Table 6.3 only some specific properties are listed.

Table 6.3 Spectral properties of quartz

Material	Dopant	UV edge 10 – 90%	Solarization
Synthetic quartz	-	160	No
Normal-quartz	-	190	Yes
Heraeus M-235	Ti:	220 – 250	No
Heraeus M-382	Ce:	330 – 390	No

Efficiency and Heat Power in Nd-YAG

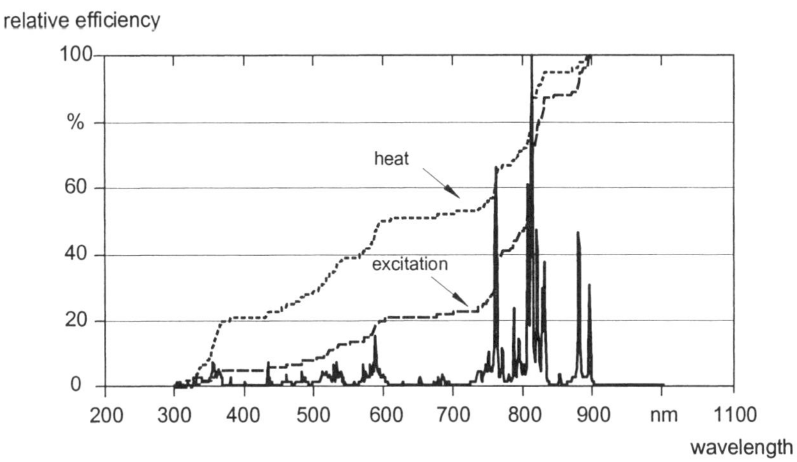

Fig. 6.28. Spectral excitation of Nd-YAG by an arc lamp

Depending on the absorption by the laser material and the emission of the excitation lamp the laser material becomes excited but also heated. In the Figs. 6.28 and 6.29 the effective excitation, spectrally resolved, is shown. The emission of an arc or a flash lamp is folded with the Nd:YAG absorption and normalized. The calculations are based on a 5 mm thick plate with a Nd doping of 1.1 at%.

The figures also show the spectrally integrated efficiency. From the quantum defect the relative amount was calculated, which contributes to the heat in the laser material. It is labeled in the illustrations as heat.

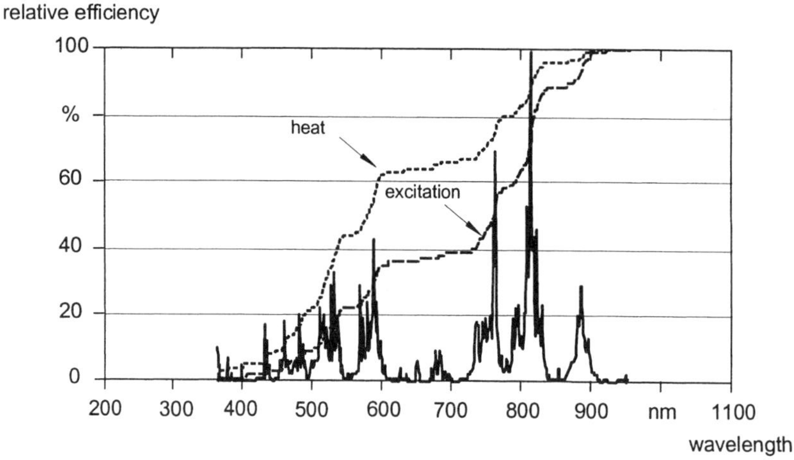

Fig. 6.29. Spectral excitation of Nd-YAG by a flash lamp

In both cases the excitation up to a wavelength of 700 nm is low, however, the heat is already considerable. By suitable filtration of the pump light the thermal lens can be reduced at the expense of the efficiency.

6.8 Radiation Characteristic

The knowledge of the radiation characteristic of an excitation lamp is necessary to design the pump light path in the cavity for the geometric arrangement and for the desired inversion profile.

The time-averaged angular distribution of the light intensity from the plasma surface can be described as a generalized cosine distribution [6.19, 6.20]

$$J(\alpha) = J_0(\cos\alpha)^{\chi}, \tag{6.30}$$

χ = distribution coefficient.

A measured angular distribution of the light intensity in axial and radial directions behind a small aperture in front of the lamp surface is shown in Fig. 6.30. From this measurement the plasma surface distribution has to be calculated.

The radial distribution coefficient χ most likely depends on the plasma temperature. In cw operation, therefore it depends on the current density and in pulsed operation on the supplied pulse energy. At high current densities (energies) the exponent χ converges towards 2, for low current densities (energies) the lamp behaves rather like a line emitter ($\chi \approx 8$).

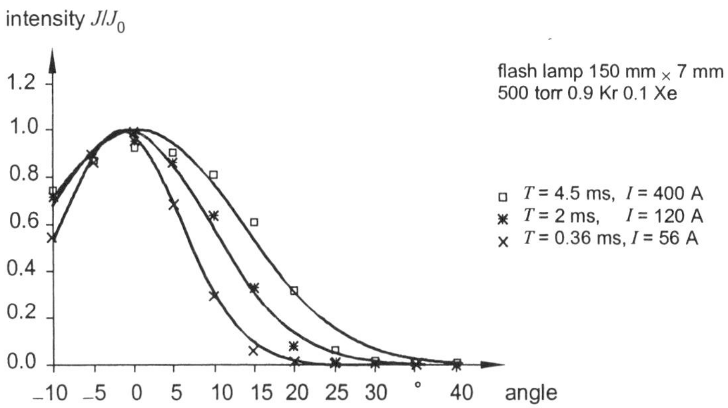

Fig. 6.30. Distribution of the flash lamp light intensity at the lamp surface. This distribution must still be corrected for the plasma surface, [6.19, 6.20]

A formulation to describe the dependence of the distribution coefficient on the current intensity and pulse duration is (see also [6.20])

$$\chi \propto F_{\mathrm{in}}^{-\alpha}, \tag{6.31}$$

$\alpha \approx 0.25.$

Figure 6.31 shows the distribution coefficient fitted to measurements of a flash lamp at different pulse energies.

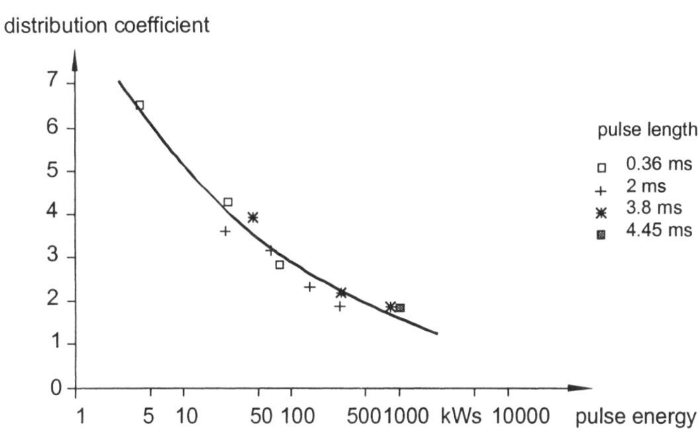

Fig. 6.31. Distribution coefficient at the plasma surface dependent on the pulse energy [6.19, 6.20]

6.9 Recommendations for Operation

The lamps have to be connected with the correct polarity. Confusing the anode and cathode leads to the lamp initially working correctly. After a short operating time the wall, however, becomes black around the original anode, the optical power of the lamp falls off and the simmer characteristic changes. In the case of vertical operation of a lamp the cathode should be down, otherwise the sputtered cathode particles fall on the anode and a similar effect occurs. The lamp has to be installed in very clean conditions. Fingerprints and dirty deposits can burn in and thereby cause an early failure of the lamp.

In Table 6.4 characteristic parameters are given that describe the arc and flash lamp function.

Table 6.4 Examples for lamp parameter and operational data

			Flash lamps		Arc lamp
Gas			Xe	Kr	Kr
Fill pressure	p_0	torr	450	700	2840
Fill temperature	T_0	K	273+20	273+20	273+20
Arc length	l	mm	150	150	75
Diameter	d	mm	7	7	5
Dead volume	V_d	mm^3	200	250	100
Power	P	kW	5	5	2.5
Current	I	A	-	-	20
Voltage	U	V	-	-	123
Pulse duration	T	ms	10	10	-
Pulse frequency	f	Hz	10	10	-
Pulse energy	E	W s	500	500	-
Trigger voltage	U_Z	kV	3.5 @dc	4.9 @dc	12 @ 20 kV/µs
Booster voltage	U_B	V	151	197	456
Simmer minimum	U_{min}	V	102	108	98
Lamp parameter	K_0	V/√A	19	21	-
Inductance	L_0	µH	150	170	-
Base voltage	U_0	V	-	-	88
Differential impedance	r	Ω	-	-	1.7

7 Excitation by Diodes

Semiconductor or diode lasers are exceptional in their high electrical/optical efficiency. Therefore, they are of increasing importance for pumping solid-state lasers. The wavelength of the diodes can be tuned to an absorption band of the laser crystals only slightly above the upper laser. The efficiency of the pump process is higher and simultaneously the thermal load is less than in the case of excitation by lamps. If the diodes can be produced at lower costs, the solid-state laser will increasingly be used in new fields, even in those which are dominated at present by CO_2 lasers.

7.1 Concepts

Depending on the geometry of the laser medium, the direction of the excitation and the coupling of the diode radiation to the crystal, several concepts can be distinguished. In Table 7.1 these concepts with their distinctive characteristics are listed. In the following they are discussed each in turn.

Table 7.1 Concepts for diode pumped solid-state lasers

Geometry of the laser medium	rod	disk	fiber	slab	plate
Direction of the pump radiation relative to the laser radiation	parallel perpendicular	parallel			
Optics for pump radiation	reflector fibers	fibers	glass plate	fibers	reflector
Pump radiation	divergent	1 axis collimated	2 axes collimated		
Temperature gradient relative to the laser radiation	perpendicular	parallel	perpendicular		
Cooling of the laser medium	liquid	contact cooling	air		

Longitudinally Pumped Rod

A longitudinally pumped laser rod is excited at the endfaces by optically or fiber-coupled diodes. An example is given in Fig. 7.1. By matching of the pump light distribution to the desired mode volume and due to the almost complete absorption of the pump light a high efficiency with a good beam quality can be reached. Until now the achieved power, however, lies only in the 10 W range.

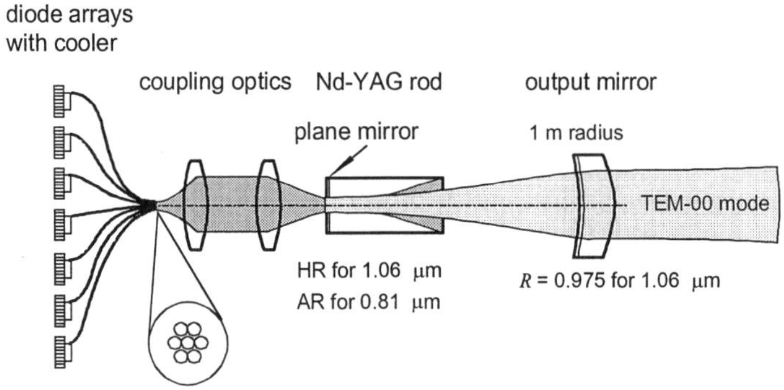

Fig. 7.1. Longitudinal diode-pumped rod laser [7.1]

Transversally Pumped Rod

By transversal pumping, the pumped volume is extended and thus the attainable total power can be increased considerably. Figures 7.2 to 7.4 provide some examples for the direct pumping and for the transfer of the diode light with fibers.

It is plausible to arrange the pump modules, the laser-diode arrays, including the cooler without further optics directly around a laser rod. In this case, the flow tube with the laser rod acts as a cylinder lens.

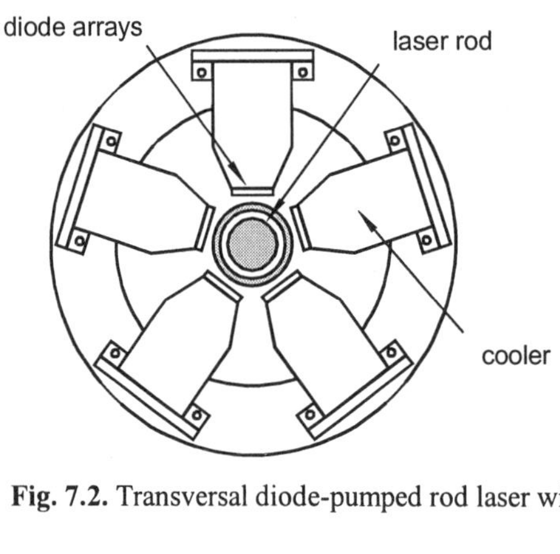

Fig. 7.2. Transversal diode-pumped rod laser without optics [7.2]

A particularly simple construction is shown in Fig. 7.3. A stack of laser arrays is arranged at one rod side. By a suitable shape of the reflector on the opposite side the desired homogeneous pump light distribution can be achieved in the laser rod.

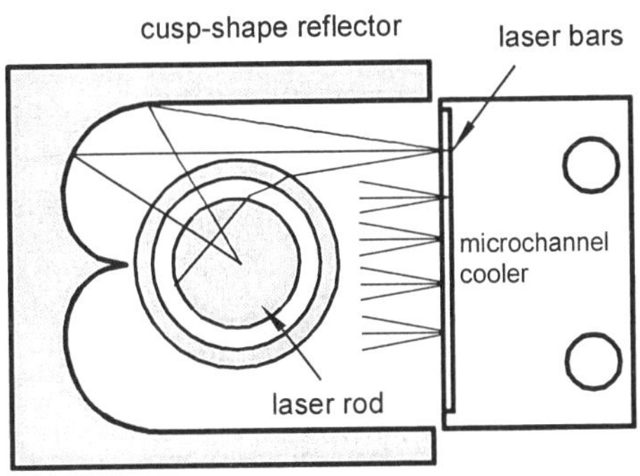

Fig. 7.3. Diode-pumped rod laser with cusp-shaped reflector [7.24]

If the pump light is transmitted to the active medium with fibers, the mechanical construction can be separated completely from the laser head and the pump diodes. With a radial odd number of pump diodes and a reflector opposite to the fibers the absorption length in the laser rod is doubled. In [7.24] the experience with three- and five-fold geometry is reported.

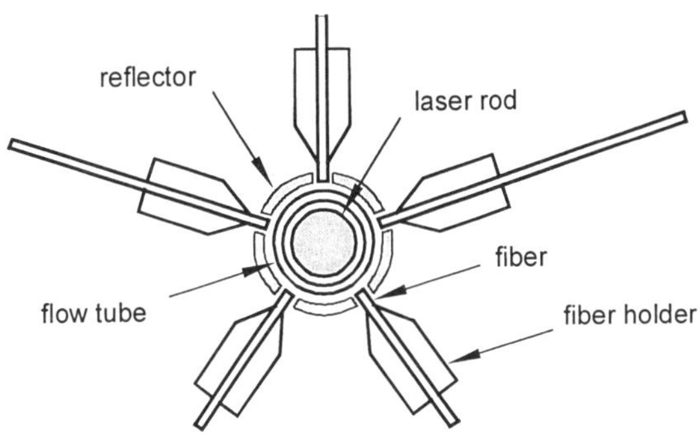

Fig. 7.4. Diode-pumped laser rod, pumping light supplied through fibers [7.24]

The cooling of transversally pumped rods is carried out perpendicular with respect to the laser radiation and therefore the thermal lens influences the laser beam. With other concepts the thermal lens can be avoided also at high powers, or at least significantly reduced.

Longitudinal Disk

Due to the geometry of a disk the cooling and the lasing take place via the faces. This by itself ensures a good beam quality. For longitudinal excitation the laser material should have a high absorption to reach a good efficiency. Ytterbium can be doped in YAG with considerably higher concentrations than neodymium, however, due to its energy level diagram it must be cooled to low temperatures [7.26]. The cooling can be carried out either in direct contact with a cooler or by a liquid. An example from [7.23] is shown in Fig. 7.5.

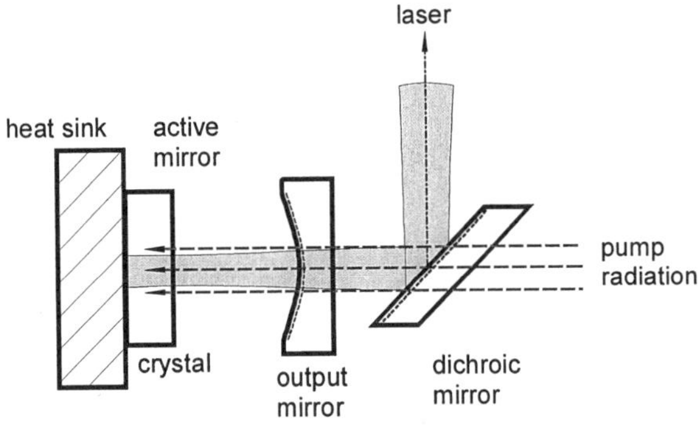

Fig. 7.5. Longitudinally pumped disk, according to [7.23]

Transversally Pumped Disk

To increase the absorption of the disk, it can be pumped from the edge, using the full diameter for absorption. To reduce the influence of the thermal lens the cooling should therefore be carried out in the direction of the radiation. Setups of this kind have not yet been realized.

Longitudinally Pumped Fiber

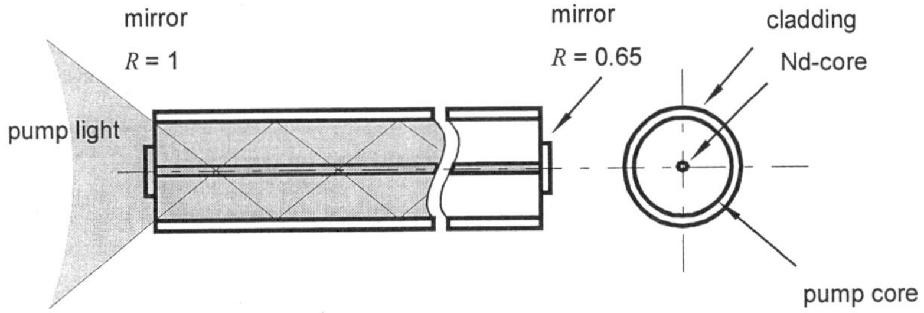

Fig. 7.6. Longitudinally pumped fiber laser [7.27]

A thin, doped fiber core in a second undoped one with relatively large diameter shows a noticeable absorption. The focused diode pump radiation at the endface is absorbed by the doped core due to multiple total reflection in the second core over the whole fiber length. With a mirror at both endfaces a laser can be realized. In [7.27] a design was reported in which a 12 μm Nd-doped core was pumped longitudinally in a 400 μm fiber. An output power of 10 W in the fundamental mode was obtained. The principle of this design is outlined in Fig. 7.6.

Transversally Pumped Fiber

Longitudinal pumping is limited by the endface diameter of the fiber. If one succeeds, by suitable preparation of the fiber cladding, to feed the pump radiation at periodic distances transversally into the doped core, the attainable laser power could be increased. Constructions of this kind are not known at present.

Slab

In [7.28] and [7.29] a slab design (Fig. 7.7) with internal total reflection is used to achieve a long path in the active medium. The slab offers a large cooling surface and also a large surface for the pump light. Due to the internal total reflection, however, the suppression of parasitic modes is difficult. A slight tilt of the polished side faces or drilling holes outside the resonator path of the laser beam has been proposed.

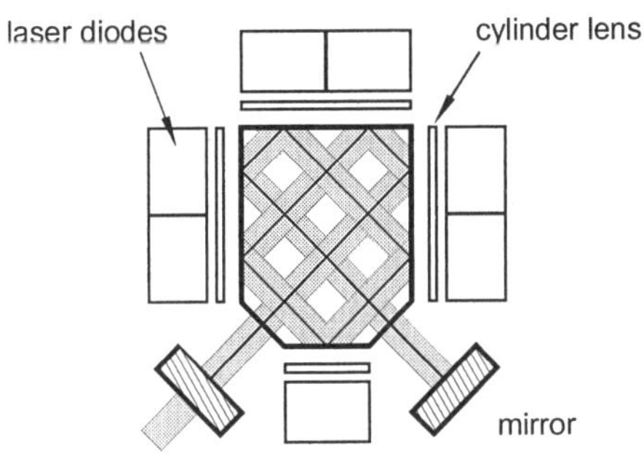

Fig. 7.7. Slab resonator [7.28, 7.29]

Transversally Pumped Slab

With the slab concept a high pump power can be supplied to the side faces without complicated optics. A good beam quality can be achieved with the slab principle. However, no high cw power with good beam quality has been reached with this approach up to now.

Cooling is essential for the obtainable output power. Since the permissible temperature range is very narrow, the cooling must be designed to be as efficient as possible. With the diodes directly assembled on a microchannel cooler (Fig. 7.8), a diode-pumped solid-state laser can be scaled to the kilowatt range.

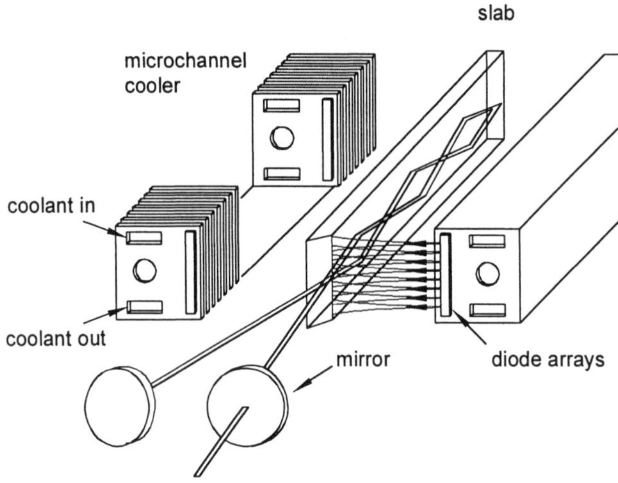

Fig. 7.8. Diode-pumped slab laser with microchannel cooling, according to [7.4]

Some details on diode pumped lasers are given in Table 7.2.

Table 7.2 Diode-pumped solid-state lasers

Type	Crystal	Efficiency η / %	Power P_{av}/W	Energy E/Ws	Remark	Ref.
Rod						
long.	Nd-YAG	4.4[a)]	0.6		00-mode, $M^2 \approx 1$	[7.1]
transv.	Nd-YAG	17.8[a)]	196	0.675	400 μs pulsed operation	[7.2]
transv.	Nd-YAG	26[b)]	390		multimode, cusp reflector	[7.24]
transv.	Nd-YAG	44[b)]	320		multimode, fiber-pumped	[7.25]
transv.	Nd-YAG	20[b)]	80		00-mode, fiber-pumped	[7.25]
Disk						
long.	Tm:Ho:YAG		0.5		Ti-sapphire pumped	[7.23]
	Yb:YAG		80			[7.26]
Fiber						
long.	Nd	25[b)]	9.2		$M^2 \approx 2$	[7.27]
Slab						
transv.	Nd-YAG	23[b)]	275	0.11	100 μs pulsed operation	[7.4]

[a)] optical /electrical, [b)] optical / optical, long. = longitudinal, transv. = transversal

Longitudinal Slab

A setup for a variant of a longitudinally pumped slab is shown in Fig. 7.9.

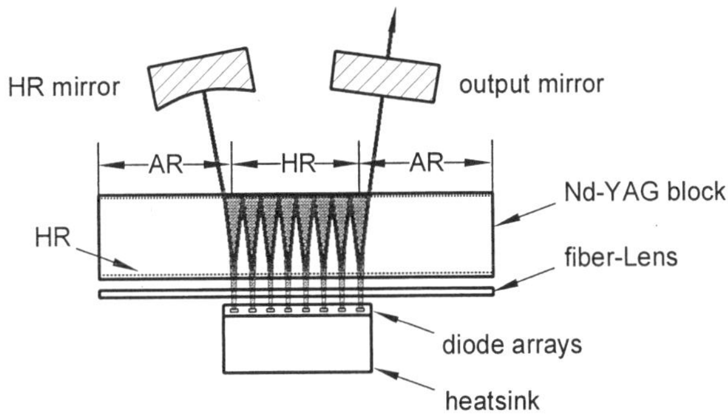

Fig. 7.9. Diode-pumped slab laser [7.3]

7.2 Layout of the Laser Diodes

The simplest laser diode, a homostructure, consists of a gallium-arsenide (GaAs) crystal with two differently doped regions, a Se-doped n-region and the Zn-doped p-region. This doping produces additional free electrons in the conduction band of the n-region and positive holes (lack of electrons) in the p-region. The active area is the transition from p to n. Applying a voltage causes a drift of electrons and holes to the p-n junction, where they recombine and emit light. This system is well known as the light emitting diode (LED). The wavelength of the emitted light is given by the band gap energy:

$$\lambda_{[\mathrm{nm}]} = \frac{hc}{\Delta E} = \frac{1240}{\Delta E_{[\mathrm{eV}]}}, \tag{7.1}$$

$c \approx 3\times10^8$ m/s = light velocity
ΔE = band gap of the semiconductor
$h \approx 4.135\times10^{-15}$ eV/Hz = Planck's constant.

The band gap is determined by the semiconductor material, it can, however, be varied in a wide range by mixing two semiconductor systems.

The p-n transition can also amplify light by induced emission. Although the system is rather complex, it is to a first approximation a four-level system, as discussed in Chap. 1. By injecting a current, a gain is produced in the transition, and laser oscillation can start, if the gain compensates the losses. The feedback in this case occurs by the two surfaces of the cleaved crystal. Due to the crystal structure, the flatness is sufficient. The refractive index of GaAs is 3.5 and yields a rather high reflection of 32%.

Figures 7.10 and 7.11 show the schematic arrangement of layers of a GaAlAs laser diode. This system is called a double heterostructure, because of the different materials GaAs and GaAlAs.

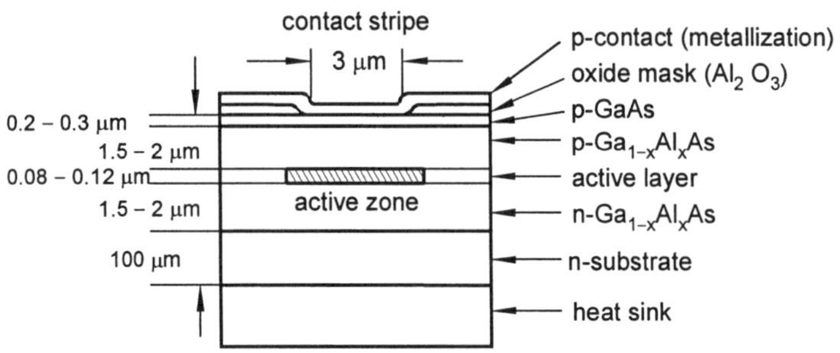

Fig. 7.10. Schematic diagram of double heterostructure [7.5]

The necessary inversion is obtained by a special semiconductor geometry. The charge carriers are confined by the two layers above and below the active layer in the vertical direction, since these have a higher band gap than the active layer due to their special doping. They act as barriers for charge carriers injected in the active volume. By the increased aluminum contents a lower refractive index in these layers with respect to the active layer is achieved at the same time. The light is guided in the vertical direction as in a waveguide. The active layer has a width of $\approx 0.1\ \mu m$.

Because of the laterally limited contact stripes, inversion is only produced in the active zone. A light wave can only be amplified in this zone (gain-guided).

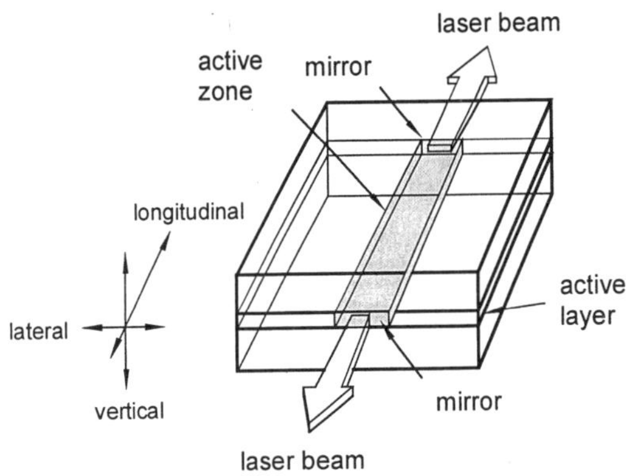

Fig. 7.11. Sketch of an oxide stripe laser diode [7.5]

The optoelectrical characteristic of a laser diode is shown in Fig. 7.12. At low current only spontaneous emission occurs and the output power is very low. Above the threshold I_{th}, self-oscillation starts. Stimulated emission exceeds spontaneous emission and the output power increases strongly.

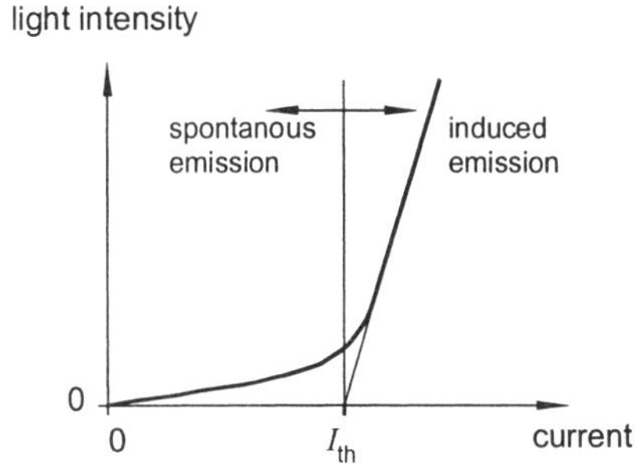

Fig. 7.12. Characteristic of a laser diode

The power, which can be generated with a single diode, is limited by the mirror load and particularly in the cw mode by the thermal load. To increase the power the active zone can be broadened (broad stripe lasers) or several active zones are arranged close together. Then the optical fields influence each other and the phases are coupled (Fig. 7.13). These arrays produce an angular spectrum of the output power as shown in Fig. 7.14.

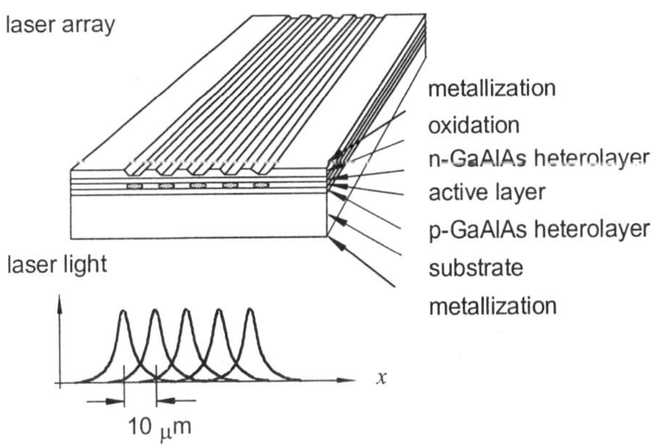

Fig. 7.13. Design of a laser array and its transverse intensity distribution [7.5]

For high power diodes several broad stripe lasers or arrays with narrow (≈ 3 μm) single stripes are arranged on one chip up to the width of a centimeter. Each laser oscillates independently and therefore emits incoherently. For the pumping of solid-state lasers this is, however, of no importance.

Figure 7.15 shows the typical dimensions of a laser diode. With a suitable heat sink the semiconductor chip was assembled on a diamond layer, optical cw-powers up to 100 W have been achieved for short times, [7.7].

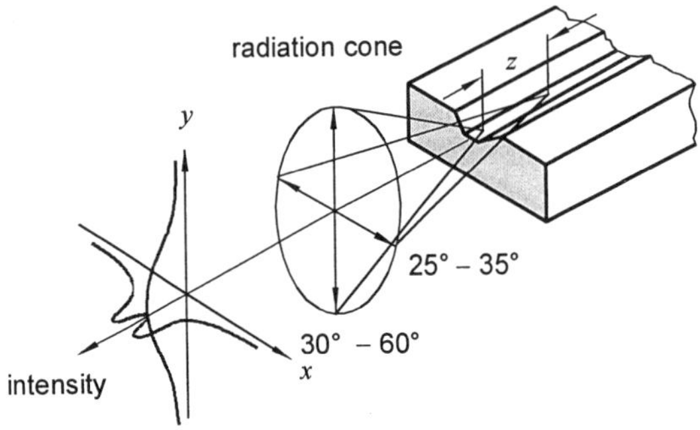

Fig. 7.14. Angular spectrum of a linear diode array [7.6]

A single laser diode completely assembled, hermetically sealed in a system unit is shown in Fig. 7.16. To obtain a good heat distribution, the diode is mounted on a copper block, which itself is cooled by a Peltier element.

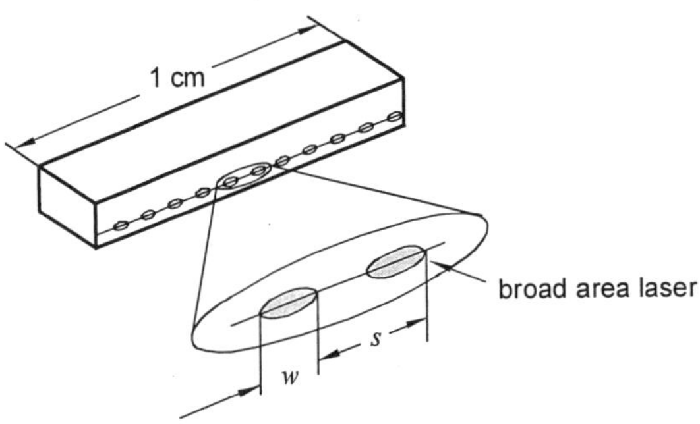

Fig. 7.15. Structure and dimension of a broad area array [7.7]

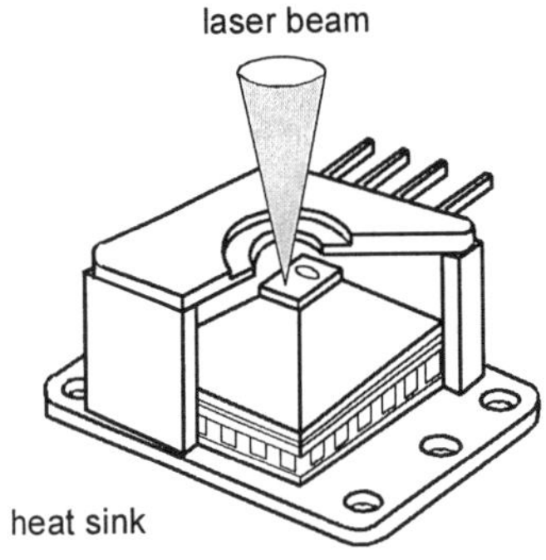

Fig. 7.16. Assembled laser diode with integrated Peltier cooler [7.8]

Production

For the fabrication of GaAs diodes the desired sequence of layers is grown on a GaAs disk, using for example the metal organic chemical vapor deposition (MOCVD) process. From the gaseous phase Ga and As form crystalline layers on the substrate. After deposition of a SiO_2 insulation layer and structuring the contacts the wafer is polished to the desired thickness. By gold-coating on both sides the electrical connections are produced. After diamond cutting, the wafer can be cleaved into the real diode bars. On the cleaved mirror facets the low or high reflective layer is coated. Figure 7.17 shows this process in a simplified form [7.15]

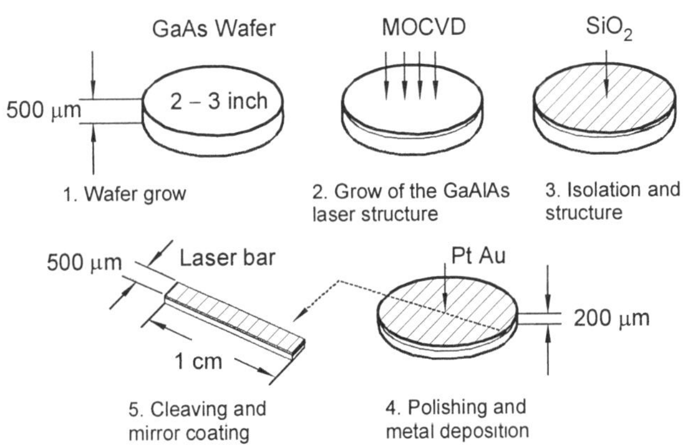

Fig. 7.17. Simplified fabrication process of laser diodes [7.15]

7.3 Semiconductor Materials

The band gap, and with it the laser wavelength, are determined by the material, the composition and the design. Figure 7.18 shows some examples. By suitable composition of binary components to ternary systems the emission wavelengths can be varied in a broad range as indicated by the solid lines in Fig. 7.18. In the case of GaAs and AlAs the variation of the gallium and aluminum contents ($Al_xGa_{1-x}As$, $0 \leq x \leq 1$) allows wavelengths between 600 nm (2.15 eV for AlAs) and 900 nm (1.42 eV) for GaAs. Since the lattice constant for all of these compositions differs very little, layers of various compositions can easily be grown without any large tensions between the layers.

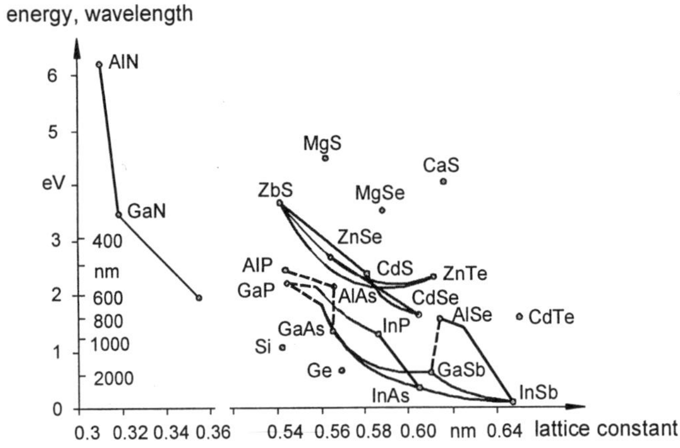

Fig. 7.18. Wavelength ranges for various material systems [7.17]

For quaternary systems, i.e. systems with four components, the wavelength range can be adjusted by the four binary components. For example in Fig. 7.19 with GaInAsP emission between 550 nm and 3500 nm can be obtained. The diode emission wavelengths match the pump bands of many rare earth ions and offer the possibility to excite solid-state lasers very efficiently.

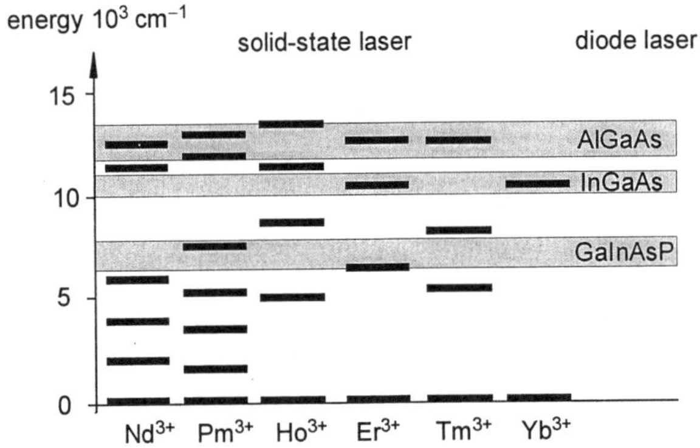

Fig. 7.19. Diode laser emission lines matching the absorption bands of rare earth ions, [7.18]

The specification of the system requires an evaluation of availability, efficiency, lifetime and facet load of the laser diodes. In Fig. 7.20 the efficiency and the safe facet load for the different wavelength ranges and systems are compiled. In Table 7.3 the band gaps of III-V systems and the refractive indices for the corresponding wavelength are listed.

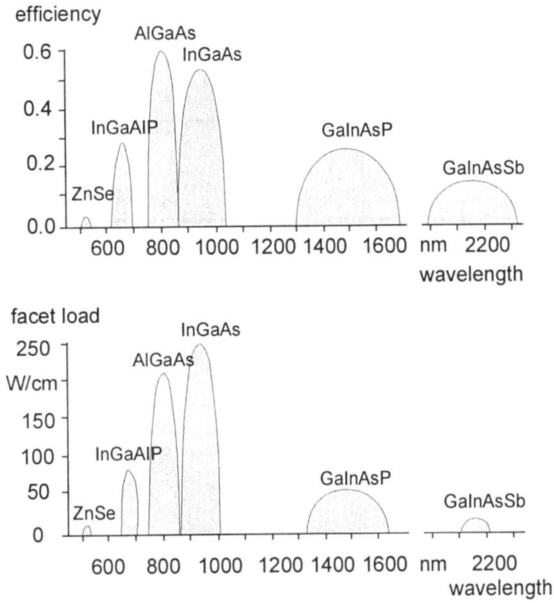

Fig. 7.20. Efficiency and facet load for various material systems as a function of the wavelength [7.19]

Table 7.3 Band gap energy and refractive index of III-V-material systems [7.16, 7.20]

Material	Energy eV	Wavelength µm	Refractive index	Lattice constant nm
Ge	0.66	1.88	4.0	0.565
Si	1.11	1.15	3.5	0.543
IV - IV				
α-SiC	2.8 – 3.2	0.39 – 0.44	2.63	*a*:0.308; *c*:1.511
III - V				
AlP	2.45	0.52	3.0	0.545
AlAs	2.16	0.57	3.2	0.587
AlSb	1.58	0.75	3.8	0.614
GaP	2.26	0.55	3.3	0.545
GaAs	1.42	0.87	3.6	0.565
GaSb	0.73	1.70	4.0	0.609
InP	1.35	0.92	3.5	0.587
InAs	0.36	3.5	3.8	
InSb	0.17	7.3	4.2	0.648
II - VI				
CdS	2.42	0.51	2.5	*a*:0.414; *c*:0.671
ß-ZnS	3.66	0.34	2.4	0.542
CdTe	1.5	0.83	2.8	0.648
IV - VI				
PbS	0.41	3.02	3.7	0.593

7.4 Technical Data of the Laser Diodes

Electrical and Optical Characteristics

The *UI* characteristic is described approximately by

$$U \approx U_0 + rI, \tag{7.2}$$

U_0 = voltage at $I = 0$,
r = differential resistance,
I = current,

and the electrical power P_{in} dissipated in the diode follows as

$$P_{in} \approx (U_0 + rI)I. \tag{7.3}$$

The example in Fig. 7.21 demonstrates the threshold and the linear dependence of the laser power P on the current.

$$P = \eta_I(I - I_{th}) \tag{7.4}$$

η_I = current efficiency,
I_{th} = threshold current.

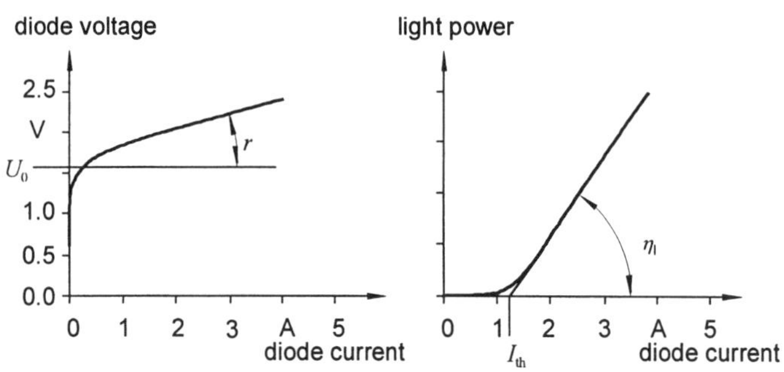

Fig. 7.21. Characteristics of a laser diode array, [7.9]

The threshold current I_{th} is temperature dependent:

$$I_{th} = I_0 e^{(T_j - 300\ \mathrm{K})/T_0}, \tag{7.5}$$

I_0 = threshold current for $T = 300$ K,
T_j = junction temperature for the current I_{th},
$T_0 = 130 - 140$ K at 300 K = temperature-dependent system constant.

With (7.2) and (7.3) follows for the laser power P as a function of the electrical power P_{in}

$$P = \eta_I \left(\sqrt{\frac{U_0^2}{4r^2} + \frac{P_{\text{in}}}{2r}} - \frac{U_0}{2r} - I_{\text{th}} \right). \tag{7.6}$$

With these approximations, the total efficiency P/P_{in} has a maximum for a pump power of

$$P_{\text{in}} = \sqrt{I_{\text{th}}^2 + I_{\text{th}} U_0 / r \left(2I_{\text{th}} r + U_0\right) + 2I_{\text{th}}\left(I_{\text{th}} r + U_0\right)}. \tag{7.7}$$

Increasing the diode current above this value will reduce the efficiency as, for example shown, in Fig. 7.22. An efficiency - electrical to optical - of more than 30%, for special structures even more than 50%, is state-of-the-art.

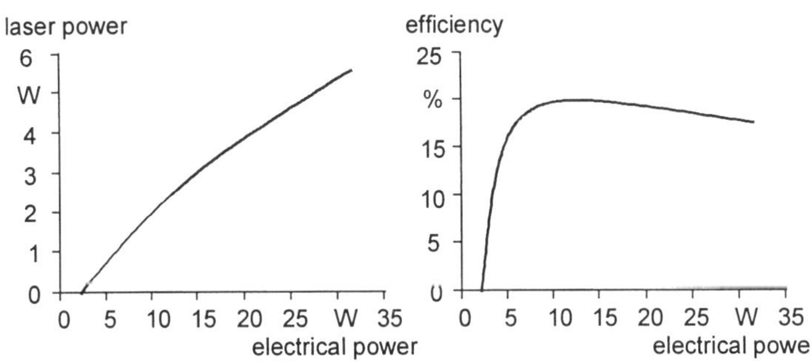

Fig. 7.22. Typical power efficiency of a laser diode

In Fig. 7.23 is shown the angular intensity spectrum of a typical diode.

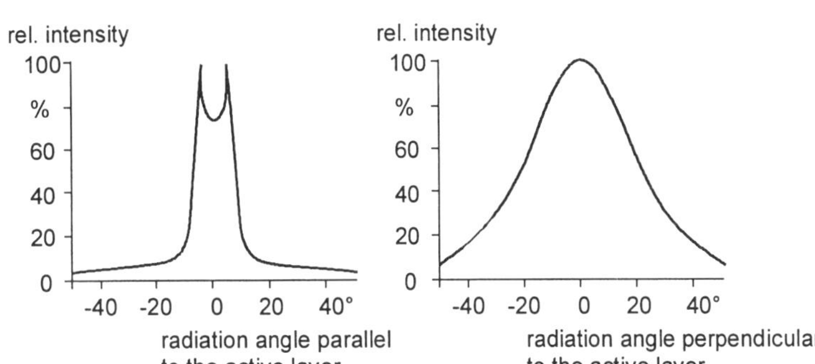

Fig. 7.23. Angular intensity spectrum of diodes

Due to diffraction, the divergence perpendicular to the plane of the thin, active layer is significantly larger than that parallel to the layer. This must be taken into account in the design of the optical system. Perpendicular to the p-n transition the beam parameter product is about 0.4 mm mrad (emitter width 1 μm), and parallel about 11 mm mrad (emitter width 200 μm).

The emission wavelength λ is dependent on the junction temperature T_j

$$\lambda(T) = \lambda_0 + K_\lambda(T_j - 300\ \mathrm{K}), \tag{7.8}$$

λ_0 = wavelength at 300 K,
K_λ = temperature coefficient of the wavelength,

and is shifted due to the dissipated electrical power in the pulsed operation as well as in the cw mode. An example is given in Fig. 7.24 with a shift of 0.3 nm/K, [7.11].

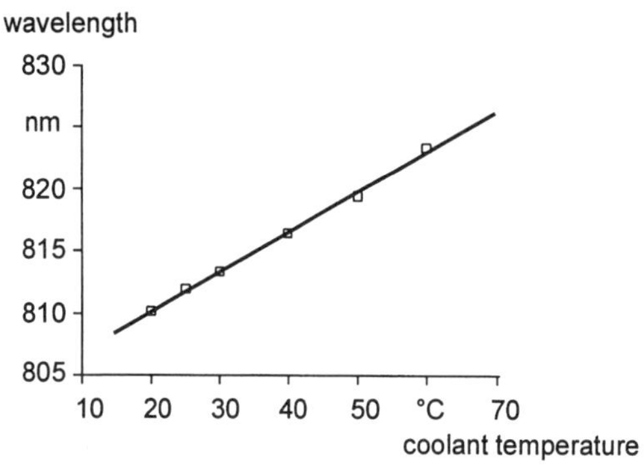

Fig. 7.24. Wavelength shift versus the coolant temperature [7.11]

Lifetime

The lifetime of a laser diode array as well as the dominant influencing parameters are not yet known. However, the lifetime decreases strongly with increasing junction temperature. Besides spontaneous failures the laser power decreases progressively with the hours of operation.

The degradation is quoted in [7.13] as 5 to 10% after 1000 h and 70°C temperature. In [7.12] a 10 W laser diode was operated with constant current and the optical power was measured. After 1000 h the laser power had decreased to approximately 8 W. Another method to indicate the degradation is to control the current for constant optical power. In [7.11] the regulation current is between 0.4 and 1.4 mA/h and so the lifetime of the used 20 W diodes lies between 4200 and 15 000 h at a laser power decrease of 20% with respect to the initial value. The lifetime τ is given in [7.14] as

$$\tau = \mathrm{f}(P_{in})\mathrm{e}^{E_\tau / kT_j}, \tag{7.9}$$

$f(P_{in})$ = power dependent coefficient,
E_τ = activation energy for ageing.

Table 7.4 Physical data of semiconductors and assembling materials

Material	ρ g/cm^3	c W s/g K	k W/(cm K)	α 10^{-6}/K	dn/dT 10^{-4}/K	Ref.
Semiconductors						
Ge				5.2		[7.20]
Si	2.30	0.75	1.4 – 1.57	4.2		[7.20, 21, 22]
GaP				5.81		[7.20]
AlAs				5.20	1.6	[7.20, 32]
AlGaAs		0.12	5.2			
GaAs	5.32	0.16	0.45	5.73 – 6.0 – 7	3.7	[7.20, 32]
Solders						
Pb		0.34	29			
Into		0.25	2.48			
Sn		0.64	2.3			
Substrates						
Cu	8.96	0.38	3.8 – 3.91	16.6 – 17.8		[7.22]
W	19.3	0.13	1.78	4.5		[7.21]
Mo	10.22	0.25	1.38	5.5		[7.22]
89W-11Cu	17	0.16	1.8	6.5		[7.22]
85W-15Cu	16.4	0.17	1.9	7.2		[7.22]
80W-20Cu	15.6	0.18	2.0	8.3		[7.22]
90Mo-10Cu	10.08	0.27	1.7	6.5		[7.22]
85Mo-15Cu	10.01	0.28	1.6	7.0		[7.22]
80Mo-Cu	9.94	0.28	1.7	8.0		[7.22]
Diamond	3.52	0.51	15 – 20	2.3		[7.22, 21]
SiC	3.20	0.73	2.72	3.7		[7.22]
BeO	2.90	0.96	2.5	7.6		[7.22]
Si_3Ni_4			0.19	0.8		[7.21]
Al_2O_3	3.60	0.80	0.17	6.7		[7.22]
SiO_2			0.014	0.55		[7.21]
H_2O			0.006			

Heat Transition

Further essential characteristic quantities are the heat-transfer resistance between the active layer and the cooling surface and the adaptation of the materials used regarding their expansion properties.

Table 7.5 Technical data of laser diodes

Type			Linear array	2-d-array
Manufacturer			Siemens	Spectra Diode Labs
Reference			[7.9]	
Optical power cw	P	W	1	5
quasi cw	P	W	1.5	800
Pulse energy	E	mW s		
Pulse duration	T	μs	<150	100
Frequency	f	Hz		15
Slope efficiency	η_{slope}	W/A	> 0.7	
Quantum efficiency	η_Q	%	> 48	48.4
Efficiency, overall	η	%	> 30	32.5
Threshold current	I_{th}	A	0.6	12
Forward voltage	U_0	V	1.6	45
Voltage	U_F	V	< 2	
Backward voltage	U_R	V	3	
Differential resistance	r	Ω	0.2	0.7
Emission wavelength	λ	nm	808	808
Spectral width	$\Delta\lambda$	nm	+ 6 −2	± 4
Temp. coefficient	K_λ	nm/° C	0.28	
Linear polarization	p	%	1	
Parallel to the p-n transition				
Operation temp.	T	°C	25	16
Max. solder temp.	T_{max}	°C	130	
Temp. range	ΔT	°C		
Temp. resistance	R_H	K/W	10	
Characteristic. temp	T_0	K	160	
Divergence	θ_x	° FWHM	14	
	θ_y	° FWHM	40 – 45	
Stripe number		20	66	
Line number		1	26	
Emission area		μm×μm	1×200	10 000×6500

The junction temperature T_j due to the thermal resistance, the coolant temperature and the heating is

$$T_j = T_k + R_H P_H, \tag{7.10}$$

T_k = temperature of the cooling surface,
R_H = thermal junction resistance to cooling surface ≈ 5°C/W [7.10].

In Table 7.4 data are listed for the heat transition and for the thermal expansion of some semiconductor materials, solders and assembly substrates. Special alloys are available which adapt to different expansion coefficients. Diamond layers between coolers and semiconductors distribute the heat due to their high thermal conductivity and the operating temperatures can be lowered considerably. This benefits the diode lifetime. In Table 7.5 typical data are given of two laser diodes.

7.5 Excitation of a Solid-State Laser

For the example of a Nd:YAG material, pumped with 808 nm laser diode arrays, the efficiency of a simple pump assembly can be estimated as follows.

Example 7.1 Excitation of Nd:YAG by a laser diode array

Pumping source:	Wavelength	$\lambda_p = 810$	nm
	Optical power	$P_p = 100$	W
	Efficiency	$P_p/P_{in} = 0.3$	
	Divergence	$\beta_x = 32$	° (full angle)
	($1/e^2$)	$\beta_y = 90$	° (full angle)
Laser medium:	Wavelength	$\lambda = 1064$	nm
	Absorption coefficient	$\alpha = 0.4$	mm^{-1} (@ 810 nm)
	Saturation intensity	$J_s - 1.6$	kW/cm^2
	Scattering losses	$-\ln V = 0.01$	1/cm
	Cross section	$f = 2{\cdot}2$	mm^2
Resonator:	Extraction efficiency.	$\eta_{extr} = 0.5$	
	Reflection	$R = 0.9$	

The excitation efficiency becomes $\eta_{excit} = 0.2$ and the laser power P_{cw} is about 10 W.

The cw power of a solid-state laser is, according to (1.61)

$$P_{cw} = \eta_{excit} \frac{\ln\sqrt{R}}{\ln(V\sqrt{R})}\left(P_p + P_s \ln(V\sqrt{R})\right), \tag{7.11}$$

J_s = saturation intensity,
P_p = optical power of the diode,
V = loss factor for internal losses,
R = reflection factor of the output coupler.

The extraction efficiency presumes complete overlap of the mode volume and the pump region. Reduced overlap is taken into account by the mode efficiency

$$\eta_{Mod} \approx 0.8 - 0.9. \tag{7.12}$$

This excitation efficiency η_{excit} is made up by the laser diode efficiency,

$$\eta_{\text{diode}} \approx P_{\text{light}} / P_{\text{in}}, \tag{7.13}$$
$\eta_{\text{diode}} \approx 0.25 - 0.4$,
P_{light} = optical power of the laser diode,

by the reflection losses of the optical transitions such as protection windows, flow tube and transition from coolant to the laser material,

$$\eta_{\text{reflec}} \approx 0.96^2 \times 0.96^2 \times 0.96^2 = 0.78, \tag{7.14}$$

by the absorption coefficient of the crystal to be pumped,

$$\eta_{\text{absorb}} = 1 - \mathrm{e}^{-\alpha l}, \tag{7.15}$$
l = path length of the pumping light in the material,
$\alpha = 4\ \text{cm}^{-1}$, absorption coefficient at the pump wavelength $\lambda_\text{p} = 807$ nm,
$\eta_{\text{absorb}} \approx 0.95$, at a material thickness of 3 times absorption length,

by the quantum efficiency,

$$\eta_\text{Q} \approx 0.8, \tag{7.16}$$

and by the *quantum defect* $\eta_{\text{QD}} = \lambda_\text{p}/\lambda$,

$$\eta_{\text{QD}} = \frac{\lambda_\text{p}}{\lambda} = \frac{810}{1064} = 0.8\,. \tag{7.17}$$

The excitation efficiency results in

$$\eta_{\text{excit}} = \eta_{\text{diode}} \eta_{\text{reflec}} \left(1 - \mathrm{e}^{-\alpha l}\right) \frac{\lambda_\text{p}}{\lambda}, \tag{7.18}$$

and the total efficiency becomes

$$\eta \approx \eta_{\text{mod}} \times \eta_{\text{reflec}} \times \eta_{\text{absorb}} \times \eta_\text{Q} \times \eta_{\text{QD}}, \tag{7.19}$$
$\eta = 0.9 \times 0.3 \times 0.78 \times 0.95 \times 0.8 \times 0.8 = 0.13$.

With respect to gas discharge lamps, the efficiency is approximately a factor of 4 higher.

Direct Use of the Laser Diodes

The direct use of the diode radiation is much more efficient than the transformation by the means of a second active medium. Possibilities to bundle the radiation of the different emitters optically, permit powers up to 10 W per bar. Since the beam quality of the emitter in the two axes is very different, the radiation of the emitters is transformed by an optical system, so that waist and divergence in the two axes are the same for the resulting beam.

In Fig 7.25 the diode beams are collimated with a cylinder lens on the axis with the superior beam quality. By the reflection at a step mirror [7.31] the geometric distances between the individual emitter beams are reduced.

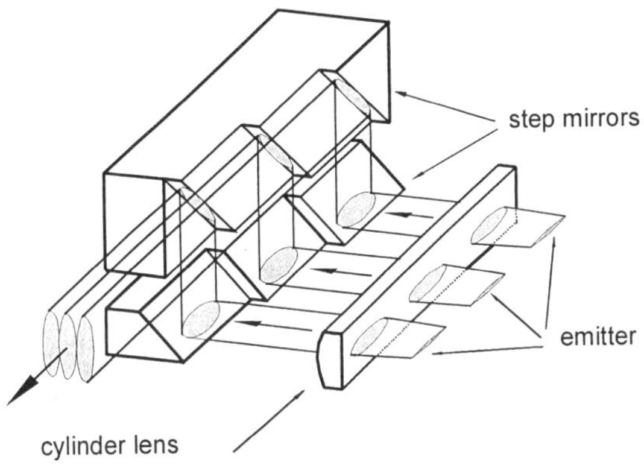

Fig. 7.25. Collimating the radiation of different emitters by means of a step mirror

With two parallel mirrors [7.30] the same effect can be achieved (Fig 7.26). In both cases, however, the diameter increases with increasing path. This affects particularly the beam diameter in the noncollimated axis.

The beam parameter product in the axis perpendicular to the p-n transition is about 40 times larger compared to the other axis. The collimation of 40 emitters produces equal beam quality in both axes. With 2 W per emitter a total power of 80 W can be reached with a beam quality of about 30 mm mrad. This is sufficient for the coupling into a 600 μm fiber. The efficiency of such an optical device is at present about 60 to 70%.

Fig. 7.26. Transformation of an array beam with two parallel glass plates

Coherent coupling of the single emitters would improve the beam quality considerably while maintaining the same output power.

8 Electrical Circuits

The principle construction of a power supply for the flash lamp is shown in Fig. 8.1. The design of a cw power supply is similar except for the capacitor to store the pumping energy.

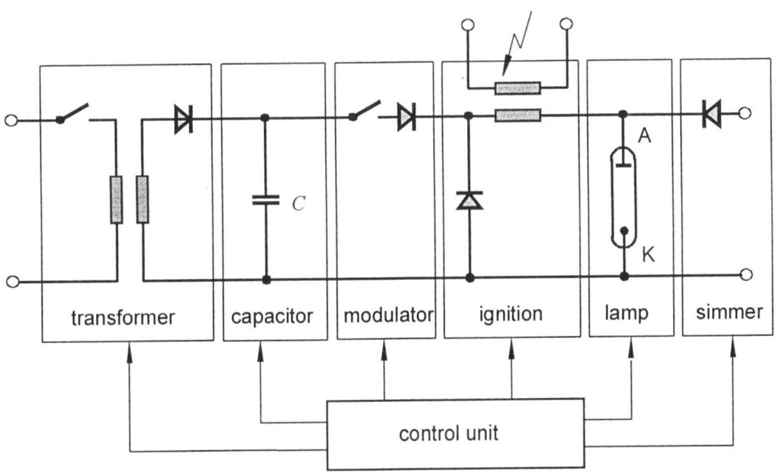

Fig. 8.1. Schematic circuit diagram of a power supply

For pulsed lasers the energy from the mains is stored intermediately by means of a suitable power supply at the desired voltage in the capacitor. After igniting the lamp an arc with small diameter is kept in the lamp by the simmer current source. The main discharging of the capacitor is made by switching on the modulator. This determines the duration and temporal shape of the lamp power and thus also the laser pulse.

8.1 Ignition Circuits

Initially the high power transient produces sufficient free charge carriers in the lamp. Standard variants are the external ignition and the series ignition.

External Triggering

Between the cathode or anode and a third electrode outside the lamp a high voltage transient is applied (Fig. 8.2). This auxiliary electrode is usually a very thin stainless steel wire which is wound around the lamp. However, it can also be the reflector or a separate wire, [8.1].

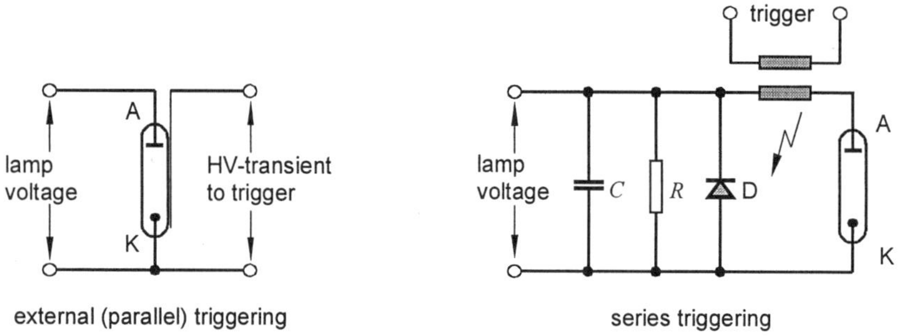

Fig. 8.2. Principle circuits to ignite gas discharge lamps

Serial Ignition

The secondary winding of the trigger transformer is in series with the main circuit of the lamp. The electrical circuit for the trigger impulse is interrupted by C_Z and R_Z.

To obtain a reproducible ionization in the lamp the electrical conditions must be well defined by the environment. Metal parts in the proximity of the lamp, such as the reflector, or assemblies which are isolated well by the deionized water may become charged and so prevent the lamp ignition. They should be connected therefore with a defined potential (cathode) by a high resistance (MΩ). An overview of possible trigger circuits is given in Fig. 8.3.

- *Circuit 1*
  The secondary winding must withstand the high trigger voltage (up to 30 kV). As the winding must, at the same time, carry the high currents and dissipate the heat the cross sections of the windings can become very large. To limit the current increase for the semiconductor modulation no additional inductance is necessary.

- *Circuit 2*
  A vacuum relay takes the main current after triggering. The contact must isolate the trigger voltage in the open condition.

- *Circuit 3, 4*
  The semiconductor elements must be able to bear the trigger voltage and, at the same time, carry the maximum main current.

- *Circuit 5*
  The beam source must be designed in such a way that a sparkover to the cavity is avoided.

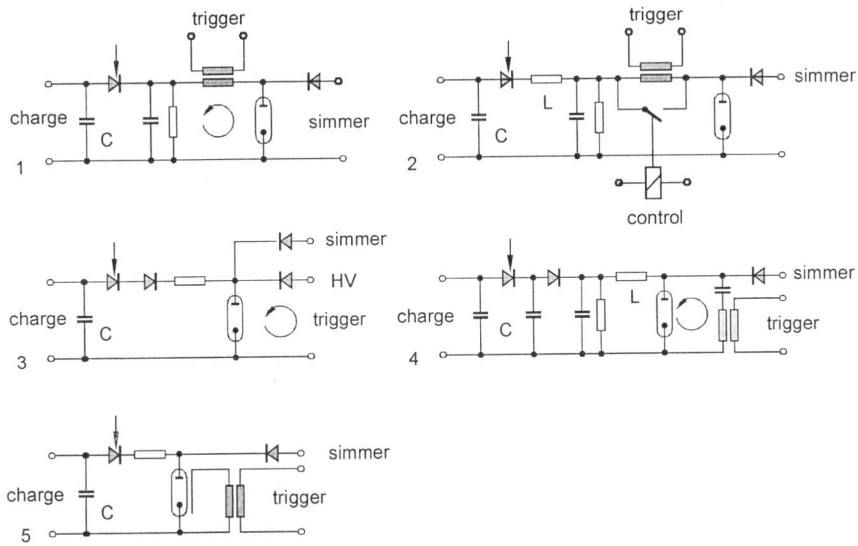

Fig. 8.3. Trigger circuits

8.2 Power Supplies

Techniques with short-circuit protected switched power supplies are the standard today. In the power range around 10 kW generally only forward converters, based on the push-pull principle, are considered because the transformer can be used in both current directions up to the magnetic saturation limit. Usually half or full bridges, as in Fig. 8.4 and Fig. 8.5 are used, with transistors or thyristors as semiconductor switches.

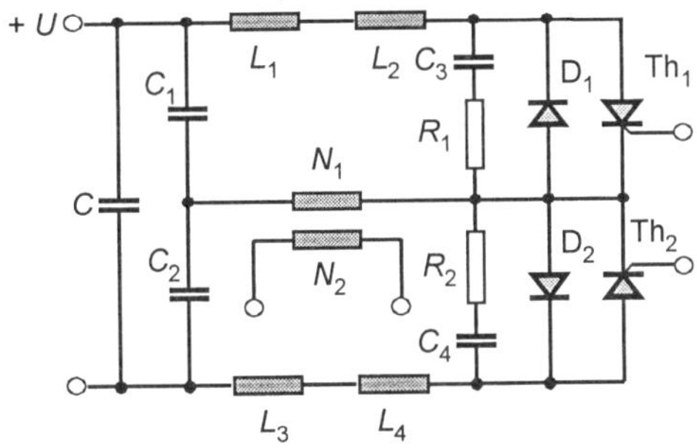

Fig. 8.4. Schematic circuit diagram of the power supply with half-bridge rectifier

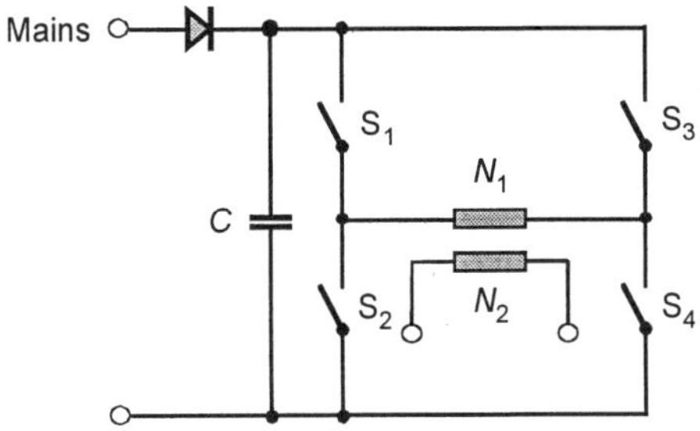

Fig. 8.5. Schematic circuit diagram of the power supply as full-bridge rectifier

Simmer Power Supply

After igniting the lamp the standby status is maintained by a relatively low current, the so-called simmer current. Since the lamp at this low current still has a negative characteristic, the simmer power supply must be an appropriate constant current source. Short-circuit proof, switched mode power supplies are an advantage. The continuous simmer current allows a failure control of the lamps even in the pulsed mode.

8.3 Energy Storage

The necessary peak power of a laser during a pulse exceeds the available average power of the power supply by at least one order of magnitude. The needed pulse energy for the laser is therefore stored intermediately in a capacitor bank and ensures the high peak power in the ms-range despite the low, limited connection power. Currently, two variants are used for pulsed lasers in this range, MP capacitors in a pulse-forming network (PFN) and electrolytic capacitors with an electronic switch.

Example 8.1 Estimation of the pump peak power for a welding point

The necessary laser pulse power to melt a metal surface is approximately 5 kW for a welding point diameter of 0.5 mm. With an efficiency of 2% this means 250 kW pulse peak power from the power supply. For an average laser power of 100 W, only 5 kW connection power is necessary.

Pulse-Forming Network

For pulsed operation at least the pulse power and pulse duration should be adjustable. The pulse power in a pulse-forming network can be changed by the appropriate voltage during charging and the duration by switching on or off one or more meshes of the PFN.

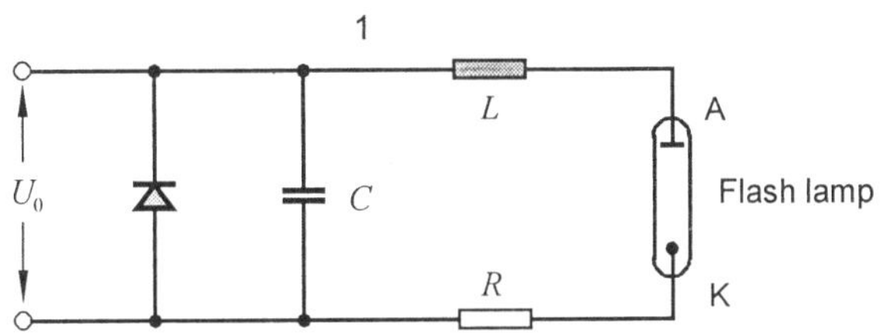

Fig. 8.6. Single mesh LC-circuit

The nonlinear differential equation for the single mesh LC-circuit (Fig. 8.6) is

$$U_0 = L\frac{dI}{dt} + RI + K_0|I|^{0.5} + \frac{1}{C}\int_0^t I dt, \tag{8.1}$$

U_0 = constant.

With the following substitutions

$$Z_0 = \sqrt{L/C},\ T = \sqrt{LC},\ \alpha = \frac{K_0}{\sqrt{U_0 Z_0}}, \tag{8.2}$$

$$I = J\frac{U_0}{Z_0},\ \tau = \frac{t}{T},\ \beta = \frac{R}{Z_0},$$

the normalized equation reads

$$I = \frac{dJ}{d\tau} + \beta J \pm \alpha|J|^{0.5} + \int_0^\tau J d\tau. \tag{8.3}$$

For different damping parameters α solutions of (8.3) were numerically calculated for $R = 0$ in [8.2] and for $R > 0$ in [5.13]. The critical damping value is $\alpha_k = 0.75$, according to [5.13] and approximately $\alpha_k = 0.84$ according to [8.1]. For larger values of α the circuit is more strongly damped, for smaller values the voltage overshoots. This overshooting can destroy the charging diodes or other components and must be prevented by suitable lamp adaptation or protection diodes.

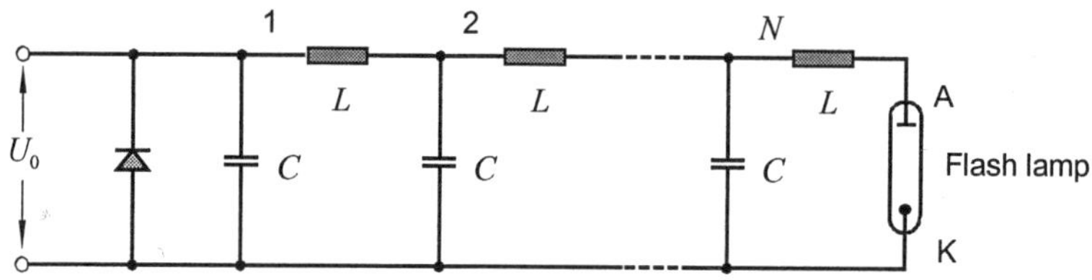

Fig. 8.7. N-mesh LC-circuit

The pulse-train duration T for the N-mesh circuit (Fig. 8.7) is

$$T = 2.2N\sqrt{LC}\,, \tag{8.4}$$

T = halfwidth of the pulse-train duration,
N = total mesh number,
L = inductance of a single mesh,
C = capacity of a single mesh.

The factor 2.2 for the halfwidth (FWHM) of the pulse has been derived empirically from Fig. 8.8.

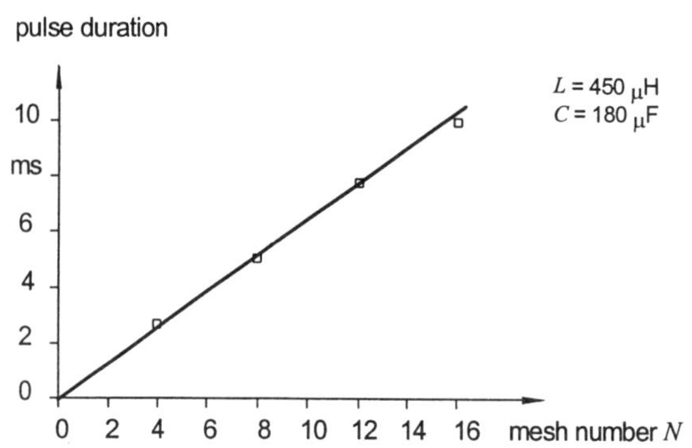

Fig. 8.8. Pulse-train duration in the N-mesh circuit

At critical damping, the energy of the capacitor chain during a pulse will be transferred completely to the lamp,

$$E = 0.5 \times NCU^2. \tag{8.5}$$

The attainable peak power for the pump pulse follows from (8.4) as

$$\hat{P} = \frac{E}{T} \approx 0.23\sqrt{C/L}\,U^2\,. \tag{8.6}$$

To determine the critical damping of an N-mesh PFN for different voltages U the overshooting at the first mesh $N = 1$ was measured (Fig. 8.9). At critical damping, $\alpha_k = 0.75$, overshooting of the pulse is just about avoided.

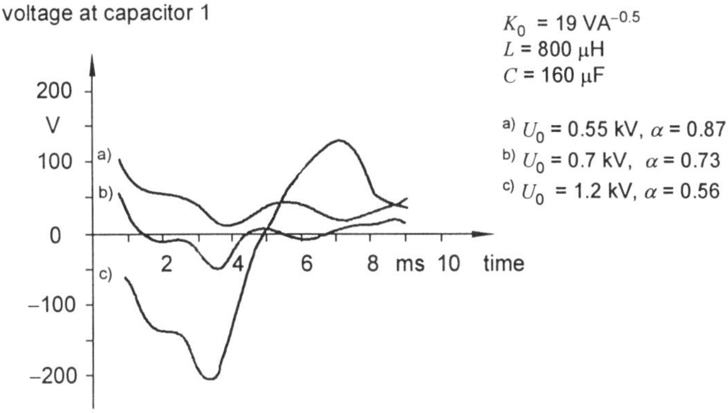

Fig. 8.9. Overshooting at various damping parameters

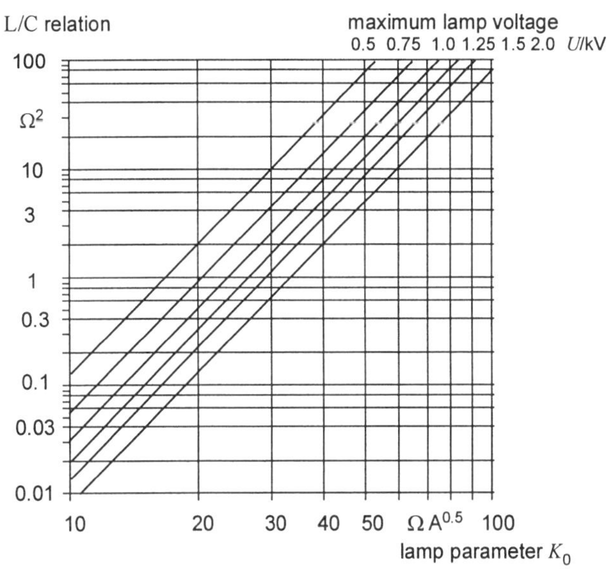

Fig. 8.10. Tolerance characteristics for a critically damped lamp in a LC-circuit

With the lamp parameter K_0 and the maximum storage voltage U for $\alpha_k = 0.75$ the critical LC-ratio can be determined at which overshooting is not yet taking place.

$$\frac{L}{C} = \frac{K_0^4}{\alpha_k^4 U^2}, \tag{8.7}$$

$\alpha_k = 0.75.$

Figure 8.10 shows the L/C ratio versus lamp parameter K_0 at critical damping parameter α_k in a double logarithmic scale with U as parameter.

Electrolytic Capacitors

The disadvantages of LC pulse-forming networks (large volume and weight, high voltages) can be avoided by the use of electrolyte capacitors and suitable electronic switches. The lamp being in standby condition due to the low simmer current, is connected to the main capacitor by a switch-on and after the derived pulse duration is switched-off. Figure 8.11 shows the current and voltage curve of the lamp when discharging an electrolytic capacitor.

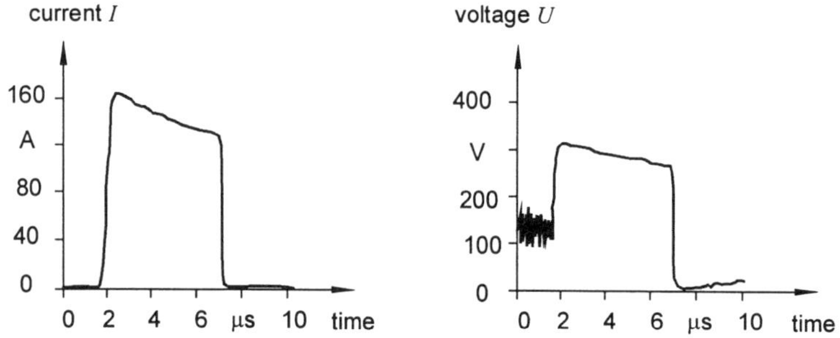

Fig. 8.11. Current and voltage for a lamp discharging an electrolytic capacitor

From the requirement that the laser peak power should not decrease by more than 30% at the highest voltage during a laser pulse, the necessary capacity C can be estimated. The laser power is proportional to the electrical power:

$$P \propto UI = \frac{U^3}{K_0^2}, \tag{8.8}$$

U = capacitor voltage,
I = lamp current.

With a laser power decrease of maximal 30% during the pulse duration the capacitor voltage drop may be less than 10%. So the maximum energy, taken from the capacitor is

$$E = 0.5C\left(U^2 - (0.9U)^2\right) \approx 0.1CU^2 . \tag{8.9}$$

For approximately constant assumed voltage U the energy taken up by the lamp during the pulse is

$$E = UIT = \frac{U^3}{K_0^2} T, \tag{8.10}$$

and hence the corresponding necessary capacity is

$$C = 8.6 \frac{UT}{K_0^2} \approx 10 \frac{UT}{K_0^2}. \tag{8.11}$$

Example 8.2 Determination of the capacitor size for a laser flash lamp

With $U = 700$ V, $K_0 = 40$ V/√A, $T = 10$ ms it follows that the capacity is $C \approx 40$ mF.

8.4 Power Modulation

By the use of special direct dc-converters [8.3] it is not only possible to switch the lamp on and off, but even, with limited resolution, to modulate the lamp current during a pulse.

The power switch is supposed to regulate the temporal shape of the laser power or even control it with a suitable transducer. In cw-mode this function is performed directly by the power supply. The settling-time constant then determines the rise and fall time of the laser power. Pulsed operation is also possible, however, the peak power of the laser is limited for long pulses (> 100 ms) to the maximum average power.

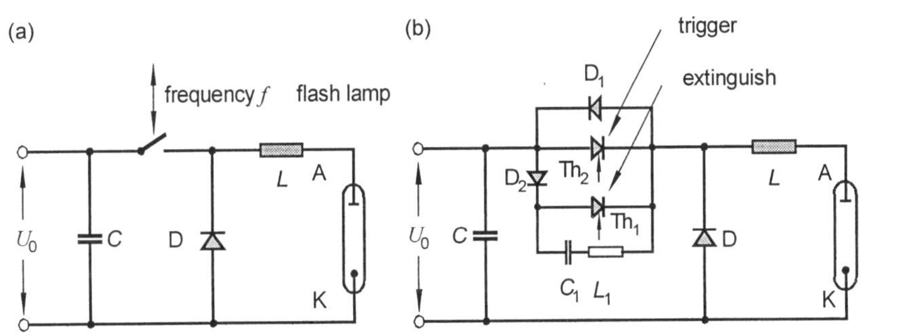

Fig. 8.12. (**a**) Schematic circuit to control the laser pulse power by a switch (**b**) setup of the switch by a thyristor

Due to the high currents in pulsed operation and the use of capacitors, special current switches are necessary which are able to switch peak currents of several 100 A in a very short time. In many cases simple setups with semiconductor switches are sufficient which connect the lamp to the storage capacitor for the desired pulse duration. By turning the switch on and off in a time interval small compared to the pulse duration, the average power can be controlled by the switching cycle and a measurement of the laser power even allows a feedback control, if the laser power is measured, (Fig. 8.12).

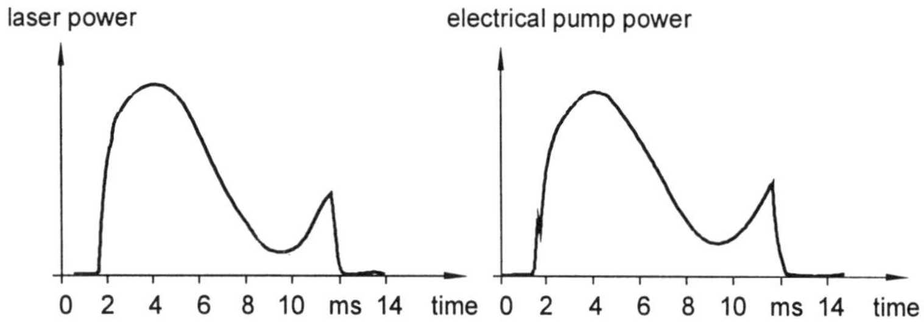

Fig. 8.13. Laser pulse which is modulated by a fast current switch

Within the resolution of the grid the laser pulse can even be programmed in its temporal behavior and adapted to the processing requirements. Figure 8.13 shows the temporal behavior of a programmed laser pulse.

9 Optics

The focus diameter, the working distance, the divergence and the beam diameter are predefined by the application problem and the beam source used. With that, the necessary optics can be completely calculated.

To minimize the geometric losses and the beam deterioration caused by diffraction at the various apertures (lens rim) the diameters of the optical components should be sufficiently large. Considering the tolerances, they should be 1.5 times the beam diameter. The reflection losses can be avoided by using antireflection coatings. For a single surface, reflections can be less than 0.2%. Achromatic lens systems should be used to keep the spherical aberrations small. Cemented achromatic lenses must be able to withstand the high intensity, otherwise achromatic lenses with air gaps have to be used.

The final lens of the objective should be protected by an antireflection-coated, interchangeable glass to avoid damage by the plasma plume or by sputtered material particles.

9.1 Focusing Optics

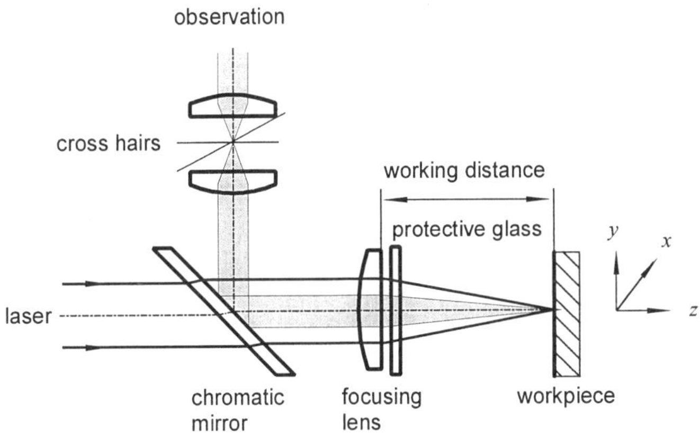

Fig. 9.1. Optical system of the processing head

The optical system focuses the laser beam, indicates the processing position and aligns the laser beam.

With a partially transmitting mirror and a monitoring telescope (Fig. 9.1) the working distance is adjusted by moving the focusing lens in the *z* direction until the workpiece surface is clearly visible in the eyepiece. The spider lines in the telescope then indicate the position of the laser beam focus in the *xy* plane.
Other methods for marking the processing position are:

- fading a light spot in the optical path,
- guiding an expanded visible He-Ne laser beam collinear to the laser, as shown in Fig. 9.2.

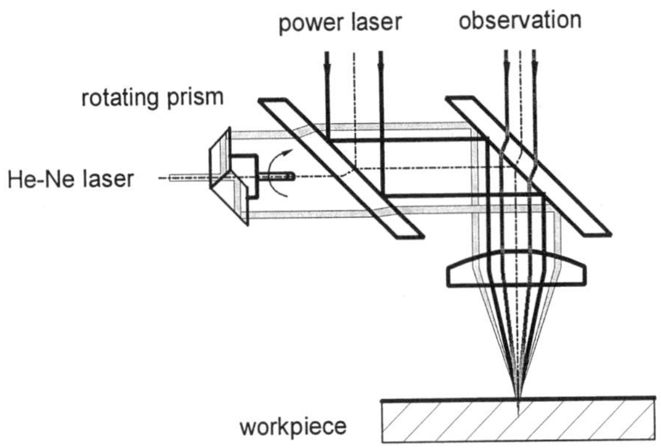

Fig. 9.2. Identification of the focus by means of a pilot laser, [9.1]

Beam Expanding Optics

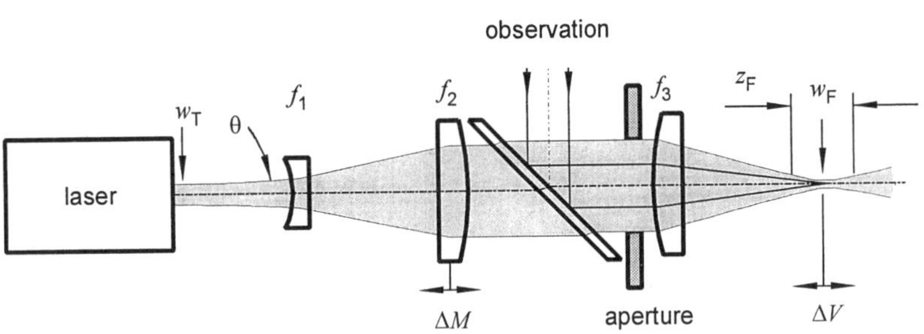

Fig. 9.3. Beam expanding optics

With the use of only one lens the desired focus diameter at the working distance can, in general, not be obtained due to the laser beam divergence. A beam expansion optics, as in Fig. 9.3, reduces the beam divergence and the beam focus diameter by the zoom factor A. However, simultaneously the focal depth is reduced and the beam diameter enlarged.

Without considering aberrations the focus diameter follows, according to (2.51)

$$2w_F = 2\frac{f_3\theta}{A}, \tag{9.1}$$

$A = \dfrac{|f_2|}{|f_1|}$ = zoom factor,

f_3 = focal length ≈ working distance,

and for the new Rayleigh length z_F in the focal region with (2.52)

$$z_F = \frac{z_R}{A^2}, \tag{9.2}$$

z_R = Rayleigh length of the initial laser beam.

In Fig. 9.4 measurements of the focus diameter for different focal lengths and zoom factors are shown.

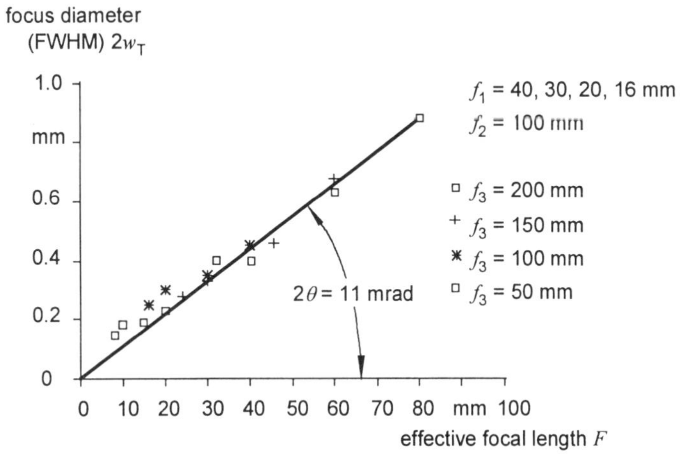

Fig. 9.4. Focus diameter as a function of the effective focal length $F = f_3/A$

The focal depth z_S is defined by a 10% decrease of the intensity when the workpiece surface is shifted by $\pm z_S/2$

$$z_S = 0.33 z_R = \frac{0.33\pi w_F^2}{\lambda M^2} = 0.33\frac{f_3^2}{A^2}\frac{\theta}{w_T}, \tag{9.3}$$

λ = wavelength,
M^2 = mode order, beam propagation factor
w_T = waist radius in the focus.

A large working distance also gives a large focal depth. On the other hand a small focus diameter and a large focal depth exclude each other.

Example 9.1 Focal depth of a Nd-glass laser

A Nd-glass laser with a mode order of about M^2=100 is focused to $2w_T \approx 0.7$ mm. The focal depth is approximately $z_S \approx 1.3$ mm.

In Fig. 9.5 the laser beam diameter near the focus is plotted for various focal lengths.

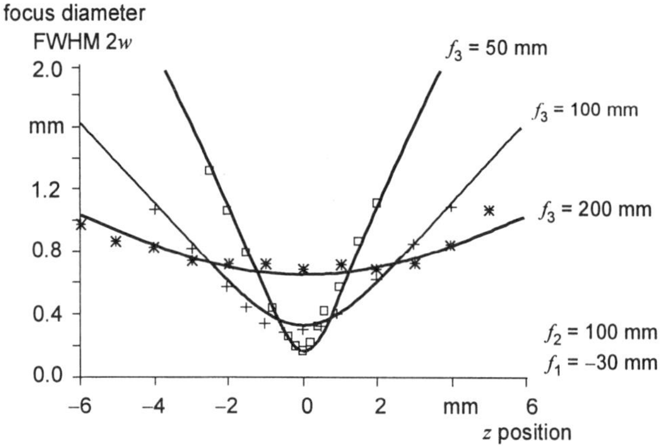

Fig. 9.5. Beam propagation in the focal region

Defocusing

Moving the lens f_2 by an amount ΔM, results in a shift ΔV of the beam waist which to a good approximation is given as:

$$\Delta V = -\left(\frac{f_3}{f_2}\right)^2 \Delta M . \tag{9.4}$$

This can be used to vary the beam intensity on the workpiece surface at constant sharp observation or fixed working distance. The sign implies that when moving the lens f_2 towards the resonator the focus moves in the opposite direction.

9.2 Lens Aberrations

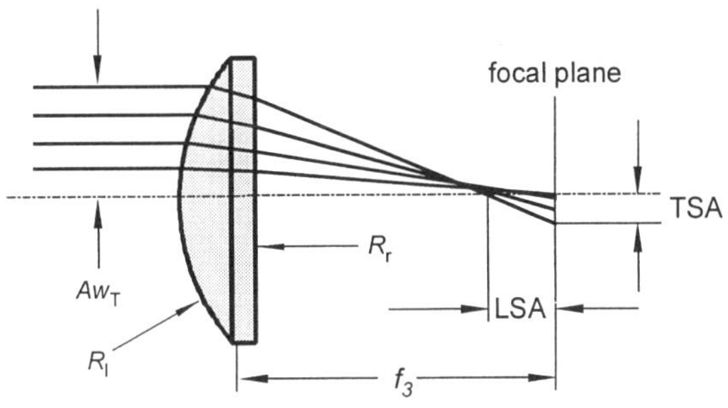

Fig. 9.6. Spherical aberration

The most relevant lens aberration of the processing optics is the spherical aberration. If a collimated beam is focused, the focal length, i.e. the intersection of the rays with the axis, is for the central ray larger than for an off-axis ray. This produces a transverse (TSA) and a longitudinal (LSA) shift as shown in Fig. 9.6.

The ratio of beam diameter to focal length is a measure for the minimum spot diameter of the lens. The ratio of lens diameter to focal length is defined as the *F*-number (*F* #) and is the relevant parameter for the spherical aberration. To a first approximation the focus diameter consists of the diffraction part and the part due to the aberration of lens f_3 [9.2]:

$$w_F = \frac{f_3\theta}{A} + f_3 K(q,n)\left(\frac{Aw}{f_3}\right)^3 . \tag{9.5}$$

$K(q, n)$ is a constant, dependent on the refractive index n and the form factor q of the lens and the imaging conditions. For focusing a parallel beam K becomes:

$$K = \frac{1}{8}\frac{1}{n(n-1)}\left[\frac{(n+2)}{(n-1)}q^2 - 4(n+1)q + (3n+2)(n-1) + \frac{n^3}{n-1}\right] \tag{9.6}$$

with

$$q = \frac{R_r + R_l}{R_r - R_l}, \tag{9.7}$$

R_l = lens radius entrance,
R_r = lens radius exit.

For an optimal focal length f_{opt}

$$f_3 = f_{\text{opt}} = Aw_{\text{T}}\left(\frac{2KA}{\theta}\right)^{1/3}, \tag{9.8}$$

there is a minimal focus diameter:

$$w_{\min} = 3Aw_{\text{T}}\left(\frac{2KA}{\theta}\right)^{1/3}. \tag{9.9}$$

In the following, the lens aberrations will be examined further only for plano-convex lenses, since they can be produced cheaply. If the spherical aberration of a single lens cannot be tolerated, it has to be replaced by an achromatic doublet. For plane-convex lenses with $q = 1$, the factor $K\,(1, n)$ becomes [9.2]:

$$K(1,n) = \frac{1}{2(n-1)^2}\left(n^2 - 2n + \frac{2}{n}\right). \tag{9.10}$$

For the common glass material BK-7 $n = 1.506$ at $\lambda = 1\,\mu\text{m}$ $K = 1.14$. In Table 9.1 the optimal focal lengths and the smallest focus diameters are listed. Parameters are the beam divergence and the diameter of the expanded laser beam.

Table 9.1 Optimal focal length and minimal focus diameter due to spherical aberration dependent on beam diameter and divergence

	$A\theta$ / mrad =	0.5	1	2	5	10
$2Aw_{\text{T}}$ / mm						
5	f_{opt} / mm	41	33	26	19	15
10		83	66	52	38	31
20		166	132	104	77	61
50		414	329	261	192	153
5	$2w_{\min}$ / mm	0.06	0.10	0.16	0.29	0.46
10	at f_{opt}	0.12	0.20	0.31	0.58	0.92
20		0.25	0.39	0.63	1.15	1.83
50		0.62	0.99	1.57	2.89	4.58

9.3 Scanning Systems

For many applications it is necessary to scan the laser beam across the workpiece (inscription, trimming, seam welding). In the following, some solutions are presented.

One-dimensional Systems

Moving a lens (Fig. 9.7a) causes the focus to be shifted as well. The aberration increases towards the lens rim, the displacement range and the velocity are restricted and the aperture angle of the beam changes with the position. The limitation of the operation area can be compensated by moving simultaneously reflector and focusing optics (*flying optics*) as shown in Fig.9.7b

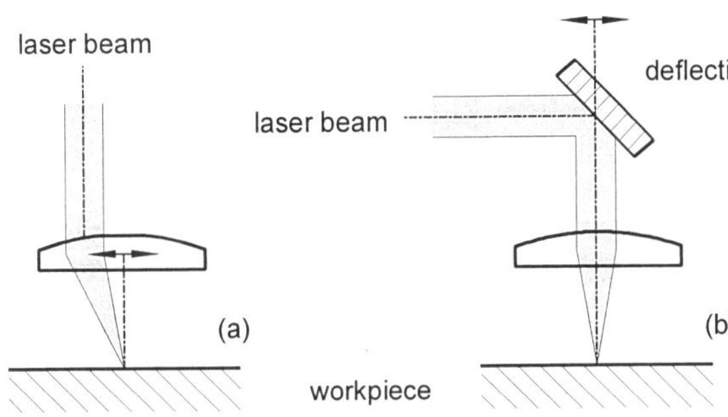

Fig. 9.7. Possibilities for one-dimensional beam deflection: (**a**) moving lens, (**b**) flying optics

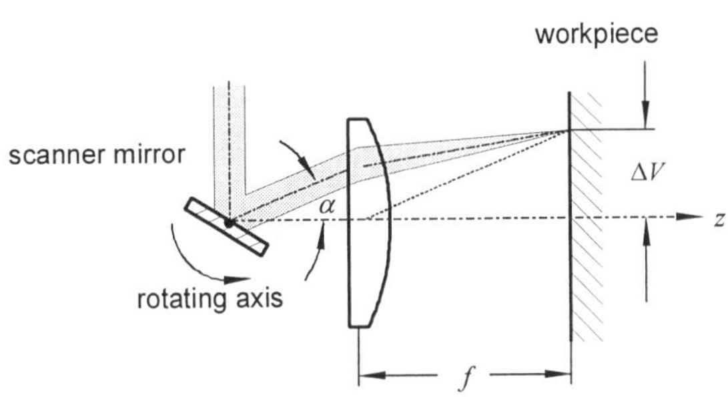

Fig. 9.8. Moving the focus position by tilting the beam with a scanner mirror

Two-dimensional Systems

These systems are usually based on an angle deflection of the beam as shown in Fig. 9.8. They require a special optical system with a flat focal plane. Only small z adjustments are possible.

1. Rotating prism

A method that preserves cylinder symmetry is the rotation of a prism, for example by means of a torque motor. Monitoring is possible using the same beam path. The deflection ΔV is very restricted, as is the deflection speed (Fig. 9.9).

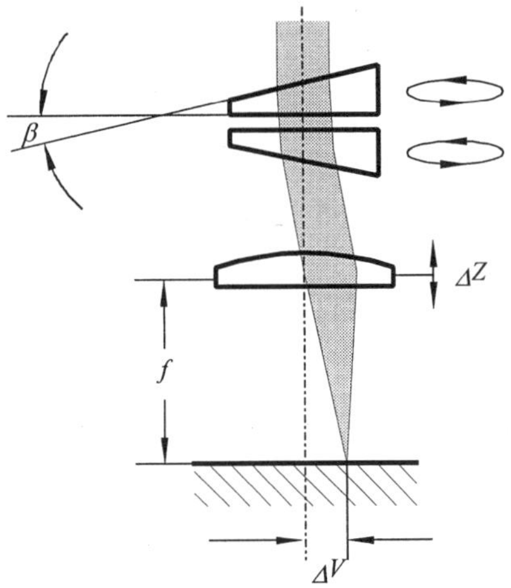

Fig. 9.9. Deflection by rotating prisms

2. Deflection by mirrors (scanners)

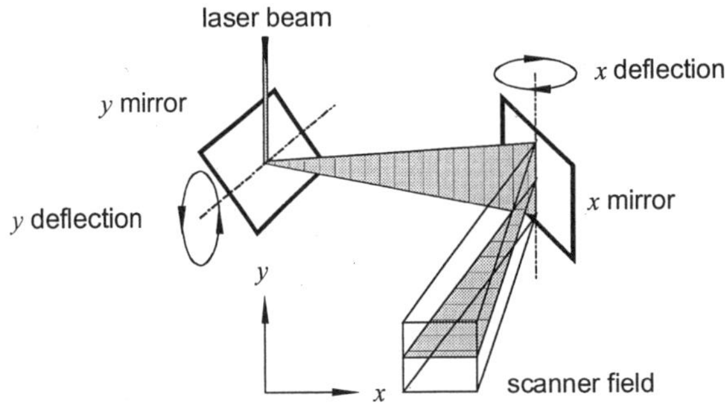

Fig. 9.10. Principle of a two-dimensional scanner

Beam deflection with mirrors (scanner principle in Fig. 9.10) and fast drives is widely used, e.g. for inscriptions. The angle (20°) and the deflection speeds are relatively high. Alternatively to inscription and marking with scanners, masks are exposed by a laser beam and then imaged to the workpiece. Most masks consist of copper and the laser works in the pulsed operation. They are used to mark components with the type number or to print the date in the packaging industry.

The mask must be replaced if the inscription is changed. Programmable masks with liquid-crystal displays have been used successfully, [9.15].

3. Multiplexers

To use a single beam source for several processing stations, the beam is multiplexed with a deflection mirror to several fibers and then guided to the corresponding processing places. An industrial design of such a multiplexer with up to six processing positions is shown in Fig. 9.11. The necessary switch time of a multiplexer is normally smaller than the temporal distance of the laser pulses.

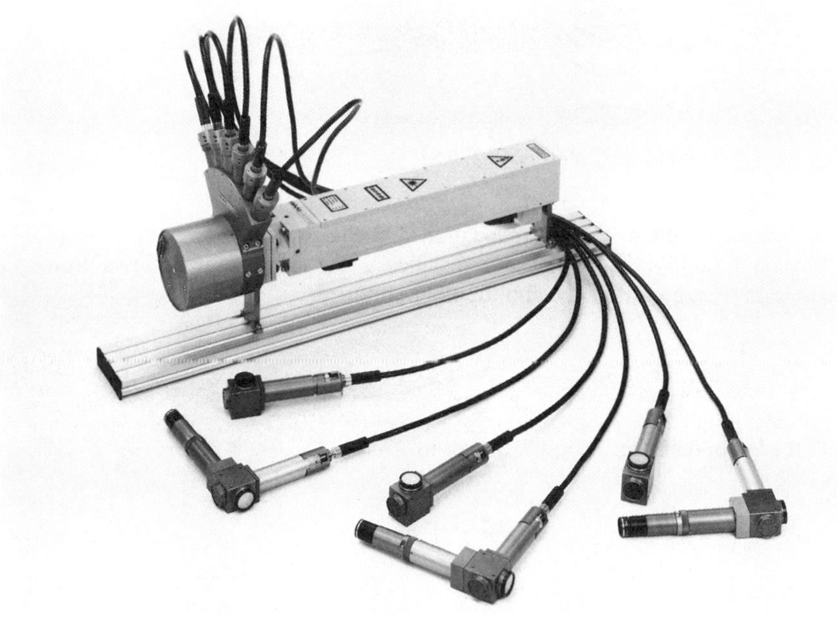

Fig. 9.11. Design of a multiplexer, Courtesy of HAAS-Laser

An arrangement to keep the focus position constant while focusing on the fiber entrance is given in [9.3]. The angle tolerance of the deflection mirror requires a somewhat larger numerical aperture of the fiber, (Fig. 9.12).
In order that the focus position on the fiber entrance is independent of small mirror rotations, three conditions have to be fulfilled simultaneously:

1. The axis of the deflecting mirror in Fig. 9.13 should be imaged by the focusing lens f_2 on the fiber.
2. The waist of the laser beam w_T should be imaged as a new waist w_{T2} on the fiber.

3. In order that the deflection can be carried out rapidly, the mirror diameter and therefore the beam diameter on the mirror must be as small as possible, without destroying the surface due to the power density.

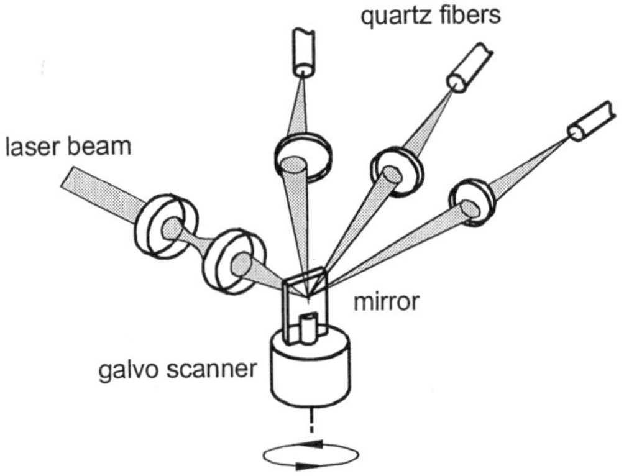

Fig. 9.12. Fast optical multiplexers, [9.3, 9.8]

The beam diameters, the waist location and the divergence result from (2.43), (2.44) and (2.45). The necessary distance z for the mirror follows from the geometric imaging of the axis on the fiber entrance

$$\frac{1}{z_3} = \frac{1}{f_2} - \frac{1}{z_4}, \tag{9.11}$$

and the beam diameter w on the mirror to

$$w^2 = \left(1 - \frac{z_1}{f_1}\right)^2 w_T^2 + \left(z_1 + z_2 - \frac{z_1 z_2}{f_1}\right)^2 \theta_T^2 . \tag{9.12}$$

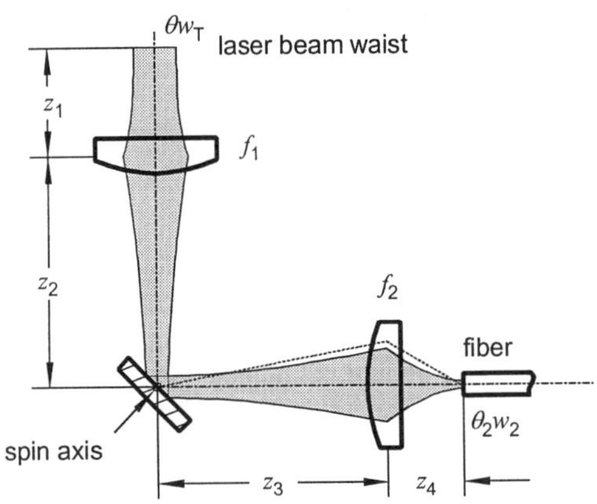

Fig. 9.13. Multiplexer deflection

9.4 Beam Splitter

In order to increase the resistance of two welded parts against twisting, one sets at the same time two welding spots at a small distance. For this purpose, the laser beam is divided by a biprism or by a simple grid [9.4]. The grid or prism in the parallel beam generates two points at a fixed distance. In the divergent beam (Fig. 9.14), the distance between the two points can be adjusted by moving the element in the *z* direction, the orientation of the points by turning the element.

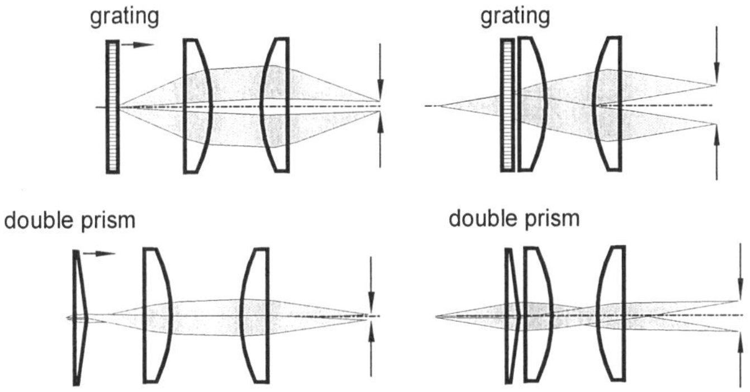

Fig. 9.14. Beam splitting by grid or biprism; distance variation by moving the grid or prism.

Precision connections of cylindrical parts often require three welding spots simultaneously. With beam splitters or beam dividers the parallel beam is split and by fibers supplied to the workpiece. The two principles currently used are shown in Fig. 9.15.

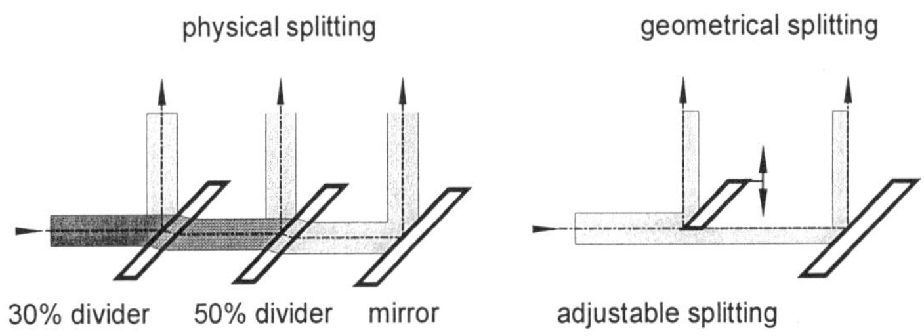

Fig. 9.15. Beam splitting with semitransparent mirrors or by an adjustable mirror

Figure 9.16 shows an industrial setup of a laser beam source with subsequent triple splitting for three fibers. In [9.5] the loss-free splitting into three and more beams with multistage structured grids is reported. Beam splitting requires pulsed lasers that have sufficient power during the pulse duration, so that the intensity in the splitted beam still has the necessary intensity at the processing position.

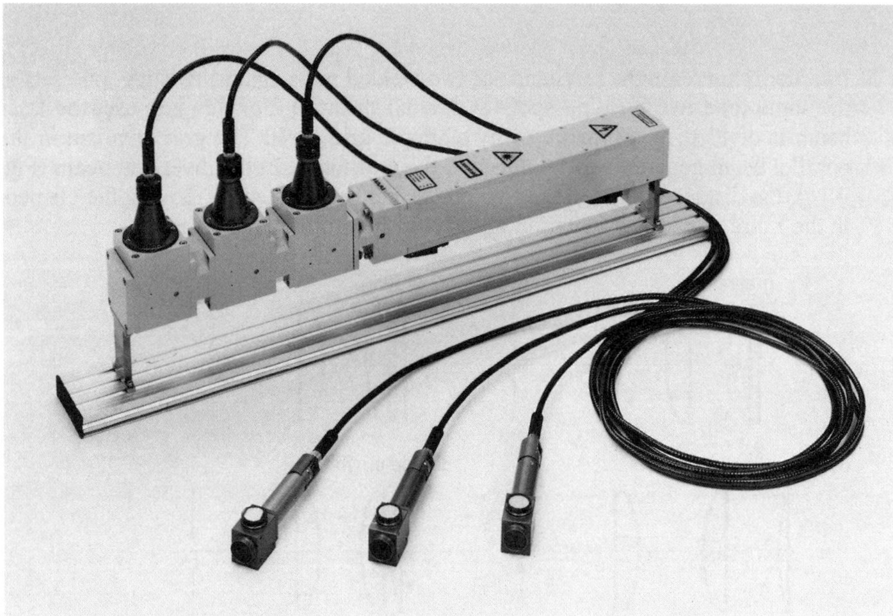

Fig. 9.16. Beam diversion in an industrial laser for three fibers, courtesy HAAS Laser

9.5 Beam Guiding by Fibers

Fibers are employed for transferring the laser beam to the workpiece [9.6]. In material processing, due to the high mean power and the high pulse peak power, mainly step index fibers with some tenths of mm in core diameter are employed. Typical realizations for different application are given in [9.7].

The advantages of transporting the laser energy by means of fibers are:

- A flexible fiber can deliver the power or energy to almost any desired place. The diameter of the protection tube is about 5 mm and is thus small and flexible.

- The necessary focusing optics are smaller than a corresponding beam source with conventional optics without light fibers.

- The distance between the laser system and the processing stations can be up to several hundred meters (Fig. 9.17).

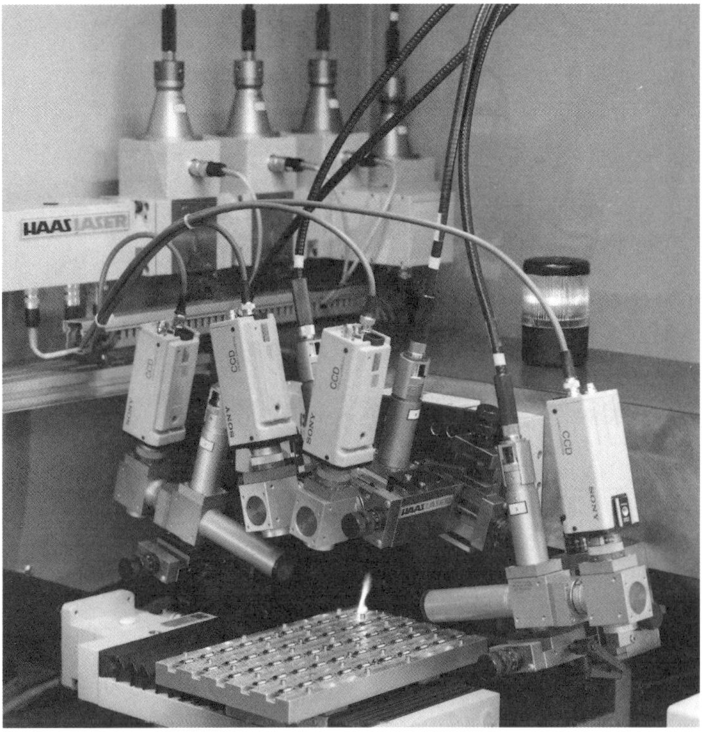

Fig. 9.17. Laser processing station, courtesy HAAS laser

Numerical Aperture NA

The NA is defined as the sine of the half acceptance angle multiplied by the refractive index of the surrounding medium [9.9]:

$$\mathrm{NA} = n \cdot \sin\alpha, \qquad \textit{numerical aperture.} \qquad (9.13)$$

Beam Fiber Coupling

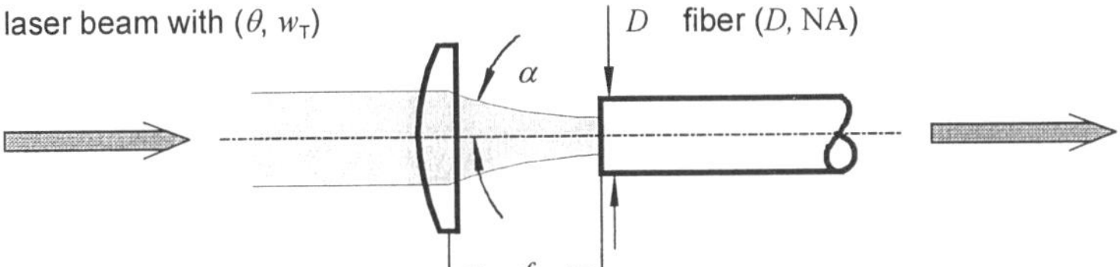

Fig. 9.18. Beam to fiber coupling

The conditions for loss-free, beam-fiber coupling at small angles are:

– the fiber diameter must be larger than the focus diameter of the laser beam

$$D > 2f\theta, \tag{9.14}$$

D = fiber diameter,
f = focal length of the coupling lens,
θ = divergence of the laser beam,

– and the acceptance angle must be larger than the focusing angle α

$$\mathrm{NA} > w_{\mathrm{T}}/f, \tag{9.15}$$

w_{T} = waist radius of the laser beam.

Example 9.2 Fiber with 2NA = 0.17 and $D = 0.6$ mm, laser beam with $2w_T = 10$ mm
Focal length $f > 2w_{\mathrm{T}}/2\mathrm{NA}$ = 58.8 mm,
Laser divergence $2\theta < D/f$ = 10 mrad.

These two conditions must, of course, be fulfilled for all operating conditions of the laser. In the limiting case one has

$$D\mathrm{NA} = 2\theta w_{\mathrm{T}}, \tag{9.16}$$

which is the condition for mode adaptation between fiber and lasers.

To check the correct focal length and the optimal mode adaptation the fiber output power P_{F} is measured as a function of the focal length. For small focal length f the beam angle α is larger than the acceptance angle NA and for constant intensity distribution $J(r)$ the power behind a fiber is reduced:

$$\frac{P_{\mathrm{F}}}{P} \approx \frac{\mathrm{NA}^2 f^2}{4w_{\mathrm{T}}^2}. \tag{9.17}$$

For large focal lengths the beam diameter exceeds the fiber core diameter and the output power is reduced again

$$\frac{P_{\mathrm{F}}}{P} \approx \frac{D^2}{4f^2\theta^2}. \tag{9.18}$$

Figure 9.19 shows the result of such a measurement.

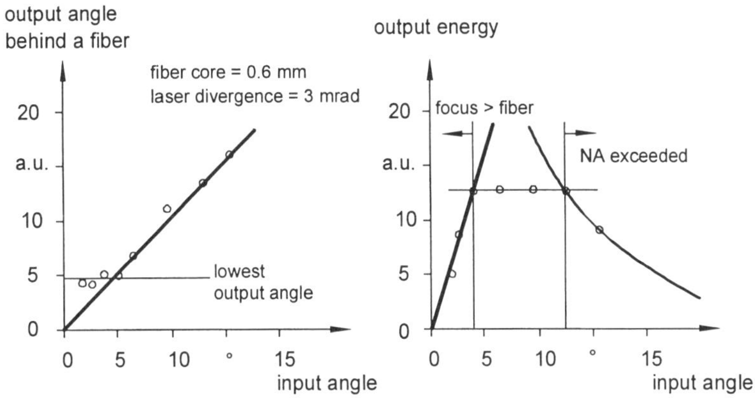

Fig. 9.19. Output divergence and output energy versus input angle.

Construction

Common fibers that are used for the beam transmission are normally quartz fibers with the following designs:

- *PC fiber (plastic-cladded-cilica, step-index)*
  A quartz core is surrounded with a silicate of lower refractive index. Both are protected by an outer plastic coating. When polishing, one must take into account small amounts of damage at the fiber rim. The full diameter cannot be used. The fibers are designed as step-index fibers.

- *Step-index quartz-quartz fiber*
  A quartz core is cladded with a doped coating of lower refractive index that itself is surrounded by a second quartz coating. For protection a plastic coating is applied.

- *Gradient-index quartz-quartz fiber*
  The refractive index decreases parabolically from the center to the fiber rim. The beam propagation is similar to the propagation in a lens line.

Step-index Fiber

The fiber is well defined by its two optical parameters, core diameter D and numerical aperture NA. While the core diameter can be easily determined with a microscope, the determination of the NA with a laser is more complicated. By tilting a fiber against a laser beam of low divergence, the angle α can be determined at which the intensity decreases by 10%.

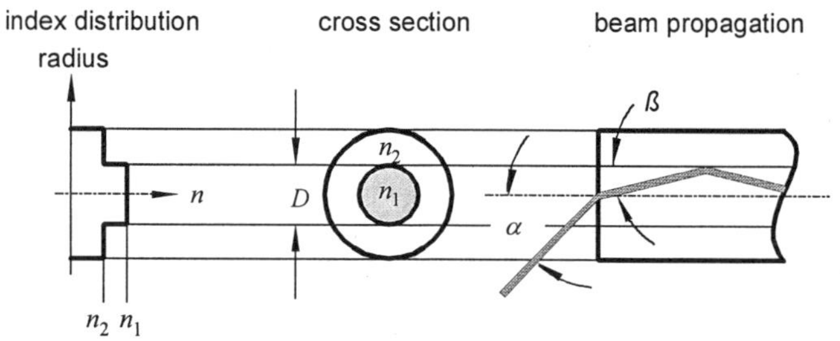

Fig. 9.20. Step-index fiber

With the refraction law and the limiting angle for total internal reflection (Fig. 9.20) one has, for the central ray, assuming $n = 1$,

$$\mathrm{NA} = \sqrt{n_1^2 - n_2^2} = \sqrt{(n_1 - n_2)(n_1 + n_2)} \approx \sqrt{\Delta n \times 2n}\,, \tag{9.19}$$

(refraction: $\sin\alpha = n_1 \sin\beta$, total internal reflection: $\sin(90° - \beta) = n_2/n_1$).

If the fiber is bent with the radius R, the effective NA, $\mathrm{NA}_{\mathrm{eff}}$, is reduced. According to [9.10]:

$$\mathrm{NA}_{\mathrm{eff}}(R) = \sqrt{\mathrm{NA}^2 - \frac{Dn^2}{R}}\,. \tag{9.20}$$

If, on the contrary, the input angle α is fixed, the bending radius should not be less than $R_{\min}$

$$R_{\min} = \frac{Dn^2}{\mathrm{NA}^2 - \mathrm{NA}_{\mathrm{eff}}^2}\,. \tag{9.21}$$

In [9.11] experimental results for the losses dependent on the fiber parameters and the bending radius are given.

Gradient Fiber

The gradient fiber [9.12] is characterized by a parabolic distribution of the core refractive index

$$n_1(r) = n_1(0)\left(1 - r^2/2\beta^2\right), \tag{9.22}$$

β = characteristic focal length.

The corresponding optical matrix is, according to Table 2.4

$$M = \begin{vmatrix} \cos(l/\beta) & \frac{\beta}{n}\sin(l/\beta) \\ -\frac{n}{\beta}\sin(l/\beta) & \cos(l/\beta) \end{vmatrix}. \tag{9.23}$$

If the waist radius is supposed to be the same as at the entrance, then with (2.29) it follows that

$$w_T^2 = A^2 w_T^2 + B^2\theta^2 = w_T^2 \cos^2(l/\beta) + \frac{\beta^2}{n^2}\theta^2 \sin^2(l/\beta). \tag{9.24}$$

This expression is length independent, if

$$w_T = \frac{\beta\theta}{n}. \tag{9.25}$$

If the focus diameter $2w_T$ approaches just the fiber diameter D, then one has for the maximum initial fiber angle

$$\theta_{max} = \frac{nD}{2\beta}. \tag{9.26}$$

A beam, focused under this angle onto the fiber, travels with constant diameter and flat wavefront through the fiber. The focusing effect of the fiber just compensates the divergence of the beam.

Example 9.3 Data of a graded core index-fiber

The quartz core of a fiber has a refractive index of 1.4498 at a wavelength of 1064 nm. Quartz doping with fluorine may result in a refractive index difference $\Delta n = n_{core} - n_{cladding}$ of about 14×10^{-3} to 20×10^{-3}. This yields a value NA ($NA^2 \approx 2n\Delta n$) of 0.20 to 0.24.
As with the step index fiber $NA \geq \theta_{max}$ and $D \geq 2w_T$ must also be fulfilled. Analogously to (9.16) it follows that $D\times NA \geq 2w_{TT}\times\theta$.
At a fiber diameter of 1 mm for $NA = \theta_{max}$ the characteristic length ($\beta = nD/NA$) is approximately 3 to 3.6 mm.

Focusing Optics Behind a Fiber

After the laser beam transmission by the fiber the beam is focused on the workpiece. A common focusing optics is shown in Fig. 9.21. The fiber endface is imaged by the two lenses f_2 and f_3 to a focus diameter of

$$2w_F = \frac{f_3}{f_2}D. \tag{9.27}$$

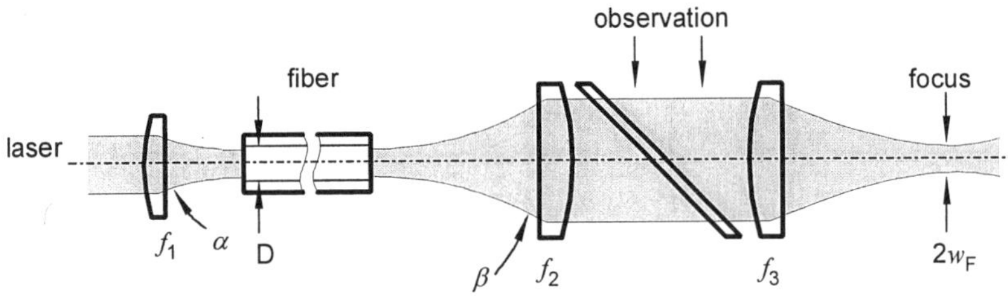

Fig. 9.21. Focusing optics behind a fiber

Fiber Losses

Three main loss mechanisms exist:

- wrong adaptation of the laser to the fiber; the aperture or the fiber diameter is exceeded,
- losses due to misalignments or due to mandatory tolerances for sufficient beam quality,
- material specific losses such as reflection and absorption; the absorption and scattering losses are normally indicated in dB/km

$$\frac{P_F}{P} = 10^{-vL/10\,\mathrm{dB}}, \tag{9.28}$$

P = laser power in,
P_F = laser power out,
v = loss coefficient in dB/km,
L = fiber length in km.

The losses due to absorption are negligible at fiber lengths of several 10 m with a loss coefficient of 5 dB/km for a wavelength of 1060 nm.

Example 9.4 Losses of a step-index fiber

Fiber length L = 100 m and a loss coefficient v = 5 dB/km yields
P_F/P = 0.89, i.e. approximately 10% loss for 100 m length.

For a laser beam with α = NA and $2w_T = D$ losses of 6% due to the beam profile and 4% due to the reflection losses at each end surface occur. The overall losses have a lower limit of about 15%.

Damage Threshold

Damages of the fiber surfaces appear due to wrong adaptation or by misalignments, in which the beam does not hit the fiber endface but neighboring parts and vaporizes or modifies the material. Imperfections in the fiber result in sudden fiber cracks at locations exposed to high pulse energies or high mean power. Low power radiation is scattered at these imperfections. A 600 μm fiber is capable of transmitting average powers around 3 to 4 kW in the cw operation and pulse peak powers up to 20 kW with 100 W s pulse energy.

9.6 Transformation of Constant Resonator Parameters

Conventional stable resonators change their beam qualities with the average pump power. Focusing with standard optics results in a power-dependent focus diameter and divergence.

With constant resonator qualities, e.g. constant aperture at the laser rod and constant phase curvature at the output coupler, a suitable design of the optics allows a constant beam angle and constant beam diameter in a fixed position behind the optics, not, however, a constant waist diameter and a fixed waist position simultaneously.

If a fiber end is placed in the plane with constant beam diameter and constant divergence, one obtains the same divergence behind the fiber and by imaging of the fiber end a constant waist on the workpiece.

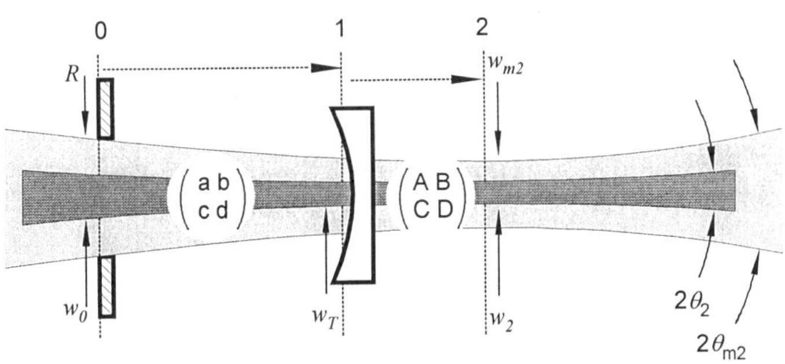

Fig. 9.22. Principle optics design to attain constant beam parameter product

We consider a stable resonator with an output coupler. Inside the resonator is a limiting aperture (rod) (Fig. 9.22). An optics is placed behind the mirror. A reference plane 1 may be placed into the mirror, where the reflected beam has a virtual waist and a virtual flat wavefront. The part of the resonator from the aperture to this plane 1 is represented by the optical matrix, [9.16],

$$M_1 = \begin{vmatrix} a & b \\ c & d \end{vmatrix}.$$

And the propagation from this plane 1 to a plane 2 behind the optics is given by the matrix

$$M_2 = \begin{vmatrix} A & B \\ C & D \end{vmatrix}.$$

The 00-mode radius of w_2 and the 00-mode divergence θ_2 in the plane 2 follow by transformation of the 00-mode waist w_T to

$$w_2{}^2 = A^2 w_T{}^2 + B^2 w_T{}^2/p^2 \text{ or } \theta_2{}^2 = C^2 w_T{}^2 + D^2 w_T{}^2/p^2. \tag{9.29}$$

p is the Rayleigh length for plane 1. The 00-mode radius w_0 at the aperture 0 is

$$w_0{}^2 = d^2 w_T{}^2 + b^2 w_T{}^2/p^2. \tag{9.30}$$

For simplicity, we assume the determinant $(ad - bc)$ to be 1, for both matrices M_1 and M_2. This is fulfilled if the reference plane lies in a medium with the same refractive index. If not, the equations have to be corrected with the corresponding refractive indices.

The multimode parameters in plane 2 follow with the relations for aperture and 00-mode radius as

$$w_{m2}^2 = \frac{R^2}{w_{02}^2}\left(A^2 w_T^2 + B^2 w_T^2 / p^2\right) \quad \text{or} \quad \theta_{m2}^2 = \frac{R^2}{w_{02}^2}\left(C^2 w_T^2 + D^2 w_T^2 / p^2\right). \tag{9.31}$$

After some algebraic transformations one obtains

$$\frac{w_{m2}^2}{R^2} = \frac{A^2 + B^2 / p^2}{d^2 + b^2 / p^2} \qquad \text{and} \qquad \frac{\theta_{m2}^2}{R^2} = \frac{C^2 + D^2 / p^2}{d^2 + b^2 / p^2}. \tag{9.32}$$

The parameters p, d and b change, in general, with the laser operation. For a constant radius and divergence or at least a small dependence, the matrix elements are chosen corresponding to the optics. This will be explained with the following two examples.

1. Aperture between output couplers and laser rod (Fig. 9.23)

The matrix elements of the resonator b and d are in this case constant, independent of the operating conditions.

Choosing the matrix elements of the optics so that

$$\frac{A^2}{B^2} = \frac{d^2}{b^2} \qquad \text{and simultaneously} \qquad \frac{C^2}{D^2} = \frac{d^2}{b^2}, \tag{9.33}$$

the beam radius and the divergence become constant in the plane 2 and independent of p

$$\frac{w_{m2}^2}{R^2} = \frac{A^2}{d^2} = \frac{B^2}{b^2} \text{ and } \qquad \frac{\theta_{m2}^2}{R^2} = \frac{C^2}{d^2} = \frac{D^2}{b^2}, \tag{9.34}$$

provided $ad - bc = 1$ and $AD - BC = 1$.

Example 9.5 Coupling with one lens (Fig. 9.23) to achieve constant beam quality

The internal matrix of the resonator is

$$\begin{vmatrix} a & b \\ c & d \end{vmatrix} = \begin{vmatrix} 1 & z \\ 0 & 1 \end{vmatrix} \tag{9.35}$$

and of the optics

$$\begin{vmatrix} 1 & x \\ 0 & 1 \end{vmatrix} \begin{vmatrix} 1 & 0 \\ -1/f & 1 \end{vmatrix} \begin{vmatrix} 1 & y \\ 0 & 1 \end{vmatrix} = \begin{vmatrix} 1-x/f & x+y-xy/f \\ -1/f & 1-y/f \end{vmatrix}.$$

With the conditions (9.34) and (9.35) it follows as a possible solution for focal length and distance at predefined fiber diameter $2w_{m2}$ and resonator (z, R)

$$f = \frac{2w_{m2}z}{R}, \tag{9.36}$$

$$x = f\left(1 + \frac{w_{m2}}{R}\right), \qquad y = f + z. \tag{9.37}$$

The numerical aperture NA of the fiber must be larger than the divergence θ_{m2}

$$\mathrm{NA} > \theta_{m2} = \frac{R_2}{2w_{m2}z}. \tag{9.38}$$

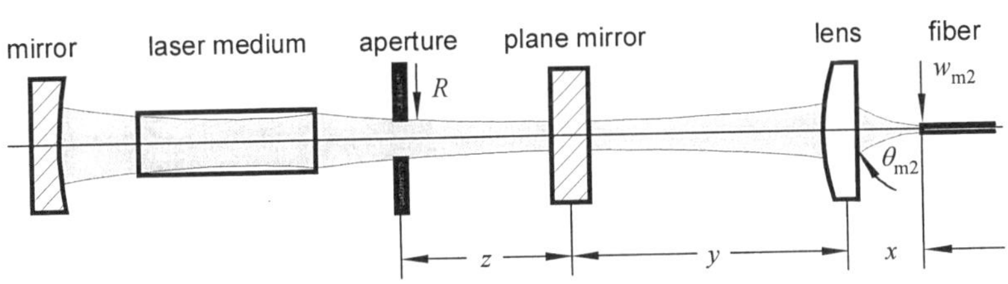

Fig. 9.23. Fiber coupling with constant beam parameters

2. Symmetrical resonator (Fig. 9.24)

If the aperture is in the center of a symmetrical resonator or just in front of the left mirror, then the resonator parameter p is

$$p^2 = -\frac{ab}{cd}. \tag{9.39}$$

Inserting this into (9.33) yields

$$\frac{w_{m2}^2}{R^2} = \frac{A^2 a/d - B^2 c/b}{ad - bc} \quad \text{and} \quad \frac{\theta_{m2}^2}{R^2} = \frac{C^2 a/d - D^2 c/b}{ad - bc}. \tag{9.40}$$

With $ad - bc = 1$ follows

$$\frac{w_{m2}^2}{R^2} = \frac{a}{d}A^2 - \frac{c}{b}B^2 \quad \text{and} \quad \frac{\theta_{m2}^2}{R^2} = \frac{a}{d}C^2 - \frac{c}{b}D^2. \tag{9.41}$$

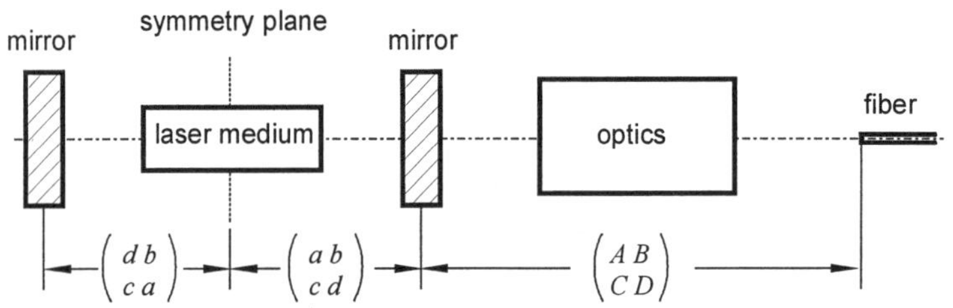

Fig. 9.24. Symmetrical resonator to achieve constant beam qualities

Radius and divergence do not change with the pump power P of the laser rod, if the derivatives of (9.41) vanish. The optics defined by the matrix elements must be chosen so that

$$\frac{\partial(c/b/\partial P)}{\partial(a/d/\partial P)} = \frac{A^2}{B^2} = \frac{C^2}{D^2} \tag{9.42}$$

is fulfilled.

Example 9.6. Symmetrical resonator with constant beam parameter (Fig. 9.25)

The thermal lens is described approximately by its refracting power D_f. The resonator matrix is

$$\begin{vmatrix} a & b \\ c & d \end{vmatrix} = \begin{vmatrix} 1 & 0 \\ -1/R & 1 \end{vmatrix} \begin{vmatrix} 1 & z \\ 0 & 1 \end{vmatrix} \begin{vmatrix} 1 & 0 \\ D_f & 1 \end{vmatrix}, \tag{9.43}$$

the derivatives are

$$\frac{\partial(c/b/\partial P)}{\partial(a/d/\partial P)}=\left(\frac{1}{z}-\frac{1}{R}\right)^2=\frac{A^2}{B^2}=\frac{C^2}{D^2}, \tag{9.44}$$

and divergence and radius follow to be

$$\frac{w_{\mathrm{m}2}^2}{R^2}=\frac{B^2}{z^2} \quad \text{and} \quad \frac{\theta_{\mathrm{m}2}^2}{R^2}=\frac{D^2}{z^2}. \tag{9.45}$$

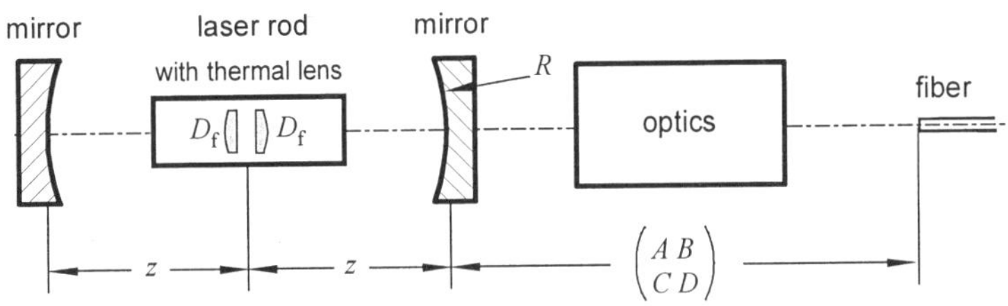

Fig. 9.25. Example for a symmetrical resonator and the optics for constant beam parameter

9.7 Laser-Induced Damage

High laser energy or peak power induce surface or bulk damage of the optical components. In general, the damage thresholds of the surface or of the coating are below the threshold of the bulk material. A defect usually modifies the surface, therefore the damage can often be detected with a microscope. A "breath test" in the laboratory provides a simple way to make qualitative modifications of the surface visible. Also, the detection of increased scattered laser light indicates damage.

In [9.13] the fundamentals and methods for damage threshold measurements are described. For the wavelength 1.06 μm at a pulse duration of 15 ns some damage thresholds are listed in Table 9.2.

At high peak intensities, particularly in the Q-switch operation, the optical material, for example a quartz fiber, can be destroyed. The fiber cracks at a certain length. This is caused by self-focusing due to the intensity-dependent refractive index. An estimation of the critical power P_c and the focusing length z_F is, according to [9.14], given by

$$P_c=\frac{1}{4\pi}\frac{\lambda^2}{n_2}, \tag{9.46}$$

$$z_{\mathrm{F}} = \frac{\pi n D^2}{4\lambda\sqrt{P / P_{\mathrm{c}} - 1}}, \tag{9.47}$$

D = is roughly the beam diameter.

Table 9.2 Damage thresholds of material surfaces and layers [9.13]

Material	Type	Method	Layer$^{(\#)}$	Energy J/cm^2	Power MW/cm^2
BK-7	HR	E	$(TiO_2/Si_2)^{12}TiO_2$	10	
BK-7	HR	IAD	$(ZrO_2/SiO_2)^{12}ZrO_2$	8	
BK-7	AR	V		2,8	
BK-7	AR			15	
BK-7	1	IAD	Ta_2O_5	2	
BK-7	1	E	Ta_2O_5	6	
BK-7	1	IAD	TiO_2	3,5	
BK-7	1	E	TiO_2	7	
BK-7	1	IAD	Al_2O_3	3,5	
BK-7	1	E	Al_2O_3	5	
Quartz				20	5
BK-7				26	
Lithium niobate				11	

λ = 1.06 µm, pulse width T = 15 ns, K_{f} = 2 Hz, 0% definition, E = electron beam, IAD = ion-assisted deposition V = vaporize. The exponent indicates the number of the layers. $^{\#}$ = number of layers

Table 9.3 Nonlinear refractive indices and focusing lengths [9.14]

Medium	λ nm	n	n_2 esu	n_2 10^{-20} m^2/W	P_{c} MW	z_{F} cm
air (100 at)	1064	1	0.2	0.8	10	7
water	1064	1.33	7	2	1.5	10
BK-7	1064	1.517	1.24	3.4	1.7	11
quartz	1064	1.456	0.95	2.73	2.3	11

z_{F} was calculated for a diameter of 1 mm and a power of 1 GW with a gaussian distribution. Further details of n_2 are given in Chap. 11.

10 Material Processing

The applications of the laser in material processing are based mainly on the local conversion of the radiation energy into heat [10.1]. At high intensities the heat is supplied to the workpiece so fast that losses due to heat conduction during welding and cutting are negligible. Nevertheless, the heat conduction is used for transformation hardening. Fig. 10.1 shows the laser cutting of a copper tube as an example.

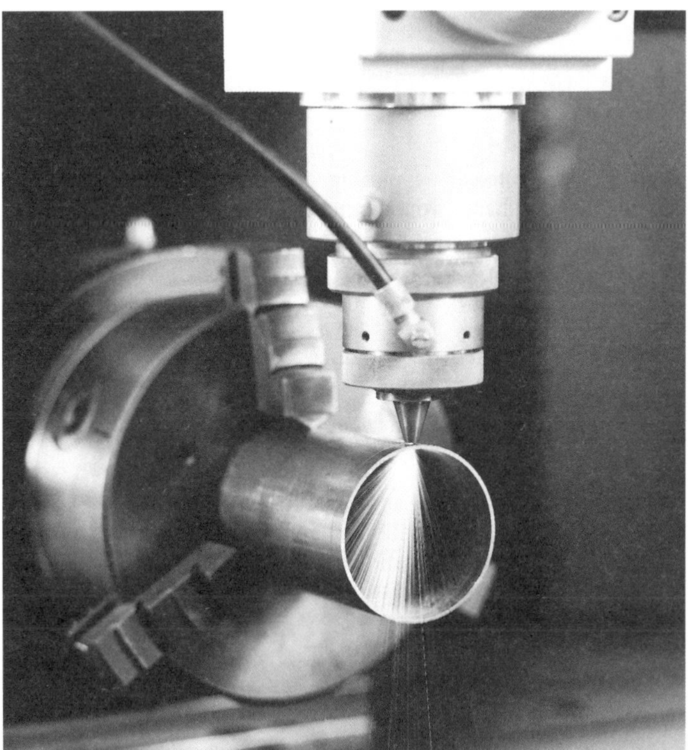

Fig. 10.1. Gas-assisted laser material processing, courtesy of HAAS

Commercial solid-state lasers in conjunction with a suitable optics allow intensities at the workpiece surface of about 10^{12} W/cm^2. Depending on the intensity and the material parameters five processing areas may be distinguished:

- **low intensity $10^4 - 10^5$ W/cm²**

  Heating: heat treatment of materials, hardening, oxidation. The material is only heated by the laser radiation. The melting temperature is not yet reached.

- **medium intensity $10^5 - 10^7$ W/cm²**

  Melting: soldering, seam- and spot-welding, marking, scribing, crystal growing, melt cutting, surface treatment.

  At the material surface the melting temperature is reached very rapidly. The intensity is then held such that the melt remains liquid until the required volume is melted.

- **high intensity $10^6 - 10^7$ W/cm²**

  Vaporization: drilling, sublimation cutting, scribing, trimming, penetration welding, carving, shock hardening.

  With high intensity, mostly in the pulsed operation, the material is vaporized rapidly.

- **higher intensity $10^8 - 10^9$ W/cm²**

  Plasma shielding:
  At further increase of the intensity a plasma is produced by the vaporized material that shields the laser beam from the material.

- **highest intensity $10^{10} - 10^{12}$ W/cm²**

  Gas breakdown:
  The high laser intensity immediately induces a gas breakdown in the atmosphere.

For the corresponding applications the parameter sets [10.2] for

- *the material,* such as heat conductivity, and absorption,
- *the laser,* such as laser power, beam quality, and pulse duration,
- *the geometry* of the workpiece, such as joint width, and material strength,
- *the processing* method, such as gas assistance, working distance, process duration, operation mode

must be adjusted with respect to each other in order to achieve or optimize the complete process.

10.1 Beam Parameters

In the following, an attempt is made to explain the various parameters, to show the interdependence in some detail and to describe the material processing quantitatively, as far as possible.

Laser Parameters

The beam parameters of the laser at the processing position are

- λ, the wavelength, for solid state lasers mostly in the range of 1 μm,
- p, the polarization, for rod cavities mostly statistical, i.e. on average there is no polarization.
- P_{av}, the average power,
- $2w_T$, the waist diameter in the focus,
- $2z_s$, the focal depth corresponding to the beam waist,
- τ, the processing time,

and especially in the pulsed operation

- $\hat{P}$, the peak power
- τ, the laser pulse-train duration
- r, thc pulsc rcpctition ratc.

The following relations are valid

$$E = \hat{P}\tau \text{ for the pulse energy,} \quad (10.1)$$

$$P_{av} = Er \text{ for the average laser power,} \quad (10.2)$$

and for the beam parameter product $w_T \times \theta$ holds with (2.33)

$$w_T\theta = \frac{w_T^2}{z_R}. \quad (10.3)$$

Optics Parameters

The optics (Fig. 10.2) is defined by two parameters:

- $A = \frac{f_2}{f_1}$ the zoom factor and
- f_3 the focal length of the objective, provided by the working distance or by the processing diameter.

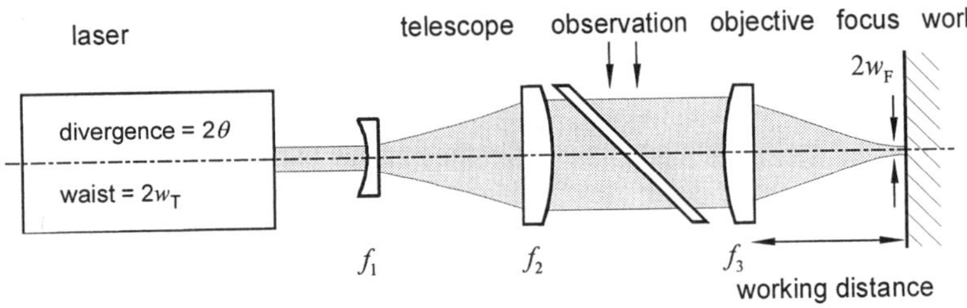

Fig. 10.2. Schematic design of the optics

An expansion of the beam can be achieved by a larger distance between laser and focusing objective (according to (2.43)).

The product of waist diameter and full divergence defines the beam parameter product of a laser beam. It is given in mm×mrad.

All of these parameters are interdependent and generally not constant in the processing range of the laser. Therefore the ranges should be noted from the corresponding data sheets.

10.2 Material Parameters

Only the intrinsic, unalterable properties of the material are considered as material parameters. They are:

- $1 - R$, the spectral and intensity-dependent absorption for a defined surface,
- k, the heat conductivity,
- c, the specific heat,
- ρ, the specific weight,
- T_m, the melting temperature,
- T_v, the evaporation temperature,
- Q_m, the melting heat,
- Q_v, the evaporation heat,
- the surface tension of boiling metals

and the derived temperature diffusivity κ

$$\kappa = \frac{k}{c\rho}. \tag{10.4}$$

With the temperature diffusivity the characteristic time (10.22) can be determined in which the temperature diffuses across a certain distance.

The absorption is dependent on the wavelength, on the temperature and thereby on the incident intensity. At intensities of about 10^8 W/cm^2, [10.1], the radiation is absorbed by the generated vapor and a laser-induced plasma arises.

The material parameters for various metals and standard materials are listed in Tables 10.1 and 10.2.

Table 10.1 Material parameters of pure metals, [10.4, 10.5,10.8]

Material		ρ g/cm^3	c J/g °C	k W/cm °C	T_m °C	T_v °C	Q_m J/g	Q_v J/g	$1-R$	κ mm^2/s
Aluminum	Al	2.70	0.90	2.22	660	2450	395	10470	0.10	91
Beryllium	Be	1.85	1.89	1.47	1277	2770	1090	24700	0.46	42
Lead	Pb	11.3	0.13	0.33	327	1750	26.4	858		22
Cadmium	Cd	8.65	0.23	0.92	321	765	55.3	879	0.30	46
Chromium	Cr	7.19	0.46	0.67	1875	2220	282	5860	0.40	20
Iron	Fe	7.87	0.46	0.75	1537	2735	274	6365	0.35	21
Germanium	Ge	5.36	0.31	0.6	937	2700	114	937		36
Gold	Au	19.3	0.13	2.97	1063	2966	67.4	1550	0.02	118
Indium	In	7.31	0.24	0.25	156	2000	28.5	1970		14
Cobalt	Co	8.85	0.42	0.71	1495	2900	245	6660	0.32	19
Copper	Cu	8.96	0.39	3.94	1083	2595	212	4770	0.06	113
Manganese	Mn	7.43	0.48	0.5	1245	2095	267	4100		14
Molybdenum	Mo	10.2	0.28	1.42	2610	5560	292	5610	0.40	50
Nickel	Ni	8.90	0.44	0.92	1453	2840	309	6450	0.28	23
Niobium	Nb	8.56	0.27	0.54	2468	3300	289	7790		23
Palladium	Pd	12.0	0.24	0.72	1552	3170	143	3475	0.26	25
Platinum	Pt	21.5	0.13	0.72	1769	3800	113	2615	0.27	26
Rhodium	Rh	12.4	0.25	0.88	1966	3880	211	5190	0.17	28
Silver	Ag	10.5	0.23	4.29	961	2212	105	2387	0.03	178
Silicon	Si	2.33	0.76	1.49	1410	2355	337	1446		84
Tantalum	Ta	16.6	0.14	0.54	2996	5425	157	4315	0.22	23
Titanium	Ti	4.51	0.52	0.17	1668	3280	403	8790	0.40	7
Vanadium	V	6.07	0.50	0.29	1900	3530	330	10260	0.40	10
Tungsten	W	19.32	0.14	1.67	3410	5930	193	3980	0.41	62
Zinc	Zn	7.13	0.39	1.13	420	907	102	1760	0.50	41
Tin	Sn	7.30	0.23	0.63	232	2270	60.7	1945	0.45	38

ρ = specific weight, T_m = melting temperature, Q_v = evaporation heat,
c = specific heat, T_v = evaporation temperature, Q_m = melting heat,
κ = temperature diffusivity, $1-R$ = absorption at 1 μm and room temperature.

Table 10.2 Material parameters of alloys [10.6, 10.7,10. 14]

Material	ρ g/m^3	c J/g °C	k W/cm °C	T_m °C	T_v °C	Q_m J/g	Q_v J/g	1−R
Al 99.5			2.7					0.059
AlMg 3			2.7					0.08
AlMgSi1Mn			2.7					0.067
AlCu6BiPb			2.8					0.073
E-Cu			8.9					0.047
CuSn8			8.8					0.128
CuZn37			8.45					0.09
CuNi10Fe1Mn			8.92					0.148
42 CrMo 4	7.85	0.477[a]	0.427[b]	1450				
90 MnV 8	7.85	0.5[a]	0.43[b]	1240				
X5CrNi 18 9	7.9	0.5	0.14 – 0. 18					
Stabilor G[d]	13.9	0.26[c]	0.131	1100	2700	49.1		0.67
Bond On 4[d]	14.4	0.65[c]	0.39	1250	3100	8625		0.7
Biosil f[d]	8.4	0.78[c]	0.1	1430	3800	162.5		0,68

a) at 20°C, value increases with increasing temperature,
b) at 20°C, value decreases with decreasing temperature,
c) temperature-dependent, mean average value in the range of 100°C to 850°C,
d) alloys of DEGUSSA.

10.3 Application Parameters

The production and application parameters are either fixed beforehand, e.g. the working distance, or their influence can be estimated only with difficulty, for example the joint gap when welding or the surface composition.

The known or required parameters for the processing are:

- cutting or welding *depth,*
- the *working distance;* i.e. the necessary, minimal free distance between workpiece surface and protection glass,
- the *process* or shielding *gas,* for protection against oxidation nitrogen or noble gases are used, for cutting support compressed air or oxygen,
- the *operating speed* (process cycle); it is provided by the application or by the processing machine,

and for the component properties:

- the *joint width*; in the case of a broad gap the laser light passes through unused, and the focus diameter and the average power must be increased correspondingly; for a small opening a total reflection can take place at the walls, which considerably increases the laser-to-material coupling,
- the *surface composition*; a shiny surface has naturally higher reflection losses, by "blackening" the surface the light coupling may improve under circumstances,
- the *material strength*,
- the *geometry,* layout; very often with a special construction the welding can be carried out in an easier way.
- the *material combination*; various materials with different geometries get combined, particularly in the case of welding.

The influences and the variety of these parameters can not be adequately defined, so that an empirical approach is normally of advantage.

10.4 Relations Between the Parameters

Simple models allow estimations of the necessary laser energy, pulse power, pulse duration and average laser power. More exact statements, however, require the one or multidimensional solution of the heat conductivity equation for the corresponding task.

$$c\rho\frac{\partial T}{\partial t} = \mathrm{div}\big(k\,\mathrm{grad}(T) + \mathrm{div}(J)\big), \qquad \textit{heat conductivity equation,} \qquad (10.5)$$

$T = T(x, y, z, t)$ temperature,
$J = J(x, y, z, t)$ absorbed laser intensity.

Minimal Energy

For a defined volume V the minimal energies for melting or vaporization can be determined. The melting energy consists of the energy to heat the volume to the melting point, and the melting heat for the phase transition [10.3],

$$E_\mathrm{m} = \rho\big(c(T_\mathrm{m} - T_0) + Q_\mathrm{m}\big), \qquad \textit{melting energy per volume,} \qquad (10.6)$$

$$E_\mathrm{v} = \rho\big(c(T_\mathrm{v} - T_0) + Q_\mathrm{m} + Q_\mathrm{v}\big), \qquad \textit{evaporation energy per volume.} \qquad (10.7)$$

The minimal energy for a cylindrical processing volume $V = \pi w_\mathrm{b}^2 z$ becomes

$$E \approx \frac{E_i}{1-R}\pi w_\mathrm{b}^2 z, \qquad (10.8)$$

w_b = processing radius,
z = material strength,
$1 - R$ = absorption coefficient,
i = m, v.

Basically, the energy balance for a spot welding is described with these equations. The losses by heat conduction can be neglected during a pulse but not, however, the losses by reflection. The main material parameter is the energy per unit volume.

Figure 10.3 reports the results of welding tests with a pulsed laser for which the laser energy was measured and the melting volume determined from the grinding pattern.

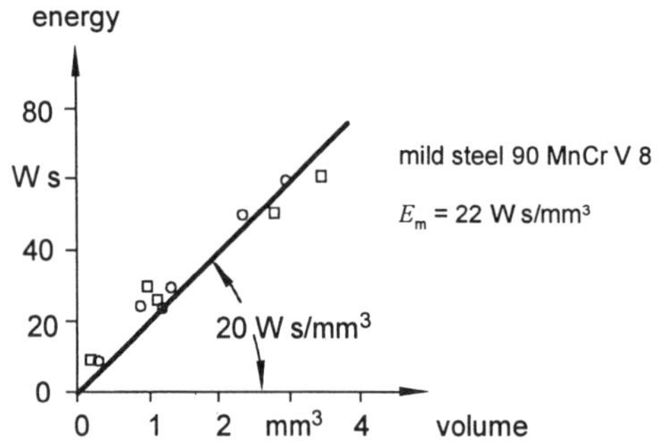

Fig. 10.3. Proportionality between laser energy and melted volume

Minimal Laser Power

To estimate the minimal laser power P_{min} that just compensates the losses due to heat conduction, the following model is used (Fig. 10.4)

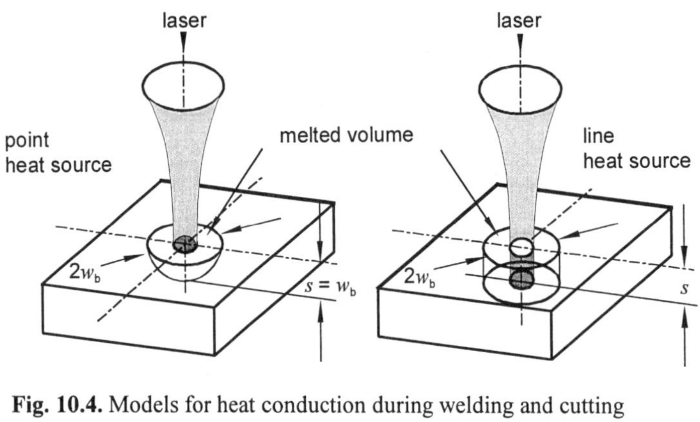

Fig. 10.4. Models for heat conduction during welding and cutting

The laser beam is assumed as a local, constant heat source with high intensity. After switching on, contours with constant temperature diffuse in the volume until a stationary condition is reached, in which the laser heats the material just to the evaporation temperature T_v at the incident point or line. From the area F at the melting temperature T_m, the heat passes to the neighboring material.

For stationary conditions, the heat dissipated per unit time through the cooling surface F is just compensated by the laser power P_{min},

$$(1-R)P_{min} \approx k\Delta T\frac{F}{w_b}, \tag{10.9}$$

$1 - R$ = absorption.

The cooling surface F is at a distance w_b from the heat source. The cooling surfaces are

$$F \approx 2\pi\, w_b^2, \tag{10.10a}$$

for a point-like heat source and

$$F \approx 2\pi\, w_b s, \tag{10.10b}$$

for a linear heat source. In order to hold the melting temperature in a hemisphere or in a cylinder of radius w_b one has the estimates

$$(1-R)P_{min} \approx 2\pi k(T_v - T_m)w_b\,, \qquad \textit{point source}, \tag{10.11}$$

$$(1-R)P_{min} \approx 2\pi k(T_v - T_m)s\,, \qquad \textit{line source}.$$

Cutting tests of a copper sheet with a pulsed laser (Fig. 10.5) allow the determination of the minimal pulse power and also the pulse power at minimal expenditure of energy per cutting length.

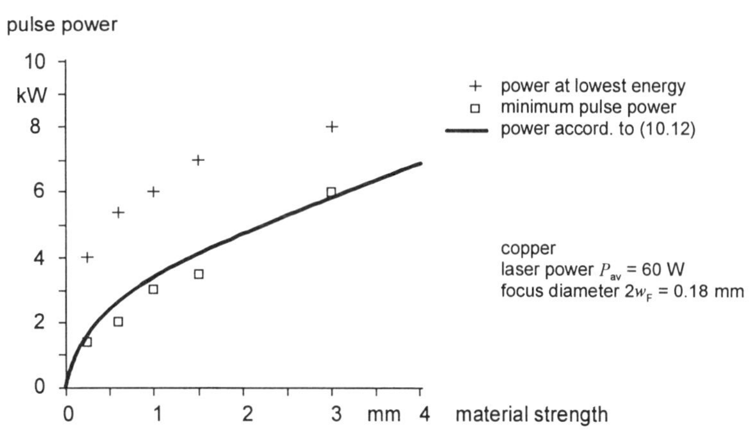

Fig. 10.5. Laser pulse power as function of the material strength for copper cutting

Depending on geometry or intensity either the point source or the line source dominates. The geometric mean value of the two takes this into account.

Three-dimensional solutions of the heat conduction equation [10.11, 10.12] give as an estimate for the minimal pulse power:

$$(1-R)P_{\min} \approx \sqrt{2/\pi}kT_i\sqrt{w_\mathrm{b}s}\ , \qquad i = \mathrm{m, v.} \tag{10.12}$$

The necessary power for given processing radius and processing depth is therefore defined by the thermal material constants and by the material reflection. For comparing different materials the power, normalized to a unit length, can be used.

Processing Performance

Besides the minimal power $P_{\min}$ that takes all losses into account, a definite volume of the material has to be melted continuously when welding or cutting. To supply the melting energy E for a certain volume to the material, an average power P_{av} must be applied during the processing time τ

$$\tau = \frac{E}{P_{\mathrm{av}}}\ . \tag{10.13}$$

This time equals, in the pulsed operation with rectangular pulses, the pulse duration. In the cw-process this time can be interpreted as the necessary exposure time for the processing diameter $2w_\mathrm{b}$. The resulting processing speed is then

$$u = \frac{2w_\mathrm{b}}{\tau} \tag{10.14}$$

in the cw-mode and

$$u = 2w_\mathrm{b}r \tag{10.15}$$

for the pulsed operation, with r = pulse repetition rate. In both cases the speed is proportional to the average laser power P_{av}

$$u \approx \frac{2w_\mathrm{b}P_{\mathrm{av}}}{E}\ . \tag{10.16}$$

Since the energy is proportional to the processing volume, the processing speed for given processing depth without further losses follows as

$$u \approx \frac{2w_\mathrm{b}\eta P_{\mathrm{av}}}{E_i\pi w_\mathrm{b}^2 s} = \frac{2\eta P_{\mathrm{av}}}{E_i\pi w_\mathrm{b}s}\ , \qquad \textit{line model}, \tag{10.17}$$

η = coupling efficiency.

In the pulsed operation the pulse duration is generally very small and the dissipated power can be neglected. Defining the line energy P_{av}/u and processing cross section $w_b s$ (10.17) then becomes

$$P_{av} \propto w_b s \, . \tag{10.18}$$

The line energy is proportional to the processing cross section.

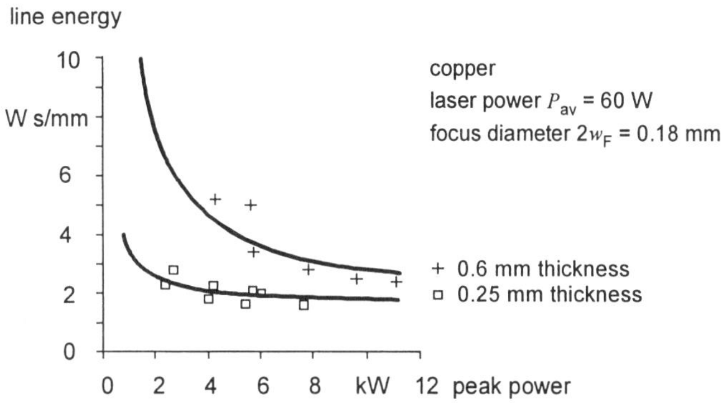

Fig. 10.6. Line energy versus laser power from cutting tests

In Fig. 10.6 the line energy decreases with increasing pulse power at first and then approaches a constant value. At a high power the heat losses are negligible.

In the cw processing the laser power is reduced by the dissipated power P_{min} and the processing speed correspondingly reduces to

$$u \approx \frac{2}{\pi} \frac{\eta}{E_i} \frac{P_{av} - P_{min}}{w_b s} \, , \qquad i = \mathrm{m, v}. \tag{10.19}$$

With P_{min} from (10.12), one then has

$$u \approx \frac{2}{\pi} \frac{\eta}{E_i} \frac{P_{av}}{w_b s} - \sqrt{\frac{2}{\pi}} \frac{kT_i}{E_i} \frac{1}{\sqrt{w_b s}} \, . \tag{10.20}$$

This can only be an approximate expression for the processing speed. It contains, however, all essential dependencies. The processing speed is linearly dependent on the laser power. A certain processing depth is reached only with a minimal laser power. For broad or deep welding seams the threshold is higher and the speed must be lowered. Essentially, (10.20) has the following form, [10.16],

$$u \approx \alpha M_1 \frac{P_{av}}{w_b s} - \beta M_2 \frac{1}{w_b} \, . \tag{10.21}$$

- The coefficients α and β are dimensionless parameters that depend on the process and on the geometry. They have to be determined empirically.
- The variables M_1 and M_2 are material parameters.
- The process speed is proportional to the average laser power per process cross section.

This principle relation between welding depth and welding speed is shown in Fig. 10.7. In the calculation the parameters are fitted to (10.20).

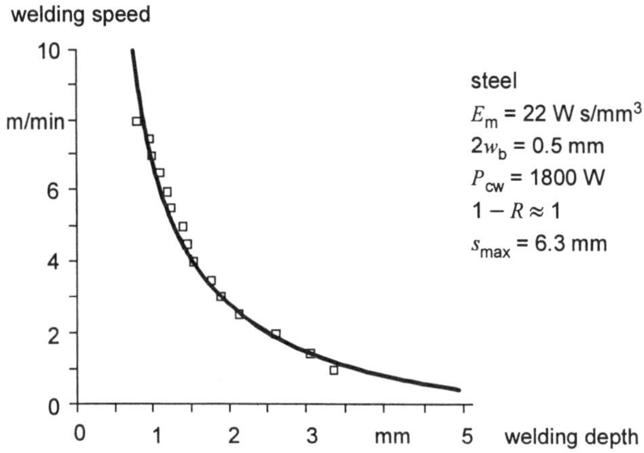

Fig. 10.7. Measured welding depth and seam width dependent on the welding speed, [10.15]

The relevant parameters are compiled in Table 10.3. These estimates only serve as approximations. In each case the specific application problem has to be optimized empirically.

The energy to melt a cubic millimeter of metals is between 1 and 10 W s, without taking into account heat losses. The evaporation energy per unit volume is approximately 10 times higher.

Processing Time, Pulse Duration τ

The temperature diffusivity from the heat conduction equation gives the characteristic distance a temperature diffuses per unit time. For a given processing width (for example the cutting width) the maximal processing time can be estimated as

$$\tau_{max} = \frac{c\rho}{k} w_b^2 . \tag{10.22}$$

Table 10.3 Thermal parameters for various materials [10.12]

Material		$1-R$	E_m J/mm^3	E_v J/mm^3	P_m/s kW/mm	P_v/s kW/mm	τ_m/F ms/mm^2	τ_v/F ms/mm^2	κ ms/mm^2
Aluminum	Al	0.1	26	309	1.8	6.8	14	45	11
Beryllium	Be	0.46	14	113	0.5	1.2	27	95	24
Lead	Pb	0.2[a]	4	52	0.07	0.36	55	145	45
Cadmium	CD	0.3	4	29	0.12	0.3	29	98	22
Chromium	Cr	0.4	20	126	0.39	0.56	52	224	49
Iron	Fe	0.35	22	165	0.41	0.74	53	223	48
Germanium	Ge	0.2[a]	11	36	0.35	1.1	30	34	28
Gold	Au	0.02[b]	195	1691	20	55	10	31	8
Indium	In	0.2	2	74	0.02	0.3	90	237	70
Cobalt	Co	0.32	24	208	0.42	0.8	57	258	52
Copper	Cu	0.06	93	806	8.9	21	10	38	9
Manganese	Mn	0.2[a]	32	184	0.39	0.7	81	280	71
Molybdenum	Mo	0.4	26	169	1.2	2.5	22	68	20
Nickel	Ni	0.28	30	235	0.60	1.2	50	201	43
Niobium	Nb	0.2[a]	41	374	0.84	1.1	49	335	43
Palladium	Pd	0.26	24	184	0.54	1.1	44	167	40
Platinum	PT	0.27	27	235	0.59	1.3	46	185	39
Rhodium	Rh	0.17	51	429	1.3	2.5	40	171	35
Silver	Ag	0.03[b]	112	948	17.2	40	7	24	6
Silicon	Si	0.2[a]	16	33	1.3	2.2	12	15	12
Tantalum	Key	0.22	43	369	0.92	1.7	47	221	43
Titanium	Ti	0.4	14	113	0.09	0.2	160	648	138
Vanadium	V	0.4	19	175	0.17	0.3	111	545	105
Tungsten	W	0.41	31	219	1.7	3.0	18	72	16
Zinc	Zn	0.5	4	29	0.12	0.3	31	112	25
Tin	Sn	0.45	2	33	0.04	0.4	43	84	27

E_i = melting-, evaporation-energy/volume,
P_i/s = stationary melting-, evaporation power/processing depth $s = \sqrt{w_b z}$
τ_i/f = pulse duration for stationary melting or evaporation/area,
$\kappa = k/(c\rho)$ = temperature diffusivity
a) The reflectivity of these metals is not known. Therefore a value of 0.8 has been assumed for this Table.
b) At high intensities the reflection gets lower, as with the use of oxygen as cutting gas [10.13].

Focus Diameter, Focal Depth

The focus diameter $2w_F$ becomes approximately the processing diameter $2w_b$ and is thus proportional to the laser divergence θ

$$w_b = w_F = \frac{f_3 \theta}{A}, \qquad \text{with (2.44).} \qquad (10.23)$$

The focal depth z_s should be approximately three times the processing depth z:

$$z \approx \frac{z_s}{3} = \frac{w_F^2}{3(w_T \theta)}, \qquad \text{with (2.46).} \qquad (10.24)$$

The intensity changes in this range are only around 10%. With the beam parameters w_T and θ follows, with (10.16), for the processing speed

$$u \propto \frac{P_{av}}{\sqrt{3 z w_T \theta}}. \qquad (10.25)$$

Beam Geometry

Assuming that the melting edges follow the lines of constant intensity J_k, the three-dimensional distribution of these lines is then important for the processing depth and also for the processing geometry. The joint width for cutting is dependent on the intensity distribution of the laser radiation. Figure 10.8 is illustrating this idea.

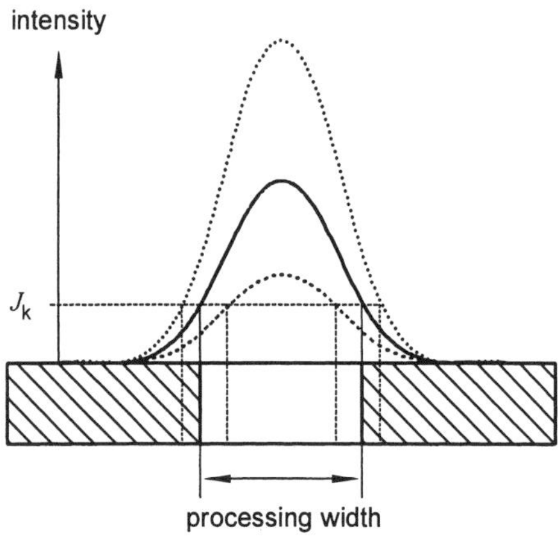

Fig. 10.8. Processing width dependence on the intensity distribution

In multimode operation the radial intensity distribution around the waist is almost Gaussian. The intensity distribution in Fig. 10.9 was recorded by a diode array.

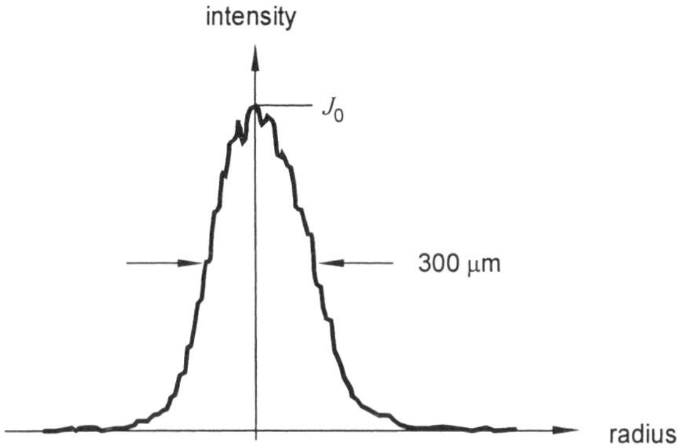

Fig. 10.9. Measured intensity distribution in a multi-mode focus

According to (2.9) one has for the Gaussian intensity distribution,

$$J_k = J_0 \frac{w_F^2}{w^2} e^{-2r^2/w^2}, \tag{10.26}$$

J_0 = peak intensity.

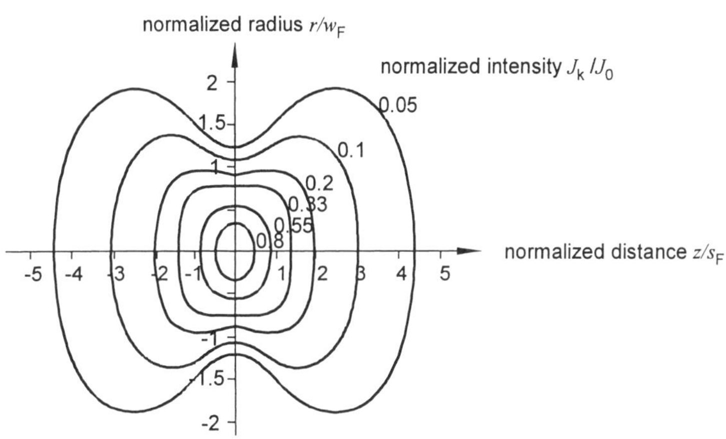

Fig. 10.10. Lines of constant intensity in the focal region

With

$$w^2 = w_F^2\left(1 + (z/z_s)^2\right) \tag{10.27}$$

follows for the *rz* dependence of the lines of constant intensity

$$\frac{r^2}{w_F^2} = -\frac{1}{2}\left(1+\frac{z^2}{z_s^2}\right)\ln\left(\frac{J_k}{J_0}\left(1+\frac{z^2}{z_s^2}\right)\right), \tag{10.28}$$

which are shown normalized in Fig. 10.10.

Figure 10.11 shows the bore diameter as a function of the z dependence. The holes in the aluminum coating of a glass substrate were caused by vaporizing the aluminum in the focal region.

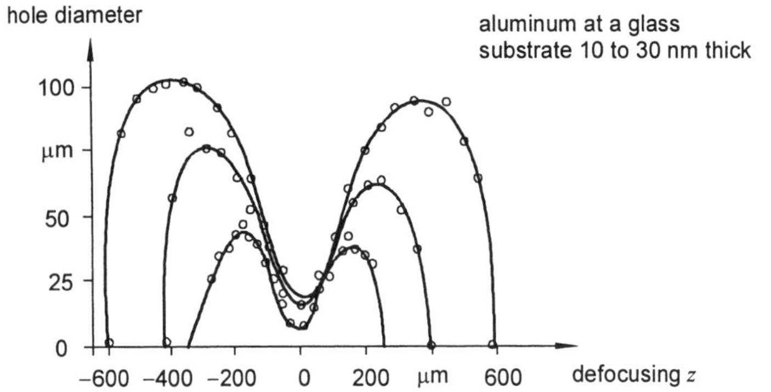

Fig. 10.11. Measured bore diameter at different distances of the focus [10.9]

Adjusting the processing intensity so that the melting front ends at

$$J_k = 0.3J_0 \tag{10.29}$$

the largest focal depth is obtained (Fig. 10.10).

The processing radius w_b – the radius, within which the melting front lies – becomes, at $z = 0$:

$$w_b^2 = -\frac{\ln(0.3)}{2} w_F^2 = 0.6 w_F^2, \tag{10.30}$$

and the processing depth s is

$$s \approx 0.69 z_s. \tag{10.31}$$

With the parameters w_T, and θ of the laser and f_3 and A of the optics these equations read

$$w_b \approx 0.78\frac{f_3\theta}{A}, \tag{10.32}$$

and

$$s \approx 0.69\frac{f_3^2\theta}{A^2 w_T}. \tag{10.33}$$

In comparison with (10.23) and (10.24) the processing diameter is insignificantly smaller and the processing depth slightly larger.

A small processing radius and a large processing depth require a low beam parameter product (high beam quality)

$$w_T\theta \approx \frac{w_b^2}{s}. \tag{10.34}$$

All relevant material processing relations are compiled once again in Table 10.4. The necessary laser parameters can be estimated from the Tables 10.1 to 10.4. This is explained with the help of the following two examples.

Example 10.1 Cutting a steel sheet

A steel sheet with a strength of s = 1 mm is to be cut. The melting cutting method assisted by compressed air is chosen. The cutting width is at most 0.25 mm. The necessary energy to melt the complete volume in a spot diameter is, according to Table 10.3

$E = E_m\pi w_b^2 = 22\ \text{W s/mm}^3 \times \pi/4 \times 0.25 \times 0.25\ \text{mm}^3 \approx 1.1\ \text{W s}.$

Due to the cutting width of 0.125 mm the maximum pulse duration follows as

$\tau_{max} = w_b^2/\kappa = 48\ \text{ms/mm}^2 \times 0.125 \times 0.125\ \text{mm}^2 = 0.75\ \text{ms},$

and the laser peak power is

$E/\tau_{max} = \hat{P}_{min} = 1.5\ \text{kW}.$

Due to this high peak power pulsed operation with an average power of P_{av} = 250 W is chosen. The pulse repetition rate follows as

$r = P_{av}/E = 250\ \text{W}/1.1\ \text{W s} = 227\ \text{Hz}.$

This leads to a maximum cutting speed of

$u = 2w_b r = 0.25\ \text{mm} \times 227\ \text{Hz} = 57\ \text{mm/s} = 3.4\ \text{m/min}.$

To receive a fairly smooth cutting surface the pulses should partly overlap. This reduces the cutting speed at least by a factor of 2 to 1.7 m/min. The necessary beam quality of the laser is

$w_T\theta = w_b^2/s = 0.125^2\ \text{mm}^2/1\ \text{mm} = 15\ \text{mm mrad}.$

Table 10.4 Relations between the processing parameters

Laser parameters		Equation	Relation with respect to the material processing
pulsed operation			
E/W s	laser energy	$E = (1-R)E_i \pi w_b^2 s$	~ to the processed volume
τ_{max}/ms	max. pulse duration	$\tau_{max} = \frac{c\rho}{k} w_b^2$	time for temperature diffusion, heat influence zone
$\hat{P}$ / kW	peak power	$P_{min} = E/\tau_{max}$	~ processing depth at constant beam radius
r/Hz	pulse repetition rate	$r = P_{av}/E$ $u = w_b r$	processing time processing speed
cw and pulsed operation			
P_{av}/W	medium power	$P_{av} \sim u$	processing speed
P_{av}/u	line energy	$P_{av}/u \sim w_F s$	processing area
quality of the beam source			
$2w_T$/mm	beam waist diameter		(focus aperture)
θ/mrad	corresponding divergence		(focus diameter)
$w_T\theta$/mm mrad	beam quality	$w_T\theta = w_b^2/s$	(w_b = processing radius), (s = material strength)
$P_{avmax} - P_{avmin}$/W		processing range	
processing sizes with focusing optics			
w_F/mm	focus radius	$w_b = \theta f/A$	$w_b \approx 0.8\ w_F$ processing diameters (f/A effective focal length)
s_F/mm	focus depth	$s \sim s_F^2/A^2$	s = processing depth
J/kW/mm^2	intensity	$J = P/(\theta w_b^2)$	processing methods processing speed material strength

Example 10.2 Cutting copper

material strength $s = 1$ mm, cutting width $2w_b = 0.2$ mm, average power $P_{av} = 100$ W

minimum energy	$E_{min} = 93 \times \pi \times 0.01$ W s ≈ 2.9 W s
pulse length	$\tau_{max} = 9 \times 0.01$ ms ≈ 0.1 ms
repetition rate	$E / \tau_{max} = \hat{P} \approx 30$ kW , $r = 100 / 2.9$ Hz $= 34$ Hz
processing speed	$u = 0.2 / 2 \times 34$ mm/s $= 3.4$ s mm/s $= 0.2$ m/min
beam parameter product	$w_T\theta = 1/0.1$ mm mrad $= 10$ mm mrad

10.5 Measurement Procedures

During processing and for reproducible adjustments the beam parameters have to be measured online during the manufacturing process. For quality control absolute values have to be maintained.

Online Power or Energy Measurements

For online measurements of energy, power or the temporal intensity distribution a preferred arrangement is shown in Fig. 10.12. A small amount of the beam is deflected and can be recorded by means of a single photodiode, diode array or camera and electronics.

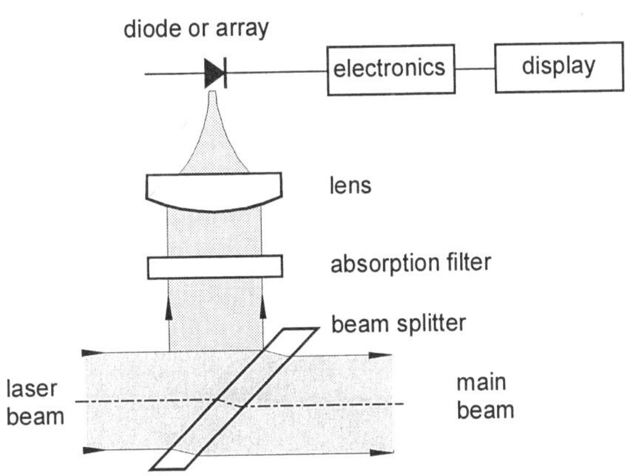

Fig. 10.12. Power or energy measurement with a beam splitter

One has to note that all the reflections at the reference surfaces, particularly of the semitransparent mirror, change with time due to dust or dirt from laser damaging. In addition, the splitter normally shows a polarization dependence. Also, one has to expect a temporal- and temperature-drift of the diode. The sensitive area of the photodetector must be large enough to compensate angle and position tolerances and the complete construction should be insensitive to reflections of other optical elements and reflections from the processing surface.

Replacing the photodiode by a diode line or an array [10.10], a two-dimensional cross section or a 3D representation of the intensity distribution in the near and far-field is possible. The waist radius and the beam diameter at different distances can be directly measured and from these the focus depth, divergence and beam quality determined.

Absolute Power Measurement

Mainly thermal methods are used for the absolute measurement of the energy or the average power. The beam is directed to a water-cooled absorber and the temperature rise or difference is measured. A flat, large absorber plate is desirable which is invariant with respect to changes in diameter, distribution and angles of the incidence beam, so that at each position the beam can be measured.

Beam Radius

The beam radius measurement is reduced to the question "what amount of energy or-power is contained in what diameter". Several definitions, depending on the beam profile, are used. For simplification, rotational symmetry is assumed:

- *1/*e *- radius*

  The beam radius at which the intensity has dropped to 1/e of the maximum intensity. This radius contains 63% of the complete energy or power for a Gaussian distribution.

- *1/*e^2 *radius*

  In this radius 86.5% of the complete energy or power is enclosed for a Gaussian distribution. These two definitions are reasonable for a Gaussian beam profile.

- *second moment*

$$\overline{r^2} = \frac{\int_0^\infty r^3 J(r)dr}{\int_0^\infty rJ(r)dr}.$$

  Here, the radial intensity distribution of the beam must be known.

– *99% energy enclosure*

In practice the beam radius is very often defined as the aperture radius at which 99% of the power (energy) can be measured behind this aperture. This is, for example, the case for the transmission with a step-index fiber. The core diameter of the fiber acts as the aperture here. This method has the advantage that the beam need not be of circular symmetry and the distribution does not need to be known.

Part III

Data and Specifications

11 Laser Materials

In the following, the specifications for the most important parameters are defined and listed for various laser materials. Only those materials are taken into account, for which laser activity has been proven for pulse or cw-operation at room temperature with an efficiency of about 1% lamp or diode with excitation and which are commercially available. Further details are provided in the subsequent chapters.

11.1 Parameter Specification

The most essential parameters used for characterizing laser materials are taken from the previous chapters or defined anew in this section. The equation numbers refer to the corresponding previous sections. These parameters are then listed in the following tables for the various laser materials.

Laser Data

λ = lasing wavelength, in general the laser wavelength of the intensity maximum. The fluorescence wavelength is smaller.
$\Delta\lambda$ = halfwidth of the laser transition
λ_p = pumping wavelength
λ_{FZ} = wavelength, below which color center generation has been observed or is possible
σ = effective cross section of stimulated emission

$$\sigma_{eff} = \frac{\lambda^4}{4\pi^2 n^2 c \Delta\lambda}, \quad \Delta\lambda = \text{halfwidth of fluorescence.}$$

The effective cross section can be determined from absorption measurements α (λ) according

$$\sigma_{eff} = \frac{\alpha(\lambda)}{N_0} \frac{Z_0}{Z_1} e^{(E-h\nu)/k_B T},$$

N_0 = concentration of the active ions,
Z_i = branching of the energy levels,

E = energy gap of the corresponding lowest branching.

The effective cross section is reduced by further excitation of the higher levels $\sigma = \sigma_e - \sigma_{ESA}$

σ_e = effective cross-section for spontaneous emission
σ_{ESA} = effective cross-section for excited state absorption

$$\sigma_e(\lambda) = \sigma_a(\lambda) e^{(E-h\nu)/k_B T}, \qquad [11.37].$$

τ_0 = lifetime of the upper laser level, extrapolated to low concentrations of the active ion

τ = lifetime of the upper level,
at standard concentration. The natural decay of the upper laser level population is approximately exponential with time. In the literature, this decay is given as

$$N_2(t) = N_2(0) e^{-t/\tau - f(t)}, \qquad [11.61].$$

For exponential depletion $\tau/\tau_0 = \eta_Q$ yields the quantum efficiency. The doping dependence becomes

$$\tau = \frac{\tau_0}{1 + \left(C / C_{0.5}\right)^2},$$

C = concentration of the active ions,
τ_0 = extrapolated lifetime for $C = 0$,
$C_{0.5}$ = concentration at which the lifetime is half τ_0.

Optical Properties

n_{rel} = refractive index (relative) at the indicated laser wavelength relative to air

$$n^2 = A + \frac{B\lambda^2}{\lambda^2 - C} + \frac{D\lambda^2}{\lambda^2 - E},$$

The special notation for selected refractive indices are given in Table 11.1. A to E are the Sellmeier coefficients, tabulated for the most common materials in Table. 11.2.

Table 11.1 Labeling of the refractive indices for various wavelengths

λ/nm	1014	852	706	656	644	589	588	546	481	480	436	405	365
Notation	n_t	n_s	n_r	n_C	$n_{C'}$	n_D	n_d	n_e	n_F	$n_{F'}$	n_g	n_h	n_i
Line				A	Cd		He			Cd			
Color				red	red	yellow		yellow		green		blue	

Table 11.2 Sellmeier coefficients for various crystals, [11.1]

Crystal	*A*	*B*	*C*	*D*	*E*
YAG	1	2.2779	0.01142	0	0
LaLuGG	1	2.3891	0.03781	0	0
GdScAG	1	2.510	0.0153781	0	0
YScAG	1	2.4118 – 2.420	0.01477– 0.01520	0	0
GdGG	2.39841	1.35724	0.03027	2.37981	165.753
GdScGG	2.11233	1.63589	0.02620	0.99161	62.87
YGG	1	2.5297	0.019694	0	0
LuGG	1	2.2573	0.022833	0	0
LuAG	1	2.4822	0.009066	0	0
GdAG	1	2.0055	0.014559	0	0

dn/dT = temperature coefficient of the refractive index
The light wave inside a material is influenced by the absolute refractive index change dn/dT in the material. However, the values of the refractive index change dn_{rel}/dT provided in the literature have been measured relative to air. With the definition of the refractive index the absolute change dn_{abs}/dT can be calculated

$$n_{rel} = \frac{n_{abs}}{n_{air}} . \tag{11.1}$$

Differentiating and setting $n_{air} = 1$ gives

$$\frac{dn_{abs}}{dT} = \frac{dn_{rel}}{dT} + n_{rel}\frac{dn_{air}}{dT} . \tag{11.2}$$

For $\lambda = 1060$ nm Fig. 11.1 gives the temperature coefficient dn_{air}/dT, [11.2].

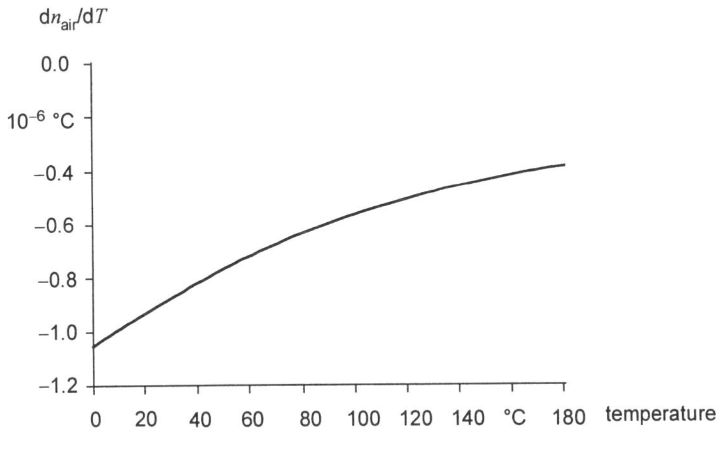

Fig. 11.1. Temperature coefficient of dn_{air}/dT, according to [11.2]

n_2 = nonlinear refractive index
The nonlinear refractive index is given either as a function of the field strength E

$$n(E) = n_0 + n_2\langle E^2\rangle = n_0 + n_2\frac{E^2}{2}, \tag{11.3}$$

$[n_2]$ = esu,

$$E[\mathrm{esu}] = 3\times 10^4 E\left[\frac{\mathrm{V}}{\mathrm{m}}\right],$$

esu = electrostatic units,

or as a function of the intensity J

$$n(J) = n_0 + \gamma_2 J,$$

$$\gamma_2 = \left[\frac{\mathrm{m}^2}{\mathrm{W}}\right], \ \gamma_2\left[\frac{\mathrm{m}^2}{\mathrm{W}}\right] = 0.9\times 10^9 n_2[\mathrm{esu}],$$

with n_2 and γ_2 being the nonlinear refractive indices.

With the help of the Abbe number ν_d and the refractive index n_d for the Fraunhofer line the nonlinear refractive index can be calculated in approximately, according to [11.6], as

$$\frac{n_2}{\left[10^{-13}\,\mathrm{esu}\right]} = \frac{n_\mathrm{d}-1}{\nu_\mathrm{d}^{5/4}}. \tag{11.4}$$

Table 11.3 provides in addition to the nonlinear refractive index the corresponding self-focusing length z_F (9.47) and the critical power P_c (9.46) for some laser materials.

The conversion from electrostatic units [esu] to SI-units can be carried out [11.3] with

$$\frac{\gamma_2}{\left[\mathrm{m}^2/\mathrm{W}\right]} = \frac{40\pi}{cn_0}\frac{n_2}{[\mathrm{esu}]} = 4.19\times 10^{-7}\frac{1}{n_0}\frac{n_2}{[\mathrm{esu}]}, \tag{11.5}$$

$$\frac{n_2}{[\mathrm{esu}]} == \frac{cn_0}{40\pi}\gamma_2 = 2.386\times 10^6\, n_0\frac{\gamma_2}{\left[\mathrm{m}^2/\mathrm{W}\right]}.$$

p_{ij} = photoelastic constants.
These are in most cases not available at the laser wavelength. In isotropic media such as glass one has

$$p_{44} = (p_{11} - p_{12})/2. \tag{11.6}$$

Table 11.3 Nonlinear refractive index, critical power P_c and self-focusing length z_F

Medium	λ nm	n	n_2 10^{-13} esu	γ_2 10^{-20} m^2/W	P_c MW	z_F cm	Ref.
Air (1atm)	1064	1	0.2	0.84	10.75	7	
Water	1064	1.33	7	22.06	0.31	10	
BK-7	1064	1.517	1.24	3.43	1.73	11	
Quartz	1064	1.456	0.95	2.73	2.26	11	
Si-Glass	1064	1.769	4.4	10.43	0.49	13	
LaSF-7	1064	1.914	6.2	13.58	0.35	14	
Alexandrite	750	1.55	0.83	2.24	1.29	16	[11.4]
BEL	1070	1.99	2.1	4.42	1.04	15	[11.4]
GGG	1064	1.943	4	8.63	0.54	14	
GSGG	1064	1.925	8	17.42	0.27	14	
Ruby	694.3	1.755	1.1	2.63	0.83	20	[11.4]
YAG	1064	1.823	3.525	8.10	0.61	14	[11.4]
YLF	1053	1.448	0.595	1.72	3.54	11	[11.3]
APG-1	1061	1.526	1.12	3.08	0.91	11	
LG-650	1061	1.511	1.415	3.93	1.51	11	
LG-670	1061	1.567	1.41	3.77	1.52	12	
LG-680	1061	1.56	1.6	4.30	1.34	12	[11.4]
LG-750	1053.5	1.516	1.08	2.99	1.95	11	
LG 760	1053	1.508	1.02	2.84	2.06	11	[11.4]
LG-810	1051	1.5	0.515	1.44	4.07	11	[11.4]
LG-812	1054	1.426	0.49	1.44	4.30	11	
LHG-5	1054	1.541	1.16	3.16	1.82	12	[11.5]
LHG-6	1054	1.532	1.01	2.76	2.09	11	[11.5]
Q-88	1054	1.536	1.14	3.11	1.85	12	[11.5]

z_F was calculated for a diameter of 1 mm and a power $P = 100 \times P_c$ using a Gaussian distribution. Additional literature is in the data tables for laser glasses.

Thermal Parameters

c_p = specific heat (at constant pressure)

k = heat conductivity
The heat conductivity is temperature dependent.

α = thermal expansion coefficient for the temperature range of 20°C to 100°C.

Mechanical Properties

ρ = specific weight

ν = Poisson number

E = modulus of elasticity (Young's modulus)

σ_{max} = fracture tension,
which may not be exceeded during laser operation. No safety margin is taken into account. This value depends strongly on the surface finish and decreases with increasing surface area. It can only be taken as a reference value.

σ_f = fracture tension due to microcracks in the material

K_c = fracture toughness is correlated with the fracture tension via

$$\sigma_f = \frac{K_c}{\sqrt{\pi a}}, \tag{11.7}$$

Y = geometry factor,
a = flaw depth [11.80, 11.81].

H_K = Knoop hardness as an indicator for polishing the material.

Crystal Properties

T_m = melting temperature
for crystals, or for glasses the transformation temperature. This temperature may not be exceeded during normal handling and processing, e.g. when coated with dielectric layers.

C [at%] = concentration in atom per cent
is the part of the respective element relative to the total molecular weight. Manufacturers of crystals normally specify only the percentage share of the doping ions, which replace a crystal ion in the host. The at% details used in the following refer to this manufacturer definition.

C [wt%] = concentration in weight percentage
is the part of the atomic weight of the respective element relative to the molecular weight of the complete chemical formula.

F_A, F_W, F_{AW} = conversion factors for the calculation of the absolute ion number

N_0 [10^{20} ions/cm^3] = $F_W \times C$[wt%], wt% = weight percentage

N_0 [10^{20} ions/cm^3] = $F_A \times C$[at%], at% = atom percentage
at% = $F_{AW} \times$wt%.

The number of the ions per volumes can be calculated for a given weight percentage C [wt%] as

$$N_0\left[\frac{1}{\text{cm}^3}\right]=\frac{N_A\rho}{M}\frac{C[\text{wt\%}]}{100},$$

$N_A = 6.022\times10^{23}\ \text{mol}^{-1}$,
M = atomic weight of the doping ion,
C = concentration in weight %,
ρ = specific weight.

For the specifications of the weight percentage one has to consider whether the part of the pure active ion or the part of the oxide is meant relative to the crystal material. In the case of Nd:YAG whether the pure neodymium or the neodymium oxide Nd_2O_3 with respect to $Y_3Al_5O_{12}$.

Example 11.1 Atom per cent of concentration of Nd in YAG

The atom % concentration C(Y) of yttrium in YAG is given as

$$C(\text{Y})=\frac{3_\text{Y}}{3_\text{Y}+5_\text{Al}+12_\text{O}}=\frac{3}{20}\times100\,\%=15\ \text{at\%}$$

If one per cent of the Y^{3+} ion is replaced by Nd^{3+} ions (manufacturer's specification at% = 1%!) then the chemical formula of $Y_3Al_5O_{12}$ changes into $Nd_{0.03}Y_{2.97}Al_5O_{12}$. The absolute atom per cent concentration of Nd in YAG becomes therefore

$$C(\text{Y})=\frac{0.03_\text{Nd}}{0.03_\text{Nd}+2.97_\text{Y}+5_\text{Al}+12_\text{O}}=\frac{0.03}{20}\times100\,\%=0.15\ \text{at\%}$$

Example 11.2 Ruby with 1 at% Chromium

Here, the aluminum ion Al^{3+} is replaced by the chromium ion Cr^{3+}, chemical formula $Cr_{0.02}Al_{1.08}O_3$ atomic weights Al = 26.981, O = 16, Cr = 52 specific weight (ruby) ρ = 3.98 g/cm^3

$$\frac{C}{\text{at\%}}=\frac{(0.02\times52_\text{Cr})}{(0.02\times52_\text{Cr})+(1.08\times26.981_\text{Al})+(3\times16_\text{O})}100\%=1.33[\text{wt\%}]$$

$$\frac{N_0}{\text{at\%}}=\frac{(6.022\times10^{23}\times3.98)}{52}\frac{1.33}{100}\frac{1}{\text{cm}^3}=6.13\times10^{20}\,\text{cm}^{-3}\ \text{or}$$

$$\frac{N_0}{\text{wt\%}}=4.61\times10^{20}\,\text{cm}^{-3}$$

Example 11.3 Nd:YAG with 1 at% Nd (Manufacturer specification)

1% of the Y^{3+} ions are replaced by Nd^{3+} ions. The chemical formula $Y_3Al_5O_{12}$ of the pure garnet is changed into $Nd_{0.03}Y_{2.97}Al_5O_{12}$. With the atomic weights Y = 88.905, Al = 26.981 O = 16 and Nd = 144.24 the weight concentration follows as

$$C_{\mathrm{Nd}} = \frac{(0.03 \times 144.24_{\mathrm{Nd}})}{(0.03 \times 144.24_{\mathrm{Nd}}) + (2.97 \times 88.9_{\mathrm{Y}}) + (5 \times 26.98_{\mathrm{Al}}) + (12 \times 16_{\mathrm{O}})} 100\%$$

$$= 0.729[\mathrm{wt\%}]$$

This means: 1 at% ≡ 0.729 wt% or 1 wt% ≡ 1.37 at%.

With the specific weight $\rho = 4.56$ g/cm³ of YAG follows for the absolute ion density

or

$$\frac{N_0}{\mathrm{wt\%}} = 1.89 \times 10^{20}\,\mathrm{cm}^{-3}$$

Table 11.4 provides the number of the ions per cubic cm for one atom percentage (F_A) and Table 11.5 for one weight percentage (F_W).

Table 11.4 Conversion factor F_A of an atom percentage into the absolute ion number

10^{20} cm^{-3}/%	Cr	He	Ho	Nd	Pr	Ti	Tm	Yb
Atom weight	52	167.3	165	122.2	140.9	47.9	169	173
Alexandrite	3.51						3.69	
Ruby	1.14							
BEL				1.94			6.061	
GGG	0.84		2.59	1.26			2.53	
GSAG	0.84		2.13	1.26			2.08	
GSGG	0.81		2.35	1.22			2.30	
Al_2O_3	4.07							
YAG	0.92	14	1.67	1.38			1.63	1.38
YAP	1.97			1.95				
YLF				1.39				
YSAG	0.86		1.67	1.30			1.63	
YSGG	0.83		1.91	1.25			1.86	
YVO_4				1.54				

Partially the values are dependent on the stoichiometry, for details see Litton (1991)
Atom per cent, according to manufacturer specifications

Table 11.5 Conversion factor F_W of a weight percentage into the absolute ion number

10^{20} cm^{-3}/%	Cr	Er	Ho	Nd	Pr	Ti	Tm	Yb
Alexandrite								
Ruby	3.24							
BEL								
GGG	8.20		1.26	2.95		1.26	7.08	
GSAG	6.74		1.26	2.43		1.26	5.82	
GSGG	7.47		1.21	2.69		1.21	6.59	
Al_2O_3								
YAG	5.28		1.39	1.90		1.39		1.59
YAP								
YLF								
YSAG	5.27		1.31	1.90		1.31		
YSGG		1.24		1.24				

Table 11.6 Crystal data for the calculation of the absolute ion numbers

	ρ g/cm^3	Z_{EZ}	$V_{EZ} = a \times b \times c$ Å×Å×Å
Alexandrite	3.69	4	9.40×5.48×4.42
BEL	6.061	4	7.53×7.35×7.44
YAP	5.35	4	5.33×7.37×5.18
YLF	4.03	4	5.17×10.74
Al_2O_3	3.98	6	4.76×13.00
GGG	7.08	8	12.383
GSAG	5.82	8	12.393
GSGG	6.59	8	12.5453
YAG	4.55	8	12.013
YSAG		8	12.283
YSGG	5.383	8	12.433

For atom per cent specifications relative to the complete chemical formula the conversion is

$$N_0 = Z_{EZ} \times Z_{FE} \times Z_{EF} \frac{\text{at\%}}{100} = \frac{Z_{EZ} \times Z_{FE}}{V_{EZ}} \frac{\text{at\%}}{100}, \tag{11.8}$$

ZEZ = number of the formula units per elementary cell EZ,
ZFE = number of the atoms per formula unit FE,
ZEF = number of the elementary cells per cm3,
VEZ = volume of the elementary cell EZ.

Also, one has to consider whether the specifications refer to the pure ion (Nd) in relation to all other atoms including (O_2) or to the ion (Nd) in relation to the substitution ion (3Y). Table 11.6 gives the necessary data for the conversion.

K = distribution coefficient
indicates the ratio of the doping concentration in the already grown crystal to the concentration in the melt. At $K = 1$ the concentration in the melt and crystal are the same. At $K < 1$ the concentration increases in the crystal with increasing grown crystal length. A distribution coefficient $K \neq 1$ results in an axial doping gradient along the pulling axis.

The concentration C of the crystal follows from the melt concentration

$$C = C_{\mathrm{m}} K (1-g)^{K-1}, \tag{11.9}$$

g = mass ratio of the material crystallized relative to the initial melt

Derived Data

The results for the derived data have been calculated according to the following equations. The equation number refers to the equation from the previous text.

$h\nu = \frac{hc}{\lambda}$, laser photon energy, $h = 6.626\times10^{-34}$ J s, $c = 2.998\times10^{8}$ m/s,

$J_{\mathrm{s}} = \frac{h\nu}{\sigma\tau}$, saturation intensity for a 4-level system (1.5),

$J_{\mathrm{s}} = \frac{h\nu}{2\sigma\tau}$, saturation intensity for 3-level system,

$H_{\mathrm{s}} = \frac{h\nu}{\sigma}$, saturation energy density for 4-level system (1.26),

$H_{\mathrm{s}} = \frac{h\nu}{2\sigma}$, saturation energy density for 3-level system,

$\phi_{\mathrm{B}} = \arctan(n)$, Brewster angle,

$R_{\mathrm{T}} = \frac{\sigma_{\max}\kappa(1-\nu)}{\alpha E}$, thermal shock parameter (4.21).

It is an indication of the stress a laser material can take and is close to the value a rod per unit length can be loaded with pumping light.

$\frac{\tau}{r_0^2} = \frac{c_p \rho}{k}$, the reciprocal temperature diffusivity (4.11) taken as the time for thermalization within an area of 1 mm^2.

The photoelastic coefficients $C_{r,\phi}$ are required for calculation of the refracting power parameters $M_{r,\phi}$.

$$C_r = \frac{(17\nu - 7)p_{11} + (31\nu - 17)p_{12} + 8(\nu + 1)p_{44}}{48(\nu - 1)} \quad (4.34),$$

$$C_\phi = \frac{(10\nu - 6)p_{11} + 2(11\nu - 57)p_{12}}{32(\nu - 1)} \quad (4.35)$$

(Only for garnets such as YAG or GSGG).

p_{ij} = elastooptical coefficients from the literature.

The absolute dependence of the refractive index on the temperature

$$\frac{dn}{dT} = \frac{dn_{rel}}{dT} + n_{rel}\frac{dn_{air}}{dT} \quad (11.2).$$

Material parameters for the thermal lens

$$M_{r,f} = \frac{1}{2k}\left(\frac{dn}{dT} + 2n^3 \alpha C_{r,\phi} + \alpha(n-1)\frac{l_x}{l}\right) \quad (4.41).$$

In the tables $dn_{air}/dT = -0.9$ is used for $T = 25°C$ and $l_x/l \approx 1/30$ is assumed.

Supplier Data

With these data the capabilities of the suppliers are listed, as far as they are known:

l_{max} = maximal deliverable length,

d_{max} = maximum diameter or thickness,

C = offered doping concentration of the laser ion or the codoping.

Measured Values

Measured values provide the best comparison of the various materials. We have attempted to find particularly favorable values for the respective material:

η_{slope} = slope efficiency, (1.62),

η_{excite} = excitation efficiency,

η_{extr} = extraction efficiency, (1.149),

η_H = heat efficiency, (4.2),

$-\ln V$ = resonator losses per unit length and round trip, (1.59),

$\kappa = \eta_H / \eta_{excite}$ = heat coefficient, (4.4),

E_{max} = maximum output energy per pulse,

T = corresponding pulse length,

P_{av} = maximum average power in the pulsed operation,

P_{cw} = maximum cw power,

$d\times l\times h$ = dimensions of the laser material used ([diameter] times [thickness] times length [height]).

11.2 Active Ions

The wavelength of the laser transition is determined by the active ion embedded in the solid. The imbedded ion can broaden the laser line, shift the line – such as in glasses – or can change the lifetime of the transition. Major properties, however, such as absorption and laser transitions, are determined by the active ion only and modified by the host crystal.

Table 11.7 gives an overview of the active elements in various solid materials and the corresponding laser wavelengths. The selection is restricted to materials that are commercially available at present. The transitions for the various ions are arranged according to the wavelength in Table 11.8.

Table 11.7 Typical laser parameter of various ions in solids

Ion	λ nm	σ 10^{-20} cm^2	τ_0 µs	ΔE cm^{-1}	Transition
Cr	700	0.7	200	100	$^4T_2 \rightarrow {}^4A_2$
Dy					
Er^{3+}	2800				$^4I_{11/2} \rightarrow {}^4I_{13/2}$
	1640	0.4	64000	400	$^4I_{13/2} \rightarrow {}^4I_{15/2}$
	850	1.4	25		$^4S_{3/2} \rightarrow {}^4S_{13/2}$
Ho^{3+}	2000	1	10000	300	$^5I_7 \rightarrow {}^5I_8$
	750	97	65		$^5S_2 \rightarrow {}^5I_7$
Nd^{3+}	1320				$^4F_{3/2} \rightarrow {}^4I_{13/2}$
	1064	50	240	2000	$^4F_{3/2} \rightarrow {}^4I_{11/2}$
	950				$^4F_{3/2} \rightarrow {}^4I_{9/2}$
Pr	640	0.17	56	522	$^3P_0 \rightarrow {}^3F_2$
Tm^{3+}	1600	1.3	830		$^3H_4 \rightarrow {}^3H_6$
	780	0.5	6	586	$^3F_4 \rightarrow {}^3H_6$
Yb^{3+}	1030	2	1100	612	$^2F_{5/2} \rightarrow {}^2F_{7/2}$

ΔE energy difference between the lower laser energy level to the ground state

Further bibliographical references can be found in the following tables.

Table 11.8 Laser wavelengths of various ions in solids, [11.18]

Wavelength $\lambda/\mu m$	Laser ion	Transition
0.544	Tb^{3+}	$^5D_4 \rightarrow {}^7F_5$
0.598 – 0.622	Pr^{3+}	$^3P_0 \rightarrow {}^3H_6$
0.66 – 1.08	Ti^{3+}	$2E \rightarrow {}^2T_2$
0.637 – 0.65	Pr^{3+}	$^3P_0 \rightarrow {}^3F_2$
0.68 – 0.704	Cr^{3+}	$^2E \rightarrow {}^4A_2$
0.694 – 0.7221	Pr^{3+}	$^3P_0 \rightarrow {}^3F_3$
0.679 – 0.745	Sm^{2+}	$^5d, {}^5D_0 \rightarrow {}^7F_1$
0.717 – 0.7496	Pr^{3+}	$^3P_0 \rightarrow {}^3F_4$
0.74 – 0.89	Cr^{3+}	$^3E, {}^4T_2 \rightarrow {}^4A_2$
0.75 – 0.761	Ho^{3+}	$^5S_2 \rightarrow {}^5I_7$
0.843 – 0.863	Er^{3+}	$^4S_{3/2} \rightarrow {}^4I_{13/2}$
0.891 – 0.947	Nd^{3+}	$^4F_{3/2} \rightarrow {}^4I_{9/2}$
0.9017 – 0.931	Pr^{3+}	$^3P_0 \rightarrow {}^1G_4$
0.979 – 1.018	Ho^{3+}	$^5F_5 \rightarrow {}^5I_7$
0.996	Pr^{3+}	$^1D_2 \rightarrow {}^3F_3$
1.014 – 1.031	Ho^{3+}	$^5S_2 \rightarrow {}^5I_6$
1.023 – 1.038	Yb^{3+}	$^5F_{5/2} \rightarrow {}^2F_{7/2}$
1.037 – 1.123	Nd^{3+}	$^4F_{3/2} \rightarrow {}^4I_{11/2}$
1.0465 – 1.0664	Pr^{3+}	$^1D_2 \rightarrow {}^3F_4$
1.229 – 1.26	Er^{3+}	$^4S_{3/2} \rightarrow {}^4I_{11/2}$
1.306 – 1.415	Nd^{3+}	$^4F_{3/2} \rightarrow {}^4I_{13/2}$
1.391 – 1.408	Ho^{3+}	$^5S_2 \rightarrow {}^5I_5$
1.462 – 1.482	Tm^{3+}	$^3F_4 \rightarrow {}^3H_4$
1.58	Tm^{3+}	$^1G_4 \rightarrow {}^3F_3$
1.645 – 1.777	Er^{3+}	$^4S_{3/2} \rightarrow {}^3I_{9/2}$
1.965 – 2.0025	Er^{3+}	$^4F_{9/2} \rightarrow {}^4I_{11/2}$
1.933 – 2.171	Ho^{3+}	$^5I_7 \rightarrow {}^5I_8$
2.274 – 2.424	Tm^{3+}	$^3F_4 \rightarrow {}^3H_5$
2.689 – 3.01	Er^{3+}	$^4I_{11/2} \rightarrow {}^4I_{13/2}$
2.845 – 3.018	Ho^{3+}	$^5I_6 \rightarrow {}^5I_7$
2.972 – 3.022	Dy^{3+}	$^6H_{13/2} \rightarrow {}^6H_{15/2}$
3.369	Ho^{3+}	$^5S_2 \rightarrow {}^5F_5$

Laser transitions demonstrated at room temperature and flash lamp excitation

For the most important ions the corresponding energy level diagram for absorption or emission and further information, as well as examples, are given in the following.

Chromium

Due to the energy band structure in the energy level diagram the Cr-doped vibronic lasers are tunable between 750 and 900 nm with a tuning range of around 100 nm, depending on the specific crystal..

Table 11.9 Lasers parameters of Cr in various solids

laser data Host	λ nm	$\Delta\lambda$ nm	σ 10^{-20} cm^2	τ µs	τ_0 µs	ΔE cm^{-1}	Transition	Ref.
Ruby	649	0.5		3000		1500	R_2	[11.35]
Alex.	750		0.7[c)]	260[c)]			$^4T_2 \rightarrow {}^4A_2$	[11.7]
Emerald	780		3.1	60				[11.7]
GGG	745		0.6	380	159			[11.8]
GSGG	770		0.8	50	115			[11.8]
GSAG	784							[11.45]
Mg_2SiO_2	1200[b)]		0.2					[11.33]
pump data Host	λ_p nm	σ_p 10^{-20} cm^2	σ_{ESA} 10^{-20} cm^2	η_H			Transition	Ref.
Alex.	410	13.6	2 – 7					
Ruby	404	22					$^4A_2 \rightarrow {}^4F_2$	[11.35]
Ruby	554	22					$^4A_2 \rightarrow {}^4F_1$	[11.35]
Emerald	650	8.26					$^4A_2 \rightarrow {}^4T_2$	[11.7]
Emerald	600	5.6					$^4A_2 \rightarrow {}^4T_1$	[11.7]
GSGG	650	3.6 ± 0.6		0.25[a)]			$^4T_2 \rightarrow$	[11.32]
measured Host	P_{av} W	T ms	E W s	η_{slope} %	$-\ln V/l$ %/cm		$d \times l \times h$ mm^3	Ref.
Alex.	105	0.16	7	3.9	0.3		6.35×100	[11.39]
Alex.			2.5	1.7	0.64		8.16×139	[11.40]
GSAG			0.11	0.12	0.54		6.35×76	[11.45]

a) without ESA, b) pumped with 1064 nm, c) temperature-dependent, at 290°C is $\sigma = 3\times10^{-20}$ cm^2

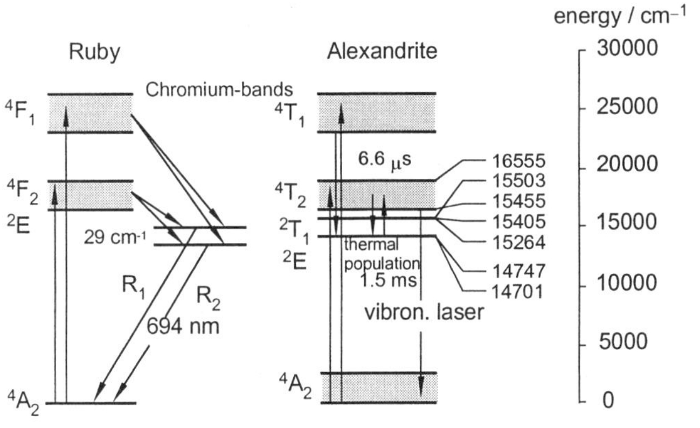

Fig. 11.2. Energy levels of chromium in ruby and alexandrite, [11.19, 11.20,11.36]

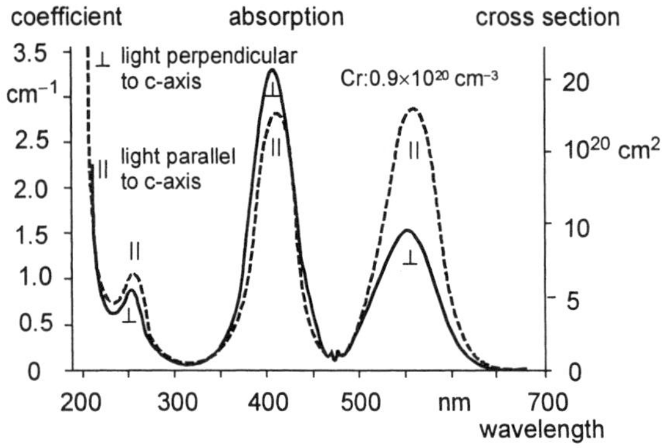

Fig. 11.3. Absorption spectrum of chromium in ruby [11.19]

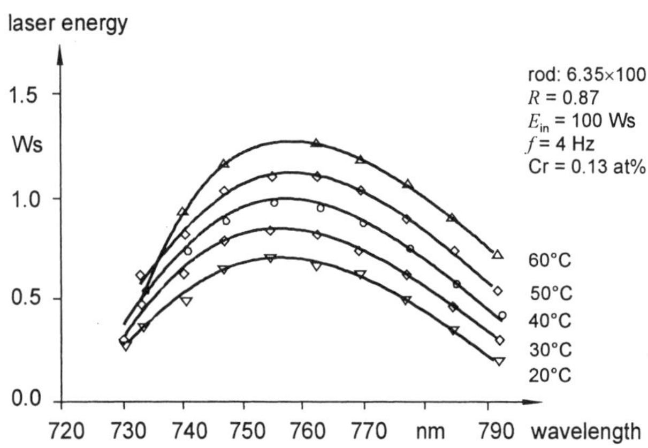

Fig. 11.4. Tuneablity and temperature dependence of chromium in alexandrite, [11.2]

Erbium Er^{3+}

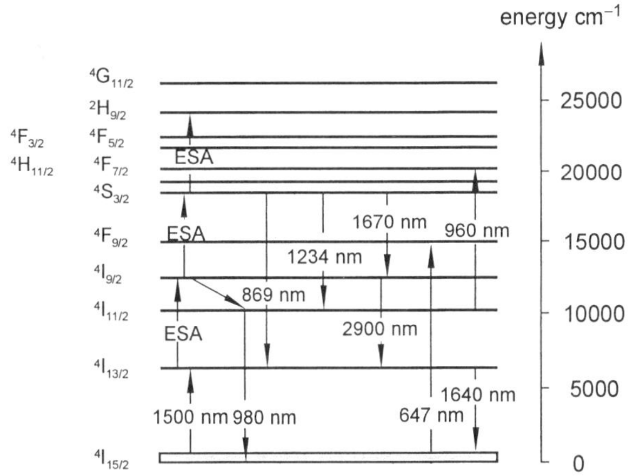

Fig. 11.5. Energy level diagram of Er^{3+}, [11.21, 11.22, 11. 23, 11. 24, 11.34, 11.38]

Table 11.10 Laser parameters of Er in various solids

Host	λ nm	σ 10^{-20} cm^2	τ_0	ΔE µs	Transition cm^{-1}	Ref.
YAG	2940	3			$^4I_{11/2}\rightarrow {}^4F_{13/2}$	
	1640	0.45 – 0.49	64000	470	$^4I_{13/2}\rightarrow {}^4I_{15/2}$	[11.9, 11.10]
YAP	1640	0.28 – 0.31	450 – 510		$^4I_{13/2}\rightarrow {}^4I_{15/2}$	[11.9, 11.10]
YLF	2900					
	1730		400		$^4S_{3/2}\rightarrow {}^4I_{9/2}$	
	1640	0.33 – 0.42	64000	346	$^4I_{13/2}\rightarrow {}^4I_{15/2}$	[11.9, 11.10]
	1230					
	850	1.4	25		$^4S_{3/2}\rightarrow {}^4S_{13/2}$	[11.11]
Glass		0.6			$^4I_{13/2}\rightarrow {}^4I_{15/2}$	[11.37]

Pump data

Host	λ_p nm	σ_p 10^{-20} cm^2	σ_{ESA} 10^{-20} cm^2	η_H	Transition
	600 – 800				

Measurements

Host	λ nm	P_{av} W	T ms	E W s	η_{slope} %	C at%	$d\times l$ mm^3	Ref.
YAG	2940	1.1		0.3	0.8	40	6.35×75	[11.56]
YAG	2940	10	0.242	1.4	2.1		6×80	[11.57]
YAG	2940	0.8	0.170	0.6	0.28	50		[11.58]

Due to the lower thermal conductivity of YAP (5.5 W m^{-1} K^{-1}), compared with YAG (10 W m^{-1} K^{-1}) and the higher Er-doping (40%) the thermal lens of Er:YAP is much stronger than of Nd:YAG.

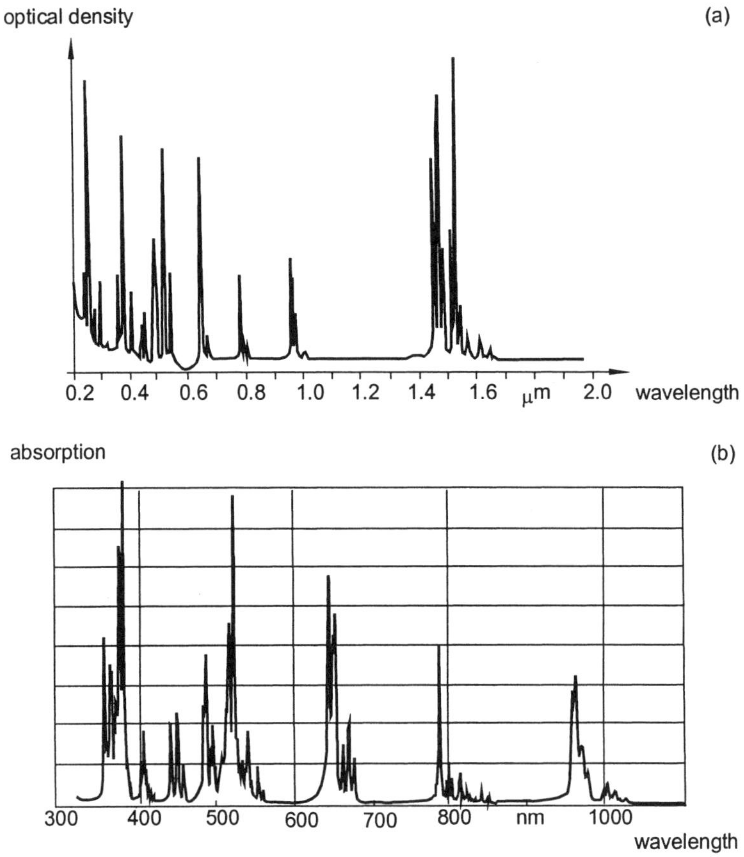

Fig. 11.6. Absorption spectrum of Erbium Er^{3+} in YAG, (**a**) [11.25], (**b**) [11.26]

Er^{3+} allows the interesting possibility via a 3-level excitation to achieve simultaneously transitions for blue $^2H_{9/2}$, green $^4S_{3/2}$ and red $^4F_{9/2}$, [11.22, 11.23].

Holmium Ho^{3+}

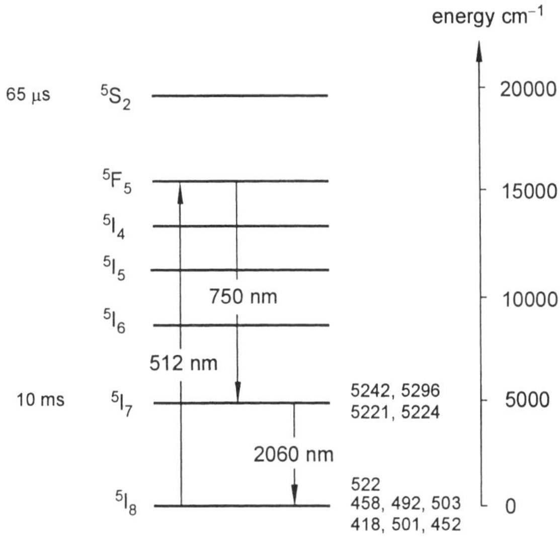

Fig. 11.7. Energy level diagram of Ho^{3+} in YLF, [11.11, 11.41]

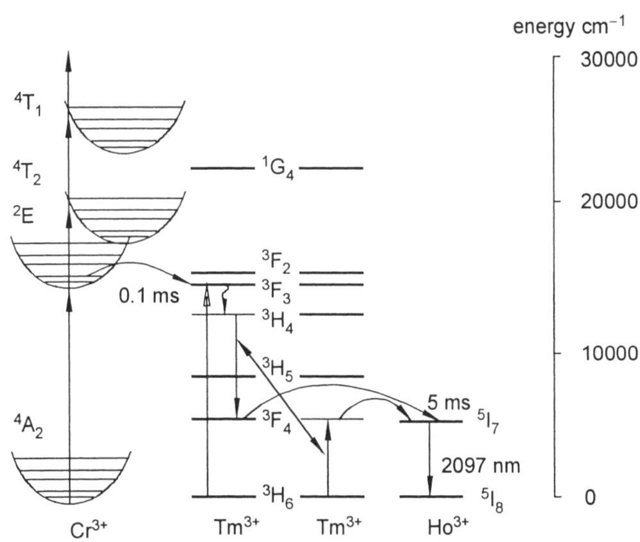

Fig. 11.8. Excitation transfer via Cr and Tm to Ho in YAG, [11.59]

Table 11.11 Typical laser parameter of holmium in solids

Host	λ nm	σ 10^{-20} cm^2	τ_0 μs	ΔE cm^{-1}	Transition	Ref.
YAG	2000	0.94 – 1.6	10000	458	$^5I_7 \rightarrow {}^5I_8$	[11.9, 11.10]
	2127	3.9	522		$^5I_7 \rightarrow {}^5I_8$	[11.41]
	2120	5.4	503		$^5I_7 \rightarrow {}^5I_8$	[11.41]
	2096	10.1	458		$^5I_7 \rightarrow {}^5I_8$	[11.41]
	2089	14.3	458		$^5I_7 \rightarrow {}^5I_8$	[11.41]
	2091	0.9	8500		$^5I_7 \rightarrow {}^5I_8$	[11.60]
YLF	2060	1.5	282		$^5I_7 \rightarrow {}^5I_8$	[11.11]
YAP	2000	0.84	316 – 460		$^5I_7 \rightarrow {}^5I_8$	[11.9, 11.10]
YLF	750	97	65		$^5S_2 \rightarrow {}^5I_7$	[11.11]

Measurements

Host	λ nm	P_{av} W	T ms	E W s	η_{slope} %	D/P 1/(m kW)	$d \times l$ mm^2	Ref.
YAG	2100	20	0.2	3	2	1.52	4×102	[11.42]
YAG	2080	7.5	1.5	17	2		5×60	[11.44]
YAG[a)]	2100		0.54	1.2	5.1		5×73	[11.59]
YLF	2060	10		0.1	3.5		5×50	[11.43]

[a)] optimal doping $Cr^{3+} = 7.7 \times 10^{19}$ cm^{-3}, $Tm^{3+} = 8.0 \times 10^{20}$ cm^{-3}, $Ho^{3+} = 5.0 \times 10^{19}$ cm^{-3}

By additional doping with other elements, for example additional Cr, and broadband pumping the absorption bands of these elements may also be used for excitation. An energy transfer takes place from the levels of the codoping into the energy levels of the real laser ion; this can increase the excitation efficiency considerably. The relationship of excitation efficiency to heat efficiency, however, becomes more unfavorable than without codoping.

Holmium must be doped low – 0.5 at% typically – to avoid absorption losses by reabsorption at the laser wavelength. Therefore crystals for holmium lasers will be codoped with chromium and thulium or erbium and thulium to increase the efficiency with the help of the energy transfer.

Neodymium Nd^{3+}

Table 11.12 Typical laser parameter of neodymium in solids

Host	λ nm	σ 10^{-20} cm^2	C_0 10^{20} cm^{-3}	τ_0 μs	ΔE cm^{-1}	σ_e 10^{-20} cm^2	Ref.
$^4F_{3/2} \rightarrow {}^4I_{11/2}$ 1.0 μm							
YAG	1064	46 – 88	2.7	240 – 272	2000	34	[11.54, 11.12]
GGG	1060	29	3.46	282			[11.66, 11.46]
GSGG	1061	31		281		11	[11.47]
KGW	1067	38		109			[11.82]
BEL	1070	21		150			[11.49]
BEL	1079	15					[11.49]
YAP	1079	37		210			[11.64]
YLFσ	1047	18		570		26	[11.52]
YLF π	1053	12		520		37	[11.52]
YVO	1063.4	120	3.85	116		62	[11.50, 11.51]
NYAB	1064	100					[11.65]
SiO2	1089	0.82		512			[11.77]
$^4F_{3/2} \rightarrow {}^4I_{13/2}$ at 1.3 μm							
YAG	1381	10					[11.12]
GGG	13311	5.5					[11.66]
GSGG	1335	7.3					[11.46]
BEL							
YLF		5.1					
YVO	1342	60 – 70					[11.50, 11.62]

Pump data Host	λ_{FZ} nm	Ref.
GGG	500	[11.48]

Measurements

Host	P_{av} W	P_{cw} W	T ms	E W s	η_{slope} %	$d\times l\times h$ mm^3	Ref.
$^4F_{3/2} \rightarrow {}^4I_{11/2}$ at 1.0 μm							
GGG	230		3	23	2.4	9×35×177	[11.48]
BEL		9	0.125		2.5	6.35×76	[11.49]
YAG	400	400	0 – 20	60	3 – 4	8×200	typical
YAP		163			3.16	5.8×111	[11.63]
SiO_2	5	5			51 (diode pumped)		[11.78]
$^4F_{3/2} \rightarrow {}^4I_{13/2}$ 1.3 μm							
YAG	165		8			8×150	[11.53]
YAP		82.8		2.02		5.8×111	[11.55]

The lifetime of the lower laser energy level $^4I_{11/2}$ is stated in [11.76] to be approximately 10 ns.

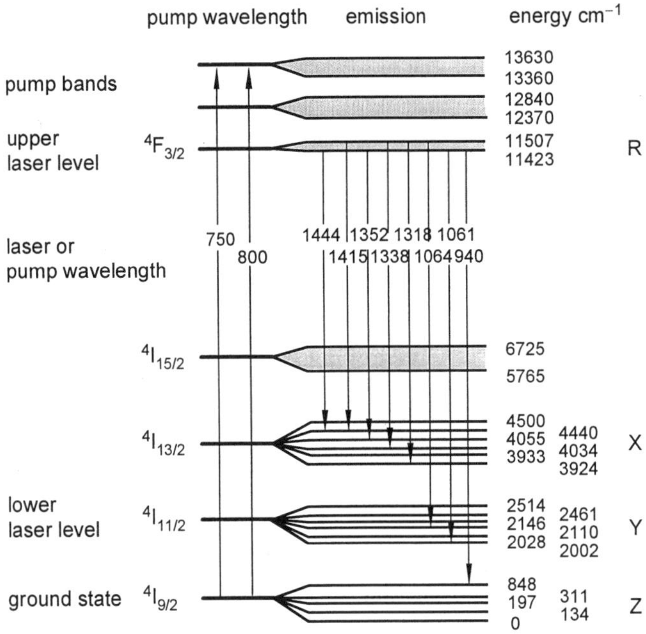

Fig. 11.9. Energy level diagram of Nd^{3+} in YAG, [11.12]

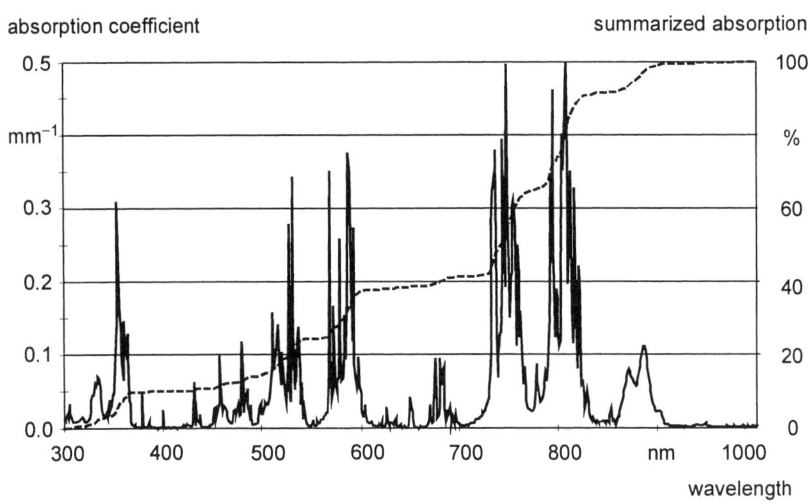

Fig. 11.10. Absorption spectrum of Nd^{3+} in YAG and integrated intensity

Figure 11.11 shows the wavelength range around 800 nm with greater resolution which is particularly interesting for pumping with laser diodes.

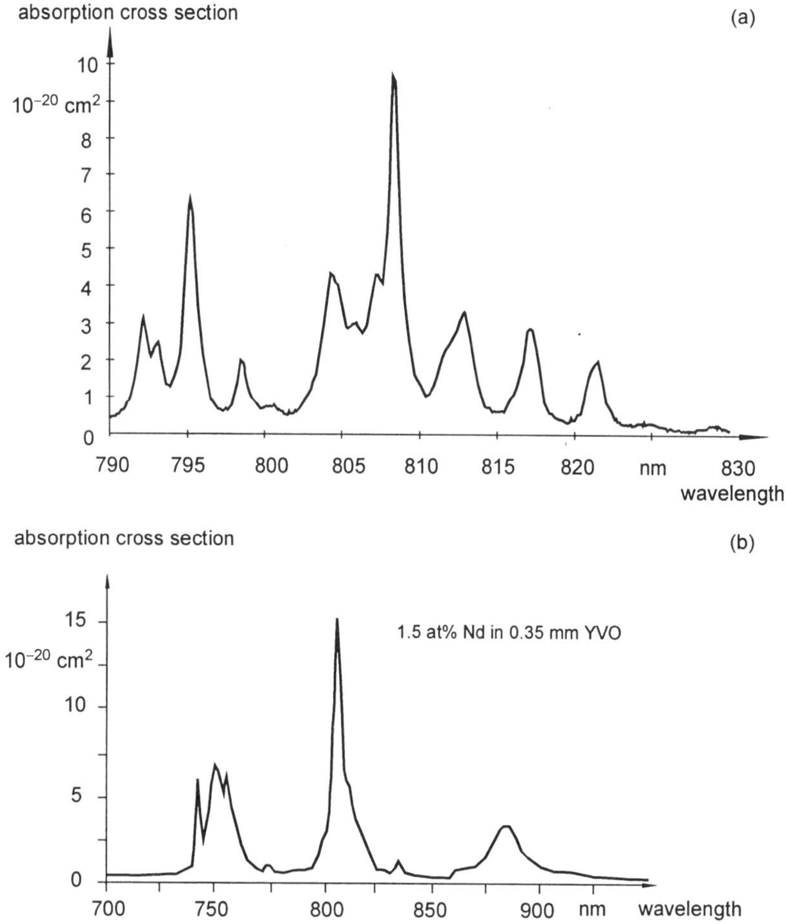

Fig. 11.11. (**a**) Absorption spectrum of Nd in YAG in the range 790 to 830 nm, [11.26] (**b**) Absorption spectrum of Nd in YVO in the range 700 to 950 nm, [11.26]

In addition to the wavelength around 1.064 μm other lasing transitions have been shown to exist for Nd^{3+}. In Table 11.13 the corresponding relative intensities of the transitions are indicated.

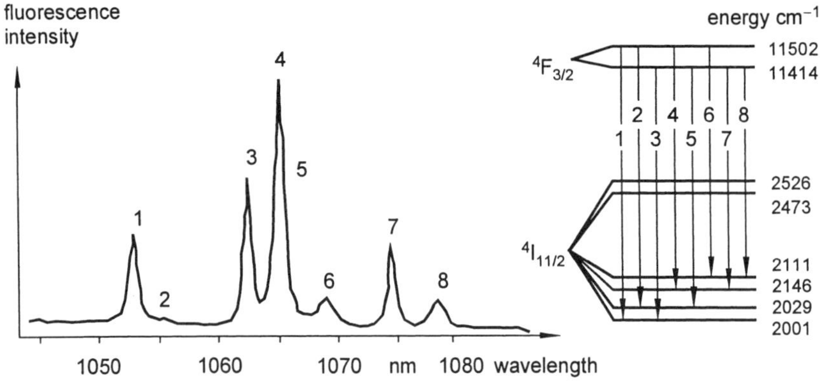

Fig. 11.12. Fluorescence spectrum of Nd^{3+} in YAG, [11.19]

Table 11.13 Laser transitions of Nd in YAG

Wavelength µm	Relative intensity	σ	J_s 10^{-20} cm^2	$R_{1.44}$ kW/cm^2	Ref.
1.05205	0.46				[11.19]
1.06152	0.92				[11.19]
1.06415	1.00	41	2	0.1	[11.27, 11.28]
1.0646	≈ 0.5				[11.19]
1.0738	0.65				[11.19]
1.0780	0.34				[11.19]
1.1054	0.09				[11.19]
1.1121	0.49				[11.19]
1.1159	0.46				[11.19]
1.12267	0.40				[11.19]
1.3187	0.25	9.5	6.7	0.63	[11.27, 11.28]
1.3200	0.09				[11.19]
1.3335	0.13	5	13.1	0.82	[11.27]
1.3351	0.15	6	10.9	0.78	[11.27]
1.3381	0.33	9	7.2	0.65	[11.27, 11.28]
1.3410	0.09				[11.19]
1.3572	0.35	7.5	8.6	0.71	[11.28]
1.4150	0.62				[11.28]
1.4444	0.54	4.5	13.4	0.85	[11.27, 11.28]

$R_{1.44}$ is the maximal permitted reflection of the respective wavelength for which pure 1.44 µm radiation with $R = 0.85$ can still be reached, [11.27].

Praseodym Pr^{3+}

Table 11.14 Typical laser parameter of Pr ions in solids

Host	λ nm	σ 10^{-20} cm^2	C_0 at%	τ_0 µs	ΔE cm^{-1}	Transition	Ref.
YLF	639.5	0.17		56	5223	$^3P_0 \rightarrow {}^3F_2$	[11.13]

Measurements $^3P_0 \rightarrow {}^3F_2$ at 0.64 µm

Host	P_{av} W	C at%	T µs	E W s	η_{slope} %	$-\ln V/l$ %/cm	$d \times l$ mm^2	
YLF	0.8	0.5	0.14	0.09	0.3		6.3 × 95	[11.13]

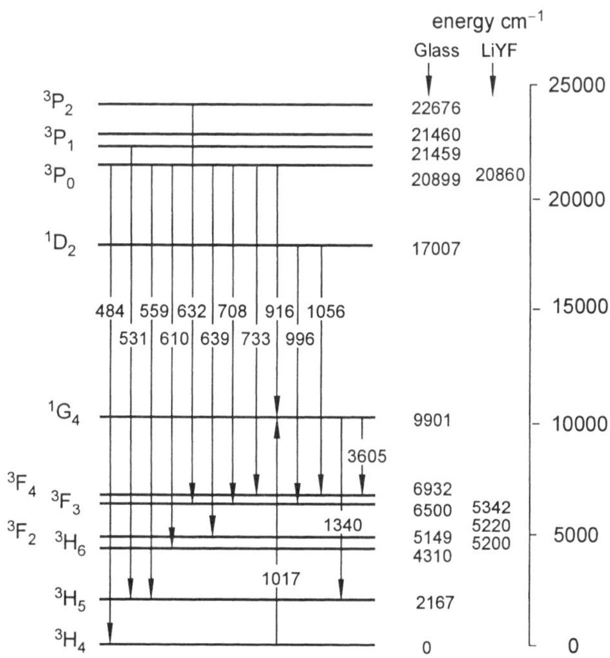

Fig. 11.13. Level scheme of Pr^{3+} in a fluoride fiber, [11.13, 11.29, 11.30]

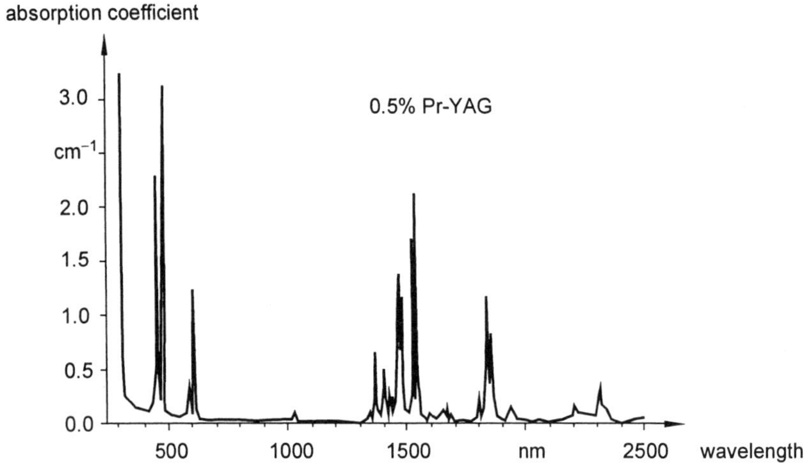

Fig. 11.14. Absorption of Pr^{3+}, [11.31]

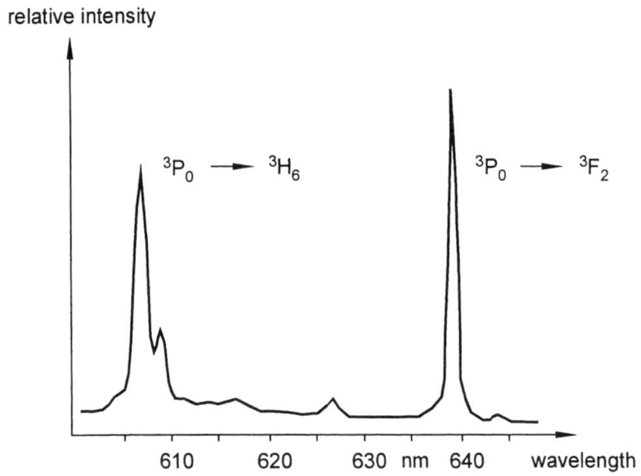

Fig. 11.15. Luminescence spectrum around the laser line of Pr^{3+}, [11.13]

Titanium

Table 11.15 Typical laser parameter of Ti ions in sapphire

λ nm	σ 10^{-20} cm^2	τ_0 µs	ΔE cm^{-1}	Transition	Ref.
720 – 920	0.17 – 0.32	3.9			[11.67]

Measurements

P_{av} W	C at%	T µs	E W s	η_{slope} %	$-\ln V/l$ %/cm	$d \times l$ mm^2	Ref.
2	0.1	0.15	0.3 – 3	0.5 – 2	0.4 – 0.7	6.3 × 100	[11.68, 11.69, 11.70]
220	0.1 wt%	5		1.9			[11.79]

Ti sapphire is tunable, cw laser operation until now is only reached by pumping with an Ar-ion laser.

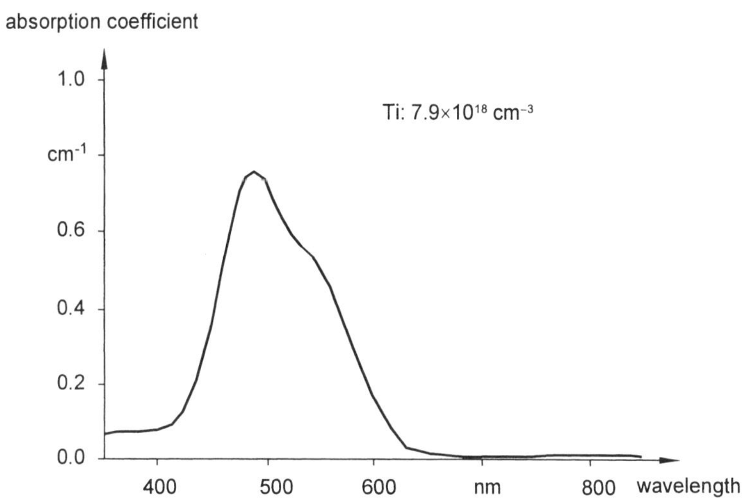

Fig. 11.16. Absorption spectrum of titanium in sapphire, [11.69]

Thulium

Table 11.16 Typical laser parameters of Tm ions in solids

Host	λ nm	σ 10^{-20} cm^2	C_0 at%	τ_0 μs	ΔE cm^{-1}	Ref.
$^3H_4 \rightarrow {}^3H_6$						
YAG	1600	1.3	1.2	740		[11.9, 11.14, 11.71]
YLF			1.06	2550		[11.71]
$^3H_4 \rightarrow {}^3F_4$						
YLF	1500					
$^3F_4 \rightarrow {}^3H_6$						
GSGG	1986	0.13				[11.73]
YAG	2014	0.16–1.3		586		[11.73, 11.9]
YAG	1965	0.1				[11.73]
YAP		0.5		6	461	[11.9]
YLF		0.32			355	[11.9]

Measurements

Host	λ nm	C(Tm) 10^{20} cm^{-3}	C(Cr)	T ms	E W s	η_{slope} %	$d \times l$ mm × mm	
$^3F_4 \rightarrow {}^3H_6$								
YAG	2014	8	0.9	1.2	2.2	4.5	5 × 76	[11.73]
YAG	2066	8	0.9	1.2	0.3	1	5 × 76	[11.73]
YAG	1965	8	0.9	1.2	0.9	0.7	5 × 76	[11.73]
GSGG	1986	8	0.9	1.2	0.3	0.4	5 × 76	[11.73]
YSGG	2010	8	0.9	1.2	0.6	0.6	5 × 76	[11.73]

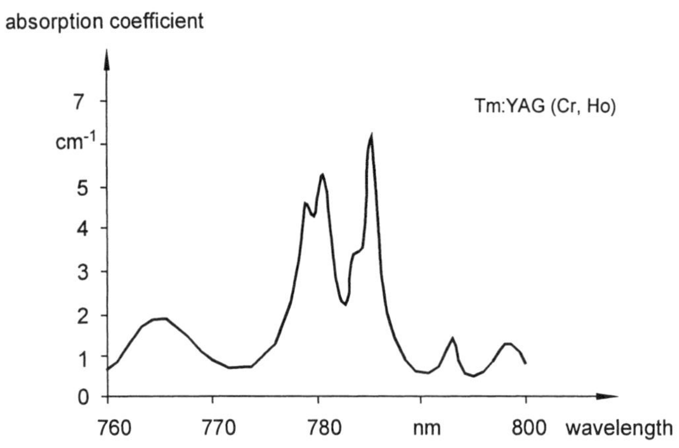

Fig. 11.17. Thulium absorption in YAG codoped with chromium and holmium, [11.60]

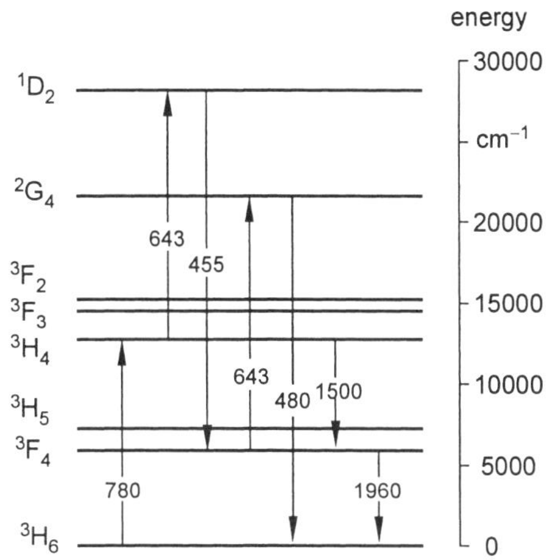

Fig. 11.18. Energy level scheme of thulium in ZBLAN, [11.72]

Ytterbium

Table 11.17 Typical laser parameter of Yb ions in solids

Host	λ nm	σ 10^{-20} cm^2	C_0 at%	τ_0 µs	ΔE cm^{-1}	Transition	Ref.
YAG	1030	2.3		950	612	$^2F_{5/2} \rightarrow {}^2F_{7/2}$	[11.15, 11.16]
FAP	1030	5.6		1100			[11.16, 11.17]
LuAG		2.7		925			[11.83]
Measurements							
Host	λ nm	C at%	P W	η_{slope} %	$d \times l \times h$ mm^2		
$^2F_{5/2} \rightarrow {}^2F_{7/2}$							
YAG	1031	19	2	72	3 × 0.7		[11.75]
	1049	19					[11.75]
Pump data							
	λ_{FZ} nm	σ_p 10^{-20} cm^2		σ_{ESA} 10^{-20} cm^2	η_H	Transition	
LuAG		0.77 Max					[11.84]
YAG		0.75					[11.84]

Diode pumped at 942 nm

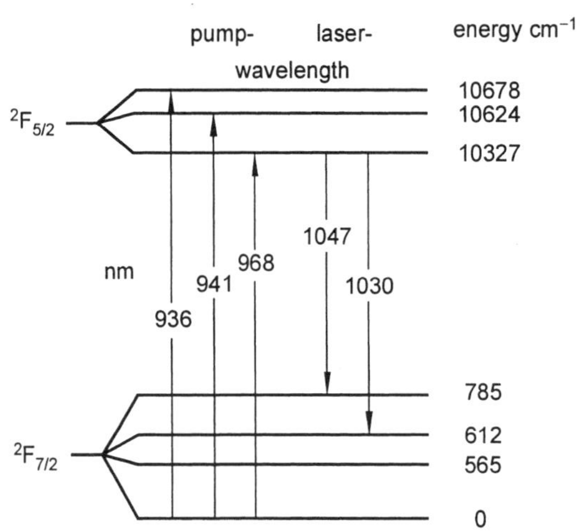

Fig. 11.19. Energy level scheme of Yb in YAG at 77 K, [11.16]

Table 11.18 Laser parameters of Yb ions in different host crystals [11.74]

Host	λ_p nm	λ nm	σ 10^{-20} cm^2	τ_0 ms	H_s Ws/cm^2	J_s kW/cm^2	E cm^{-1}	ΔE 10^{-20} W s	ΔE/K 10^{-20} W s/K
$LiYF_4$	962	1020	0.81	2.21	24	10	480	0.95	690
LaF_3	975	1009	0.36	2.16	54	25	353	0.70	508
SrF_2	967	1025	0.16	9.20	121	13	593	1.18	853
BaF_2	967	1024	0.14	8.00	138	17	578	1.15	831
$KCaF_3$	973	1031	0.22	2.70	87	32	593	1.18	853
KY_3F_{10}	976	1011	0.44	1.87	44	23	369	0.73	531
$Rb2NaYF_6$	962	1012	0.10	10.8	196	18	372	0.74	535
BaY_2F_8	961	1018	0.67	2.04	29	14	458	0.91	659
Y_2SiO_5	979	1042	0.33	1.04	57	55	617	1.23	887
$Y_3Al_5O_{12}$	942	1031	2.03	1.01	9.5	9.4	628	1.25	903
$YAlO_3$	962	1014	1.31	0.43	15	35	353	0.70	508
$Ca_5(PO_4)_3F$	905	1043	5.90	1.20	3.23	2.69	603	1.20	867
$LuPO_4$	975	1011	0.53	0.83	37.1	44.7	349	0.69	502
$LiYO_2$	908	1020	0.56	1.13	34.8	30.8	474	0.94	682
$ScBO_3$	966	1022	0.19	4.80	102	21.3	472	0.94	679

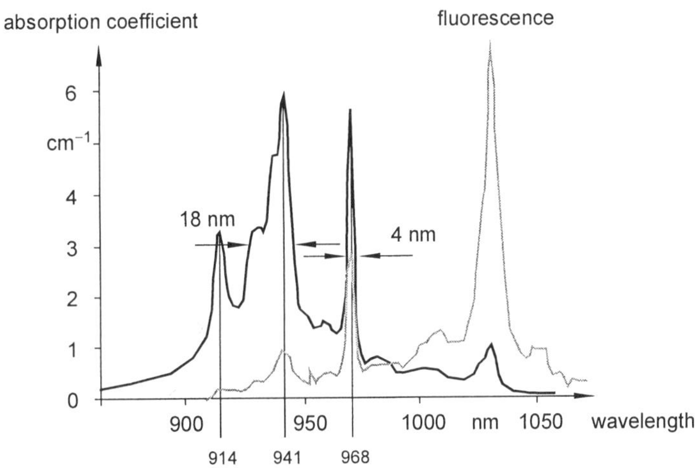

Fig. 11.20. Absorption spectrum of Yb in YAG in the range 850 to 1050 nm, doping 6.5 at% Yb, [11.16]

12 Laser Crystals

Alexandrite

Material		$BeAl_2O_4$		Active Ion	Cr^{3+}	Ref.
Optical properties			$E \parallel a$	$E \parallel b$	$E \parallel c$	
n		1.74	1.736	1.742	1.735	[12.1, 12.2, 12.3]
dn_{rel}/dT	10^{-6}/°C	8.3 – 9.41	9.4	8.3		[12.3, 12.1, 12.38]
n_2	10^{-20} m^2/W	2				[12.3]
Thermal properties			$\parallel a$	$\parallel b$	$\parallel c$	
k	W/(m °C)	23				[12.1]
α	10^{-6} °C		5.9 – 6	6.1 – 6.8	6.7 – 6.9	[12.2, 12.38]
Mechanical properties						
ρ	g/cm^3	3.69				[12.3]
ν		0.3				[12.4]
E	GPa	469				[12.3]
P_{max}	MPa	400 – 900	457 – 948	520		[12.1, 12.3]
H_K	MPa	19.6				[12.2]
γ	MPa$\sqrt{m}$	2.6				[12.4]
Crystal properties						
Structure		orthorhombic				[12.3]
Space group		D_{2h}^{16} - Pnma				[12.3]
Lattice	a/nm	0.9404				[12.3]
constant	b/nm	0.5476				[12.3]
	c/nm	0.4427				[12.3]
T_m	°C	1870				
F_A at%	10^{20} cm^{-3}	3.51				[12.3]
Derived data			literature values			
R_T	kW/m	2 – 5	3.5 – 7.4	2.35		[12.3, 12.4]
dn_{abs}/dT	10^{-6}/°C	6.7 – 7.8				
Supplier data		Cr				
l_{max}	mm	120				[12.1]
d_{max}	mm	15				[12.1]
C(ion)	at%	0.05 – 0.2				[12.1]

Alexandrite with Cr is tunable from 710 to 800 nm. Efficiency and wavelength increase linearly with the temperature to about 200°C. (750 nm at 20°C to 790 nm at 300°C). However, the lifetime decreases with temperature (260 µs at 20°C to 100 µs at 150°C).

BEL

Material		$La_2Be_2O_5$	Lanthanum beryllate			Ref.
Active ion		Nd^{3+}				
Codoping						
Optical properties		$E \parallel x$	$E \parallel y$	$E \parallel z$		
n		1.964	1.997	2.035		[12.5]
dn_{rel}/dT	$10^{-6}/°C$	2.86	1.53	−6.23		[12.5]
n_2	10^{-13} esu	6				
Thermal properties			a	b	c	
c_p	W s/g °C					
k	W/m °C	4.7 – 5				[12.7, 12.5]
α	10^{-6} °C	8	7.0	7.9	9.5	[12.5, 12.6]
Mechanical properties						
ρ	g/cm³					
ν						
E	GPa					
P_{max}	MPa					
Crystal properties						
Structure		monoclinic		biaxial		
Space group		C_{2h}^6-C2/c				[12.6]
Lattice const.	a/nm	0.75356				
	b/nm	0.73476				
	c/nm	0.74387				
Orientation						
T_m	°C		1360			[12.5]
F_A	10^{20} cm^{-3}/at%		1.94			[12.6]
Distribution coefficient			0.66 (Nd)			[12.5]
Derived data			literature values			
R_T	kW/m		0.22			[12.5]
Supplier data						
l_{max}	mm	> 76				
d_{max}	mm	> 7				
C(ion)	wt%					

In a suitable direction (z - 62° y) the crystal behaves almost athermally [12.5]. The damage threshold is about 500 W/cm per rod length in cw operation, according to [12.6]. The laser radiation is polarized due to the crystal structure [12.7].

Emerald

Material		$Be_3Al_2Si_6O_{18}$			Ref.
Active ion		Cr^{3+}			
Co-doping					
Optical properties		n_o	n_e		
n		1.57 – 1.60	1.56 – 1.59	dependent on Cr-concentration	
dn_{rel}/dT	$10^{-6}/°C$				
n_2	10^{-13} esu				
Thermal properties					
c_p	W s/g °C				
k	W/m °C				
α	10^{-6} °C				
Mechanical properties					
ρ	g/cm^3				
ν					
E	GPa				
P_{max}	MPa				
H_K	MPa				
γ	MP√m				
Crystal properties					
Structure		hexagonal			
Space group					
Lattice const.	a/nm				
	c/nm				
Orientation					
T_m	°C				
F_W	10^{20} cm^{-3}/wt%				
F_A	10^{20} cm^{-3}/at%				
Distribution coefficient					
Derived data					Ref. values
R_T	kW/m				
τ/f	s/mm^2				
dn_{abs}/dT	10^{-6} /°C				
Supplier data					
l_{max}	mm				
d_{max}	mm				
C(ion)	at%	3			[11.7]

Tunable in the range from 720 nm to 842 nm [11.7]

GGG

Material	$Gd_2Ga_5O_{12}$	Gallium-Gadolinium-Garnet			Ref.
Active ion		Nd	Codoping	Cr	

Optical properties				
n		1.943		[12.8]
dn_{rel}/dT	10^{-6}/°C	20.5		[12.8]
n_2	10^{-13} esu			
p_{11}		−0.086		[12.9]
p_{12}		−0.027		[12.9]
p_{44}		−0.078		[12.9]
λ_{FZ}	nm	500		[11.48]

Thermal properties				
c_p	W s/g °C	0.38		[12.4]
k	W/m °C	7.4	7.05	[12.9, 12.4]
α	10^{-6} °C	8.96		[12.8]

Mechanical properties			
ρ	g/cm^3	7.088	[12.4]
ν		0.28	[12.9]
E	GPa	225	[12.9]
P_{max}	MPa	440	

Crystal properties			Nd	Cr	
Structure			cubic		
Space group					
Lattice const.		a/nm	1.238		[12.9]
Orientation			<111>		[12.10]
T_m	°C		2000		[12.9]
F_A	10^{20} cm^{-3}/at%		1.265	0.84	[12.11]
Distribution coefficient			0.4	3.0	[12.10]

Derived data			
R_T	kW/m		
τ/f	s/mm^2		
C_r		0.011	
C_ϕ		−0.016	
dn_{abs}/dT	10^{-6} /°C	18.75	
M_r	10^{-6} m/W	1.375	
M_ϕ	10^{-6} m/W	1.49	

Supplier data	Nd		Cr	
l_{max}	mm		200	[12.10]
d_{max}	mm		80	[12.10]
C	10^{-20}/cm^3	2.4	0.4	[12.10]

GSAG

Material		$Gd_3Sc_2Al_3O_{12}$	Ref.
Active ion		Cr^{3+}	
Optical properties			
n		1.88	[12.8]
dn_{rel}/dT	$10^{-6}/°C$	5.20	[12.8]
n_2	10^{-13} esu		
p_{11}			
p_{12}			
p_{44}			
λ_{FZ}	nm		
Thermal properties			
c_p	W s/g °C		
k	W/m °C		
α	10^{-6} °C	6.92	[12.8]
Mechanical properties			
ρ	g/cm^3	5.822	[11.45]
ν			
E	GPa		
P_{max}	MPa		
Crystal properties		Cr	
Structure		cubic	
Space group			
Lattice const.	a/nm	1.242	[11.45]
Orientation		<111>	
T_m	°C	1815	[11.45]
Distribution coefficient		≈ 1	[11.45]
Derived data			
R_T	kW/m		
τ/f	s/mm^2		
C_r			
C_ϕ			
dn_{abs}/dT	$10^{-6}/°C$		
M_r	10^{-6} m/W		
M_ϕ	10^{-6} m/W		
Supplier data			
l_{max}	mm		
d_{max}	mm		
C(ion)	wt%		

GSGG

Material		$Gd_3Sc_2Ga_3O_{12}$			Ref.
Active ion		Nd^{3+}	Cr^{3+}	Codoping Cr	
Optical properties		Nd	Nd(Cr)		
n		1.94	1.942		[12.8, 12.12]
dn_{rel}/dT	$10^{-6}/°C$	10.45	10.9		[12.8, 12.12]
n_2	10^{-13} esu	8			[12.4]
p_{11}		−0.012	−0.097		[12.9]
p_{12}		0.019	−0.040		[12.9]
p_{44}		−0.066	−0.066		[12.9]
λ_{FZ}	nm	550 – 590			[12.13]
Thermal properties		Nd	Nd(Cr)	Cr	
c_p	W s/g °C	0.40	0.40	0.40	[12.9, 12.4]
k	W/m °C	4.86	6.02	5.63	[12.9, 12.4]
α	10^{-6} °C	7.32 – 7.39	7.5		[12.8, 12.4]
Mechanical properties					
ρ	g/cm^3	6.46			[12.14]
ν		0.28			[12.4]
E	GPa	210			[12.4]
P_{max}	MPa	144	164[a)]		[12.4, 4.11]
γ	MP√m	1.2			[12.4]
Crystal properties			Nd^{3+}	Cr^{3+}	
Structure		cubic			[12.14]
Lattice const.	a/nm	1.26			[12.14]
Orientation		<111>			
Debye T	K	503			
T_m	°C	1800 – 2133			[14.9]
F_A	10^{20} cm^{-3}/at%		1.265	0.84	
Distribution coefficient			0.65		[12.4]
Derived data	Nd	Nd(Cr)	literature values		
R_T	kW/m	0.33 – 0.45	0.66		[12.4, 12.14]
τ/f	s/mm^2	0.43 – 0.53	0.435		[12.14]
C_r		0.02	0.004	0.024	[12.4]
C_ϕ		0.001	−0.02	0.0015	[12.4]
dn_{abs}/dT	$10^{-6}/°C$	8.75 – 9.15			
M_r	10^{-6} m/W	1.18	0.81		
M_ϕ	10^{-6} m/W	0.94	0.60		
Supplier data		Nd		Cr	
l_{max}	mm	152.4			[12.14]
d_{max}	mm	12.7			[12.14]
C(ion)	10^{20} cm^{-3}	2		1.2 – 2	[12.14]

a) depends on surface finish

GVO

Material		$GdVO_4$				Ref.
Active ion		Nd^{3+}	Tm^{3+}	Tm^{3+}	Yb^{3+}	
Codoping				Ho		
Optical properties						
n						
dn_{rel}/dT	10^{-6}/°C					
n_2	10^{-13} esu					
λ_{FZ}	nm					
Thermal properties						
c_p	W s/g °C					
k	W/m °C					
α	10^{-6} °C					
Mechanical properties						
ρ	g/cm^3	5.47				[12.39]
ν						
E	GPa					
P_{max}	MPa					
HK	MPa					
γ	MP√m					
Crystal properties						
Structure						
Lattice const.	a/nm	0.721				[12.39]
	b/nm	0.721				[12.39]
	c/nm	0.635				[12.39]
Orientation						
T_m	°C					
F_A	10^{20} cm^{-3}/at%		1.2			[12.39]
Distribution coefficient						
Derived data						
R_T	kW/m					
τ/f	s/mm^2					
dn_{abs}/dT	10^{-6}/°C					
Supplier data		Nd	Yb	Tm		
l_{max}	mm	10				[12.39]
d_{max}	mm	5				
C(ion)	at%	1	2	4. 5– 8.5		[12.39]

KGW

Material		$KGd(WO_4)_2$				Ref.
Active ion		Nd^{3+}				
Optical properties			x	y	z	
n	@ 1.067 µm		1.937	1.986	2.033	[12.40]
n	@ 1.180 µm		1.928	1.979	2.026	[12.40]
dn_{rel}/dT	10^{-6}/°C	4 – 10				[12.43]
n_2	10^{-13} esu					
λ_{FZ}	nm					
Thermal properties			<100>	<010>	<001>	
c_p	W s/g °C					
k	W/m °C		2.8	2.2	3.5	[12.40]
α	10^{-6} °C		4.0	3.6	8.5	[12.40]
Mechanical properties			a	b	c	
ρ	g/cm³	7.27				[12.40]
ν						
E	GPa		115.8	152.5	92.4	[12.43]
P_{max}	MPa					
H_K	MPa					
γ	MP√m					
Crystal properties						
Structure		monoclinic				[12.40]
Space group		C^6_{2h}				[12.40]
Lattice const.	a/nm	0.8098				[12.40]
	b/nm	1.0417				[12.40]
	c/nm	0.7583				[12.40]
Orientation		<010>				[12.40]
T_m	°C	1075				[12.40]
F_A	10^{20} cm^{-3}/at%		0.633			[12.40]
F_W	10^{20} cm^{-3}/wt%		3.064			[12.40]
Mol	g		647.2			[12.40]
Formula units	ZEZ	4				[12.40]
Distribution coefficient						
Derived data						
R_T	kW/m					
τ/f	s/mm²					
dn_{abs}/dT	10^{-6}/°C					
Supplier data		Nd	Nd			
l_{max}	mm		80			[12.43]
d_{max}	mm	5	8			[12.43]
C(ion)	at%	3	1 – 8			[12.40, 12.43]

LICAF

Material		$LiCaAlF_6$			Ref.
Active ion		Nd^{3+}	Cr^{3+}		
Optical properties			‖ a	‖ c	
n					
dn_{rel}/dT	$10^{-6}/°C$		−4.2	−4.6	[12.38]
n_2	10^{-13} esu				
p_{11}					
p_{12}					
p_{44}					
λ_{FZ}	nm				
Thermal properties			‖ a	‖ c	
c_p	W s/g °C				
k	W/m °C		4.6	5.1	[12.38]
α	10^{-6} °C		22	3.6	[12.38]
Mechanical properties					
ρ	g/cm^3				
ν					
E	GPa				
P_{max}	MPa				
H_K	MPa				
γ	MP√m				
Crystal properties					
Structure					
Lattice const.	a/nm				
Orientation					
T_m	°C				
F_A	10^{20} cm^{-3}/at%				
Distribution coefficient					
Derived data			literature values		
R_T	kW/m				
τ/f	s/mm^2				
C_r					
C_ϕ					
dn_{abs}/dT	$10^{-6}/°C$				
Supplier data					
l_{max}	mm				
d_{max}	m				
C(ion)	10^{20} cm^{-3}				

Thermal lensing is very low [12.38]

LNA, LMA

Material		$LaMgAl_{11}O_{19}$				Ref.
Active ion		Nd^{3+}	Codoping		Cr	
Optical properties			o	e		
n			1.769	1.764		
dn_{rel}/dT	$10^{-6}/°C$		12.5	11.5		[12.15]
n_2	10^{-13} esu					
λ_{FZ}	nm					
Thermal properties			⊥c	‖c		
c_p	W s/g °C	0.715				[12.15]
k	W/m °C		4.58	2.68		[12.15]
α	10^{-6} °C		8	12		[12.15]
Mechanical properties			⊥c	‖c		
ρ	g/cm^3	4.18				[12.15]
ν						
E	GPa	38				[12.15]
P_{max}	MPa	20				
H_K	MPa		14.2	18.2		[12.15]
Crystal properties			Nd	Cr		
Structure		hexagonal				
Space group		$P6_3/mmc$				[12.16]
Lattice const.	a/nm	0.558				[12.16]
	c/nm	2.202				[12.16]
Orientation		a axis				[12.17]
T_m	°C	1760				[12.16]
F_A at%	10^{20} cm^{-3}	3.4				[12.16]
Distribution coefficient			≈ 1	≈ 1.3		
Derived data		⊥c	‖c			
ϕ_B	°	60.5				
R_T	kW/m	≈ 0.226	≈ 0.09			
τ/f	s/mm^2	0.653	1.115			
dn_{abs}/dT	$10^{-6}/°C$	≈ 10				
Supplier data	Nd	Nd(Cr)				
l_{max}	mm					
d_{max}	mm					
C(ion)	wt%	1.5	1.5(1)			[12.18, 12.17]

Tunable up to 1083 nm [12.18]. Can be doped very highly by Nd.
The wavelength depends on the Nd concentration [12.17].

LSB

Material		$LaSc_3(BO_3)_4$				Ref.
Active ion		Nd^{3+}	Codoping		Cr	

Optical properties			*x*	*y*	*z*	
n	@ 1 µm		1.828	1.8272	1.828	
dn_{rel}/dT	10^{-6}/°C @ 633 nm		4.37			
n_2	10^{-13} esu					
λ_{FZ}	nm					

Thermal properties			‖ z			
c_v	W s/g °C	0.62 ±0.02				[12.45]
k	W/m °C		2.8			[12.45]
α	10^{-6} °C	12.3				[12.45]
T_m	°C	1600				[12.45]

Mechanical properties			*a*	*b*	*c*	
ρ	g/cm^3	3.8				[12.45]
ν						
E	GPa					
P_{max}	MPa					
H_K	MPa					
γ	MP√m					

Crystal properties						
Structure		monoclinic for $x \leq 0.5$ trigonal for $x > 0.5$				[12.45]
Space group						
Lattice const.	*a*/nm	0.9778				[12.45]
	b/nm					
	c/nm	0.7929				[12.45]
Orientation						
T_m	°C					
F_A	10^{20} cm^{-3}/at%					
F_W	10^{20} cm^{-3}/wt%					
Mol weight						
Distribution coefficient						

Derived data						
R_T	kW/m					
τ/f	s/mm^2					
dn_{abs}/dT	10^{-6}/°C					

Supplier data		Nd	Nd(Cr)			
l_{max}	mm	1	50			[12.45]
d_{max}	mm	3	4			[12.45]
C(ion)	at%	5 – 25	$7.3(0.5)\times10^{20}cm^{-3}$			[12.45]

NYAB Neodym-Yttrium-Aluminum-Borate

Material		$YAl_3(BO_3)_4$			Ref.
Active ion		Nd^{3+}			
Optical properties			n_o	n_e	
n	@1062 nm		1.7553	1.6869	[12.36, 12.37]
n	@531 nm		1.7808	17075	[12.36, 12.37]
dn_{rel}/dT	10^{-6}/°C				
n_2	10^{-13} esu				
Frequency doubling			Type I	Type II	
d_{11}	pm/V	1.7			
d_{eff}	pm/V		1.43	0.67	[12.36]
Phase angle	°		32.9	51	[12.36]
Thermal properties					
c_p	W s/g °C				
k	W/m °C				
α	10^{-6} °C				
Mechanical properties					
ρ	g/cm^3				
ν					
E	GPa				
P_{max}	MP				
γ	MP√m				
Crystal properties					
Structure					
Lattice const.	a/nm	0.9293			[12.37]
	b/nm	0.9293			[12.37]
	c/nm	0.7245			[12.37]
Orientation					
T_m	°C				
F_A	10^{20} cm^{-3}/at%		0.557		[12.36]
Distribution coefficient					
Derived data		Literature values			
R_T	kW/m				
τ/f	s/mm^2				
dn_{abs}/dT	10^{-6}/°C				
Supplier data					
l_{max}	mm				
d_{max}	mm				
C(ion)	10^{20} cm^{-3}				
Self-frequency doubled					

Quartz

Material		SiO_2		Ref.
Active ion		Nd^{3+}		
Optical properties				
n	@1064 nm	1.44963		
dn_{rel}/dT	$10^{-6}/°C$	10		
n_2	10^{-13} esu			
ν_d		67.8		
λ_{FZ}	nm			
Frequency doubled				
d_{11}	pm/V			
d_{eff}	pm/V			
Phase angle	°			
Thermal properties				
c_p	W s/g °C	0.772		
k	W/m °C	1.5		
α	10^{-6} °C	0.55		
Mechanical properties				
ρ	g/cm^3	2.201		
ν		0.17		
E	GPa	70		
P_{max}	MPa			
γ	MP√m			
Crystal properties				
Structure			amorphous	
Lattice const.	a/nm			
	b/nm			
	c/nm			
Orientation				
T_g	°C		1600	
F_W	10^{20} cm^{-3}/wt%		0.925	
Distribution coefficient				
Derived data		Literature values		
R_T	kW/m			
τ/f	s/mm^2			
dn_{abs}/dT	$10^{-6}/°C$			
Supplier data				
l_{max}	mm			
d_{max}	mm			
C(ion)	10^{20} cm^{-3}			

Sapphire

Material		Al_2O_3	Ruby	Sapphire	Ref.
Active ion		Cr	Cr	Ti	
Optical properties			n_e π	n_o σ	
n		1.7638	1.7556	1.76	[12.19]
dn_{rel}/dT	10^{-6}/°C	12.6			
p_{11}		−0.25			[12.20]
p_{12}		−0.038			[12.20]
p_{44}		−0.1			[12.20]
Thermal properties					
c_p	W s/g °C	3.98			
k	W/m °C	28			[12.4]
α	10^{-6} °C	6.7			[12.4]
Mechanical properties					
ρ	g/cm^3	3.98			
ν		0.25			[12.4]
E	GPa	405			[12.4]
P_{max}	MPa	440			[12.4]
γ	MP√m	2.2			[12.4]
Crystal properties					
Structure		hexagonal	rhombohedral		
Lattice const.	a/nm				
	c/nm				
Orientation					
T_m	°C		2040		[12.19]
F_W	10^{20} cm^{-3}/wt%		3.24	3.29	
F_A	10^{20} cm^{-3}/at%				
Derived data		Literature value			
R_T	kW/m	3.4	3.4		[12.4]
τ/f	s/mm^2	0.56			
C_r		−0.026			
C_ϕ		−0.044			
dn_{abs}/dT	10^{-6} /°C	11.02			
M_r	10^{-6} m/W	0.16			
M_ϕ	10^{-6} m/W	0.14			
Supplier data	Cr_2O_3	TiO_2			
l_{max}	mm	100	100		[12.19]
d_{max}	mm	20	6.35		[12.19]
C	wt%	0.03 – 0.05	0.02 – 0.15		[12.19]

YAG

Material		$Y_3Al_5O_{12}$				Ref.
Active Ions		Er^{3+}	Ho^{3+}(Cr, Tm)		Nd^{3+}	Yb^{3+}
Optical properties						
n		1.82	1.818			[12.21, 12.22]
dn_{rel}/dT	10^{-6}/°C	7.3 – 8.9				[12.22, 12.38]
n_2	10^{-13} esu	3				[12.18]
p_{11}		–0.0290				[12.20]
p_{12}		0.0091				[12.20]
p_{44}		–0.0615				[12.20]
λ_{FZ}	nm	475				
Thermal properties			<100>	<110>	<111>	
c_p	W s/g°C	0.6				[12.21]
k	W/m°C	10 – 14				[12.4, 12.21]
α	10^{-6}°C	6.9 – 7.5	8.2	7.7	7.8	[12.22, 12.21,23]
Mechanical properties						
ρ	g/cm^3	4.56				[12.21]
ν		0.25 – 0.28				[12.22, 12.4]
E	GPa	282 – 300				[12.4, 12.21]
P_{max}	MPa	130 – 280				[12.21, 12.4]
H_K	MPa	11.9				
γ	MP√m	1.4				[12.4]
Crystal properties		Nd	Cr	Tm	Ho	
Structure		cubic, isotropic				[12.8]
Space group		O_h^{10}-Ia3d				
Lattice constant	a/nm	1.201				[12.21]
Orientation		<111>				
T_m	°C	1950				[12.22]
F_W	10^{20}cm^{-3}/wt%	1.903				
F_A	10^{20}cm^{-3}/at%	1.38			[12.22]	
Distribution coefficient		0.18	2.6	1.06	1.0	[12.5]
mol	g	593.6				[12.22]
Derived data				Literature values		
R_T	kW/m	0.516 – 1.254		0.7–1.4 7.9		[12.4, 12.21, 12.22]
τ/f	s/mm^2	0.27				
C_r		0.017				
C_ϕ		–0.003				
dn_{abs}/dT	10^{-6}/°C	5.66				
M_r	10^{-6} m/W	0.362				
M_ϕ	10^{-6} m/W	0.282				

Supplier data		Nd	Ho(Cr | Tm)	Er	Yb
l_{max}	mm	250	152		[12.22]
d_{max}	mm	20	9.5		[12.22]
C	at%	0.6 – 1.4	0.36(1 | 5.7)	50	[12.22, 12.23]

YAG is the most essential crystal for material processing.

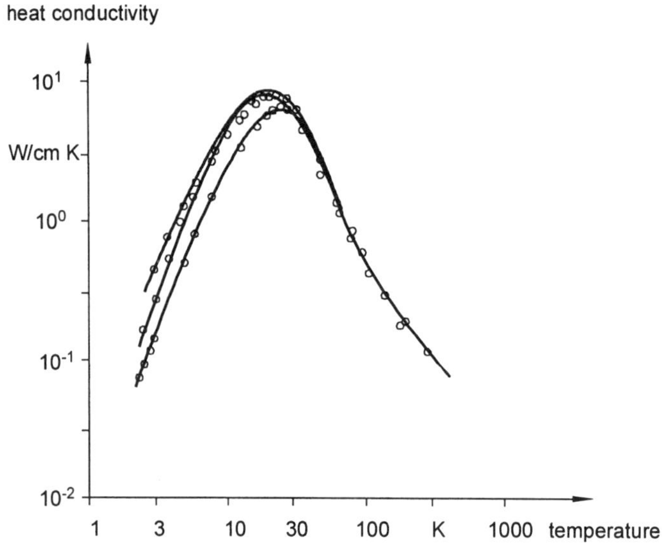

Fig. 12.1. Temperature dependence of the heat conductivity of YAG, after [12.24]

The quantum efficiency of Nd in YAG depends on the concentration and is typically between 0.6 at 1.6 at% and 0.8 at 0.9 at% [42]. Some single rods attain 0.87 [41], even at high concentration. The relative dependence on the concentration is given in [42] with η_r© = 1 – 18.2 C. Heating the crystal to 1400°C in I_2 and O_2 (atmosphere) for 8 h reduces the absorption by color centers and improves the efficiency [41].

The thermal lens in the Ho laser is about three times stronger in the Cr co-doped crystal than in Nd without Cr.

YAP (YALO)

Material		$YAlO_3$ yttrium aluminum perovskite				Ref.
Active ion		Nd^{3+}	Er^{3+}	Tm^{3+}		
Codoping				Cr		
Optical properties			*a*	*b*	*c*	
n			1.932	1.923	1.909	[12.25]
dn_{rel}/dT	10^{-6}/°C		8.5 – 9.75	8.1 – 9.86	12.8 – 14.5	[12.25, 12.29]
n_2						
λ_{FZ}	nm	280				[12.26]
Thermal properties			*a*	*b*	*c*	
c_p	W s/g °C	0.418				[12.20]
k	W/m °C	11				[12.20]
α	10^{-6} °C		9.5	4.3	10.8	[12.20]
Mechanical properties						
ρ	g/cm³	4.88 – 5.35				[12.27, 12.20]
ν						
E	GPa	220				[12.20]
P_{max}	MPa					
Crystal properties						
Structure		orthorombic				
Space group		D_{2h}^{16} – Pbnm				[12.20]
Lattice	*a*/nm	0.5176				
constant				[27]		
	b/nm	0.5307				[12.27]
	c/nm	0.7355				[12.27]
Orientation		*b* axis				
T_m	°C	1850				
F_A	10^{20} cm^{-3}/at%		1.95			[12.28]
Distribution coefficient			0.82			[12.27]
Derived data		Literature values				
R_T	kW/m					
τ/f	s/mm²	0.185	0.2			[12.20]
dn_{abs}/dT	10^{-6}/°C	6.4 – 11.1				
Supplier data		Nd				
l_{max}	mm	150				
d_{max}	mm	10				
C	wt%	0.7				[12.28]

The emission wavelength is dependent on the polarization direction and the rod orientation
λ = 1079 nm, rod axis || *a* axis and polarization || *b* axis
λ = 1064 nm, rod axis || *a* axis and polarization || *c* axis
λ = 1064 nm, rod axis || *b* axis independent of polarization direction
λ = 1064 nm, rod axis || *c* axis independent of polarization. Without polarization || to *a* axis

YLF

Material		$LiYF_4$			Ref.
Active ion		Nd	Er	Ho	
Codoping				Er, Tm	
Optical properties			n_e	n_o	
n			1.470	1.448	[12.30]
dn_{rel}/dT	$10^{-6}/°C$		−4.1	−2.0	[12.12]
γ_2	10^{-20} m²/W				
λ_{FZ}	nm				
Thermal properties			*a* axis	*c* axis	
c_p	W s/g °C		0.79		[12.30]
k	W/m °C	4.3	6		[12.38, 12.30]
α	$10^{-6}/°C$	14	13	8	[12.4, 12.30]
Mechanical properties					
ρ	g/cm³	3.95			[12.30]
ν		0.33			[12.4]
E	GPa	75 – 77			[12.30, 12.4]
P_{max}	MPa	33 – 54			[12.30, 12.4]
H_K	MPa				
γ	MP√m	0.27			[12.4]
Crystal properties					
Structure		tetragonal	uniaxial		
Space group					
Lattice const.	*a*/nm				
	c/nm				
Orientation		<100>	*a* axis		[12.30]
T_m	°C	825			[12.30]
F_W	10^{20} cm⁻³/wt%				
F_A	10^{20} cm⁻³/at%		1.3		
Distribution coefficient			0.3		[12.5]
mol		171.844			[12.30]
Derived data		literature values			
R_T	kW/m		0.14		[12.4]
τ/f	s/mm²				
dn_{abs}/dT	10^{-6} /°C				
Supplier data	Nd	Er	Ho (Er)		
l_{max}	mm	127			[12.30]
d_{max}	mm	9.5			[12.30]
C	wt%	1	5	2 (35) (10)	[12.30]

The thermal lens is assumed to be smaller than for YAG, due to the negative dn/dT. The focal length is smaller by a factor of 6 for the same pump power [4.19].

YOS

Material		Y_2SiO_5				Ref.
Active ion		Nd				
Codoping						
Optical properties			n_x	n_y	n_z	
n			1.772	1.773	1.793	[12.35]
dn_{rel}/dT	$10^{-6}/°C$					
γ_2	$10^{-20}\ m^2/W$					
λ_{FZ}	nm					
Thermal properties						
c_p	W s/g °C					
k	W/m °C	4.5				[12.35]
α	$10^{-6}/°C$					
Mechanical properties						
ρ	g/cm^3					
ν						
E	GPa					
P_{max}	MPa					
H_K	MPa					
γ	MP√m	0.54				[12.35]
Crystal properties						
Structure			monoclinic		biaxial	
Space group		c2/2				
Lattice const.	a/nm					
	c/nm					
Orientation	20° relative to		<010>			[12.35]
T_m	°C					
F_W	$10^{20}\ cm^{-3}/wt\%$					
F_A	$10^{20}\ cm^{-3}/at\%$					
Distribution coefficient						
mol	g					
Derived data		Literature values				
R_T	kW/m		0.38			[12.35]
τ/f	s/mm^2					
dn_{abs}/dT	$10^{-6}/°C$					
Supplier data						
l_{max}	mm	150				[12.35]
d_{max}	mm	25				[12.35]
C	wt%					

Birefringent, polarized laser

YSAG

Material		$Y_3Sc_2Al_3O_{12}$		Ref.
Active ion		Nd^{3+}		
Codoping		Cr		

Optical properties				
n		1.86		[31]
dn_{rel}/dT	10^{-6}/°C			
n_2	10^{-13} esu			
p_{11}				
p_{12}				
p_{44}				

Thermal properties				
c_p	W s/g °C			
k	W/m °C	10.8		[32]
α	10^{-6} °C			

Mechanical properties				
ρ	g/cm^3			
ν				
E	GPa			
P_{max}	MPa			
γ	MP√m			

Crystal properties				
Structure			cubic	
Space group				
Lattice const.	a/nm		1.2271	[31]
Orientation			<111>	[31]
T_m	°C		1900	[31]
F_W	10^{20} cm^{-3}/wt%		0.189	[31]
F_A	10^{20} cm^{-3}/at%			
Distribution coefficient			0.4	[31]

Derived data		Literature value		
R_T	kW/m			
τ/f	s/mm^2			
C_r				
C_ϕ				
dn_{abs}/dT	10^{-6}/°C			
M_r	10^{-6} m/W			
M_ϕ	10^{-6} m/W			

Supplier data				
l_{max}	mm			
d_{max}	mm			
C(ion)	wt%			

YSGG

Material		$Y_3ScGa_3O_{12}$			Ref.
Active ion		Nd^{3+}	Ho^{3+}	Er^{3+}	
Codoping		Cr^{3+}	Cr, Tm	Cr^{3+}	
Optical properties					
n		1.9626			[12.33]
dn_{rel}/dT	10^{-6}/°C	12.3			[12.33]
p_{11}		0.40			[12.33]
p_{12}		0.42			[12.33]
p_{44}		0.77			[12.33]
Thermal properties					
c_p	W s/g °C	0.534			[12.33]
k	W/m °C	7.9	8.3		[12.33, 12.32]
α	10^{-6}/°C	8.1			[12.33]
Mechanical properties					
ρ	g/cm^3	5.36			[12.33]
ν					
E	GPa	218			[12.33]
P_{max}	MPa				
H_K	MPa				
Crystal properties					
Structure		cubic			
Space group		Ia3d-O_h^{10}			
Lattice const.	a/nm				
	c/nm				
Orientation		<001>	<111>		
T_m	°C	1810			[12.33]
F_W	10^{20} cm^{-3}/wt%				
F_A	10^{20} cm^{-3}/at%				
Distribution coefficient					
Derived data		Literature values			
R_T	kW/m				
τ/f	s/mm^2	0.362			
C_r		0.079			
C_ϕ		0.206			
dn_{abs}/dT	10^{-6}/°C	10.57			
M_r	10^{-6} m/W	1.26			
M_ϕ	10^{-6} m/W	2.2			
Supplier data		Nd(Cr)	Ho(Tm, Cr)	Er(Cr)	
l_{max}	mm	152.4	101	101	[12.33]
d_{max}	mm	12.7	5	5	[12.33]
C	10^{20} cm^{-3}	2 – 3(1– 2)	0.5(8, 2.5)	30 – 40(1 – 2)	[12.33]

YVO

Material		YVO_4	yttrium orthovanadate		Ref.
Active ion		Nd^{3+}	Nd^{3+}	Ho^{3+}	
Codoping					
Optical properties			$o = a = b$	$e = c$	
n (1064 nm)			1.9573	2.16252	[34, 44]
dn_{rel}/dT	10^{-6}/°C		8.5	3.0	
n (808 nm)			1.9721	2.1858	
n(532 nm)			2.0210	2.2560	
$n = A + B(\lambda^2 - C) - D\lambda^2$					
A			3.77834	4.59905	[44]
B			0.069736	0.110534	[44]
C			0.04724	0.04813	[44]
D			0.0108133	0.0122676	[44]
n_2	10^{-13} esu				
λ_{FZ}	nm				
Thermal properties			a	c	
c_p	W s/g °C				
k	W/m °C	5.2	5.23 ∥ c	5.10 ⊥ c	[34, 44]
α	10^{-6} °C		4.49	11.4	[34]
Mechanical properties					
ρ	g/cm³	4.22			[44]
ν					
E	GPa				
P_{max}	MPa				
Crystal properties					
Structure		tetragonal	uniaxial		[34]
Space group		D_{4h}			
Lattice const.	$a = b$/nm	0.712			[34]
	c/nm	0.629			[34]
Orientation		a axis	(c axis)		[34]
T_m	°C	1825			[34]
F_W	10^{20} cm^{-3}/wt%				
F_A	10^{20} cm^{-3}/at%		1.54		[34]
Distribution coefficient					
Supplier data					
l_{max}	mm	20			[34]
d_{max}	mm	5			[34]
C	wt%	1.5 – 3			[34]

The absorption coefficient at 809 nm is $\alpha = 40$ cm^{-1} for polarization parallel to the c axis (π). The absorption range is from 801 to 821 nm.

13 Laser Glasses

The emission spectrum of doped glass is generally much broader than for crystals and depends on the special glasses (Fig. 13.1).

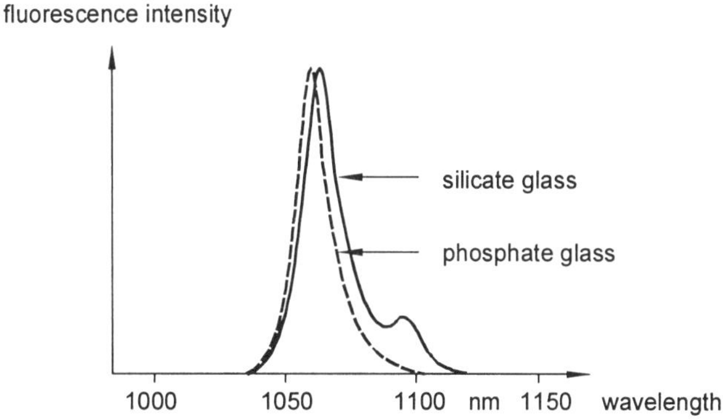

Fig. 13.1. Fluorescence of glass, [13.1]

The lifetime is dependent on the doping concentration and the type of glass, as shown in Fig. 13.2. A typical absorption spectrum of neodymium in glass is given in Fig. 13.3.

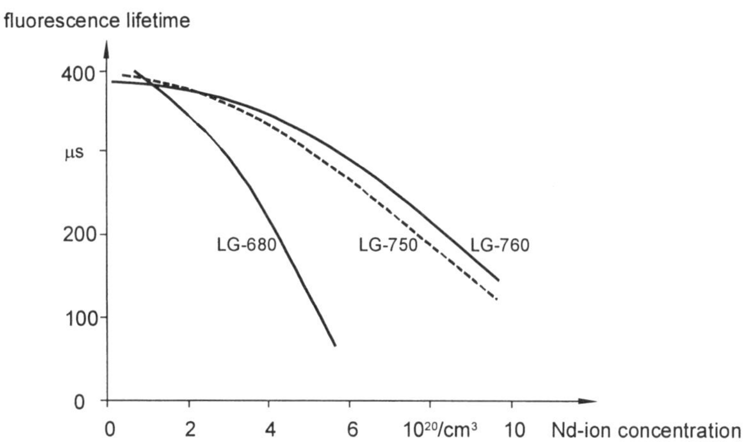

Fig. 13.2. Fluorescence lifetime versus doping concentration, [13.2, 13.3]

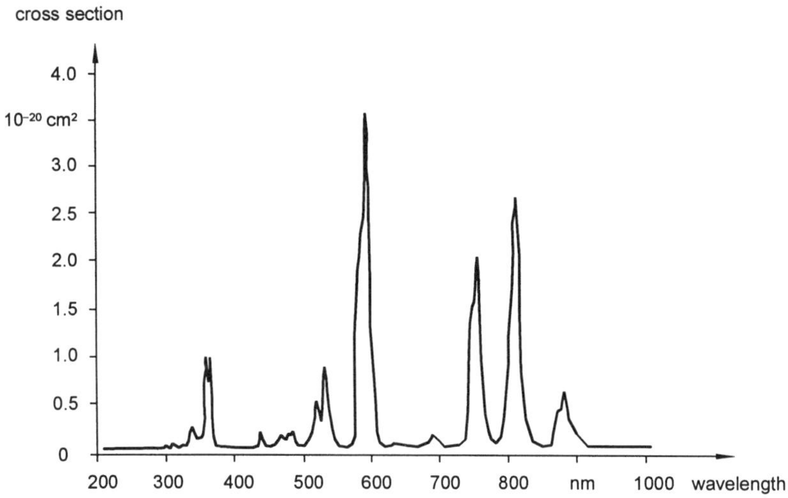

Fig. 13.3. Absorption spectrum of Nd-doped glass, [13.4]]

The average laser power that can be achieved with laser glass rods is very low compared to crystal materials. It is typically around 20 W for a rod length of 20 cm. Due to the possibility to dope glasses with very high concentrations, approximately about 10 wt% neodymium, high energies in the pulsed operation can be obtained, e.g. 100 W s at a pulse length of 10 to 20 ms. In [13.13] an average power of 180 W for a glass plate with dimensions of $0.42 \times 9 \times 15$ cm^3 is reported.

HOYA Laser Glasses

Material		LHG-80	HAP-4	LHG-8	LHG-5	LSG-91H	LEG-30
Doping		Nd^{3+}	Nd^{3+}	Nd^{3+}	Nd^{3+}	Nd^{3+}	Er^{3+}
Glass type		phos.	phos.			phos.	silicate
$C_{0.5}$	wt%		10.4	8.5 [13.10]	8.5 [13.6]		
Laser data							
λ	nm	1054	1054	1054	1054	1062	1535
$\Delta\lambda$	nm	20.2	27	21.8	22	27.4	32
σ	10^{-20} cm^2	4.8	3.6	4.2	4.1	2.7	0.77
τ_0	μs	337 [13.18]	390 [13.10]	390 [13.10]	350 [13.6]		
τ(3%)	μs	320	350	315	290	300	8700
ln$V$	%	0.15	0.15	0.15	0.15	0.15	
Optical properties							
n		1.533	1.533	1.520	1.531	1.55	1.527
dn_r/dT	10^{-6}/ °C	−3.8	1.8	−5.3	−0.4	1.6	−3
n_2	10^{-13} esu	1.24	1.25	1.13	1.28	1.58	1.22
n_2	10^{-20} m^2/W	3.36	3.32 [13.18]	3.08	3.15 [13.11]		
ν_d		64.7 [13.18]	64.6 [13.18]	66.5 [13.18]			
p_{11}		0.107	0.1835	0.1929 [13.9]	0.19 [13.8]		
p_{12}		0.1829	0.241	0.2358 [13.9]	0.26 [13.8]		
p_{44} [a)]		−0.0379	−0.029	−0.021	−0.035 [13.8]		
λ_{FZ}	nm						
Thermal properties							
c_p	W s/g °C	0.63	0.71	0.75	0.71	0.63	
k	W/m °C	0.59–0.63	1.02	0.58	0.77	1.04	0.55
α	10^{-6}/ °C	13	8.5	12.7	9.8	10.5	10.9
Mechanical properties							
ρ	g/cm^3	2.924	2.7	2.83	2.68	2.81	3.09
ν		0.267	0.236	0.266	0.237	0.237	0.261
E	GPa	50.0	68.9	50.1	67.8	87.2	55.3
σ_{max}	MPa				24 [13.7]	176	
H_K	MPa	3.36	4.61	3.43	4.21	5.79	3.99
E	MPa	50.1 [13.18]	70	68.1			
K_c	MPa√m	0.46 [13.18]	0.79		0.50 [13.6]		
R_T^*	W/√m				0.52 [13.6]		
Glass properties							
T_g	°C	402	486	485	455	465	515
F_W	10^{20}/cm^3/wt%		1.03	0.97	1.03	0.97	0.97
Weight loss in water	%/100°C/h	0.18	0.07	0.13	0.08	0.036	0.06

a) $p_{44} = (p_{11} - p_{12})/2$, All data from [13.5], unless stated otherwise.

HOYA Laser Glasses

Derived data		LHG-80	HAP-4	LHG-8	LHG-5	LSG-91H	LEG-30
$h\nu$	10^{-19} W s	1.88	1.889	1.889	1.889	1.87	1.29
J_s	kW/cm^2	12.3	15	14.3	15.8	23.1	1.93
H_s	W s/cm^2	3.93	5.24	4.49	4.6	6.93	16.8
ϕ_B	°	57	57	57	57	58	57
R_T	kW/m				0.02	0.15	
τ/mm^2	s/mm^2	2.94	1.88	3.67	2.49	1.71	
C_r		0.064	0.086	0.783	0.0934		
C_ϕ		0.047	0.075	0.0691	0.0792		
dn_{abs}/dT		–5.18	0.42	–6.67	–1.78	0.178	–4.37
M_r		0.81	2.87	0.46	3.23		
M_ϕ		–0.41	2.52	–0.25	2.58		
Supplier data							
l_{max}	mm						
d_{max}	mm						
C(Nd)	wt%	3.0	3.3	3.0	3.3	3.1	
Measured data							
η_{slope}	%		1.83	1.83	1.15	1.5	
η_{excit}	%						
η_{extr}	%						
η_H	%						
–ln$V$	%/cm						
E_{max}	W s						0.4
T	ms						
P_{av}	W						
d×l	mm^2		10×160	10×160	10×160	3×50	

KIGRE Laser Glasses

Material		Q-88	Q-98	Q-100	Q-246	QE-7	Q89
Doping		Nd^{3+}	Nd^{3+}	Nd^{3+}	Nd^{3+}	Er^{3+}	Nd^{3+}
Glass type		phos.	phos.	phos.	silicate	phos.	phos.
$C_{0.5}$	wt%	8					
Laser data							
λ	nm	1054	1053	1054	1062	1535	1054
$\Delta\lambda$	nm	22	21	21	28	30	21
σ	10^{-20} cm^2	4	4.5	4.4	2.9	0.8	3.8
τ_0	µs	350		350	370		350
τ	µs	326	308	190	330	8000	308
V	%/cm					0.02	
Optical properties							
n		1.536	1.546	1.562	1.561	1.532	≈ 1.54
dn_{rel}/dT	10^{-6}/ °C	−0.5	−4.5	−4.6	2.9	−6.3	
n_2	10^{-13} esu	1.1	1.2	1.2	1.4		
n_2	10^{-20} m^2/W	$3.11_{[13.11]}$					
ν_d		$64.8_{[13.18]}$					
p_{11}		$0.72_{[13.12]}$					
p_{12}		$0.186_{[13.12]}$					
p_{44}		−0.57					
λ_{FZ}	nm						
Thermal properties							
c_p	W s/g °C	0.81	0.8	0.8	0.93	0.80	
k	W/m °C	0.84	0.82	0.82	1.3		0.82
α	10^{-6}/ °C	10.4	9.9	9.6	9.0	11.4	8.8
Mechanical properties							
ρ	g/cm^3	2.71	3.1	3.2	2.55	2.94	3.14
ν		0.24	0.24	0.24	0.24	0.24	[13.9]
E	GPa	69.9	70.7	70.1	84.1	70.7	[13.9]
σ_{max}	MPa						$20_{[13.7]}$
H_K	MPa	4.18	5.56	5.47	5.89	5.47	
R_T^*	W/√m	$0.39_{[13.10]}$					
Glass properties							
T_g	°C	366	416	413	470	462	440
F_W	10^{20}/cm^3/wt%						
Weight loss in water	%/100°C/h	0.2	0.08	0.08	0.04		

Reference [13.14, 13.15], Q-98 and Q-100 are athermal.

KIGRE Laser Glasses

Derived data		Q-88	Q-98	Q-100	Q-246	QE-7	Q89
$h\nu$	10^{-19} W s	1.88	1.89	1.88	1.87	1.29	1.88
J_s	kW/cm^2	14.3	11.9	22.5	19.6	2.02	14.2
H_s	W s/cm^2	4.71	4.19	4.28	6.45	16.2	4.96
ϕ_B	°	57	57	57	57	58	
R_T	kW/m						0.02
τ/mm^2	s/mm^2	2.61	3.02	3.13	1.82		
C_r		0.70					
C_ϕ		0.47					
dn_{abs}/dT		−1.88	−5.89	−6.01	1.50	−7.68	
M_r		30.4					
M_ϕ		19.97					
Supplier data							
l_{max}	mm						
d_{max}	mm						
C(Nd)	wt%		4 – 6		3		
Measured data							
η_{slope}	%	4.5	4.9	8			
η_{excit}	%						
η_{extr}	%						
η_H	%						
$-\ln V$	%/cm	0.0008	0.0008	0.0008	0.002	0.02	0.0008
E_{max}	W s						
T	ms						
P_{av}	W						
$d\times l$	mm^2	6×75		6×400			

SCHOTT Laser Glasses

Material		LG-680	LG-700	LG-750	LG-760	APG-1	APG-t
Doping		Nd^{3+}	Nd^{3+}	Nd^{3+}	Nd^{3+}	Nd^{3+}	Nd^{3+}
Glass type		sil.	phos.	phos.	phos.	phos.	phos.
$C_{0.5}$	$10^{-20}/cm^3$	4.2	>8	6.6[13.10]	>8	8.0[13.10]	
Laser data							
λ	nm	1061	1053	1053	1053		
	1052[13.2]1054.6[13.18]						
$\Delta\lambda$	nm	28.2	22.3	21.5	19.6	23[13.2]	31.5
σ	10^{-20} cm^2	2.9	3.7	4.0	4.2	3.5[13.2]	2.39[13.18]
τ_0	µs	380	380	390[13.10]	360	370[13.10]	464
τ	µs	350	290	270	290	270	
Optical properties							
n		1.560	1.504	1.516	1.508	1.526[13.2]	
dn_{rel}/dT	10^{-6}/ °C	2.9	−3.2	−5.1	−6.8	1.2[13.2]	4.0
n_2	10^{-13} esu	1.6	1.08	1.08	1.02	1.13[13.2]	1.02[13.18]
n_2	10^{-20} m^2/W		3.51[13.11]			3.1[13.18]	2.85[13.18]
ν_d		57.7	67.5	68.2	69.2	67.7[13.18]	67.8[13.18]
p_{11}		0.107	0.1835	0.196	0.1725		
p_{12}		0.1829	0.241	0.238	0.2226		
p_{44}		−0.0379	−0.029	−0.021	−0.025		
λ_{FZ}	nm						
Thermal properties							
c_p	W s/g °C	0.92	0.84	0.72	0.57	0.84[13.2]	
k	W/m °C	1.35	0.69	0.52	0.60	0.83[13.2]	0.86
α	10^{-6}/ °C	9.3	10.9	11.4	12.5	7.6[13.2]	6.38
Mechanical properties							
ρ	g/cm^3	2.54	2.63	2.83	2.60	2.64[13.2]	2.56[13.18]
ν		0.242	0.243	0.256	0.267	0.239[13.2]	0.229
E	GPa	90.1	60.6	50.1	53.7	71.0[13.2]	64.4
σ_{max}	MPa	166		90			
H_K	MPa	5.88	3.14	2.85	3.04	3.09[13.2]	
E	MPa					71	64
K_c	MPa√m					0.60	0.80[13.18]
R_T^*	W/√m			0.32[13.10]		0.75[13.10]	
Glass properties							
T_g	°C	468	458	450	350	450[13.2]	549[13.18]
F_W	$10^{20}/cm^3$/wt%		0.91	0.94	1.01	0.93	0.94[13.2]
weight loss in water	mg/cm^2/day	0.05	0.05	0.13	0.15	0.008	

Reference [13.2, 13.16, 13.17]

SCHOTT Laser Glasses

Derived data		LG-680	LG-700	LG-750	LG-760	APG-1
$h\nu$	10^{-19} W s	1.87	1.89	1.89	1.89	1.89
J_s	kW/cm^2	129	18	17	15	20
H_s	W s/cm^2	6.45	5.1	4.72	4.5	5.4
ϕ_B	°	57.36	56.23	56.59	56.45	56.75
R_T	kW/m	0.2		0.06		
τ/mm^2	s/mm^2	1.73	3.2	3.92	2.47	2.67
C_r		0.067	0.085	0.081	0.075	
C_ϕ		0.051	0.073	0.072	0.064	
dn_{abs}/dT		1.50	−4.55	−6.46	−8.16	−0.17
M_r		2.36	1.41	0.15	−1.24	
M_ϕ		1.95	0.77	−0.53	−2.03	
Supplier data						
l_{max}	mm		60			60[13.2]
d_{max}	mm					
C(Nd)	wt%		1– 3.5	3.6		1.3.5[13.2]
Measured data						
η_{slope}	%		3.1			
η_{excit}	%					
η_{extr}	%					
η_H	%					
$-\ln V$	%/cm	≤ 0.2	≤ 0.2	≤ 0.2	≤ 0.2	
E_{max}	W s		45			
T	ms		6			
P_{av}	W		10 – 20			[Author]
d×l	mm^2		8×180			

Some glasses were formerly manufactured by Owens Illinois under different names:

Schott	LG-670	LG-680	LG-760
Owens-Illinois	ED-2	ED-3	EV-4

14 Material Data

In this chapter the material parameters and spectra are given for the more common materials and substances which find use in construction and design.

14.1 Optical Transparent Media

The spectral qualities of the media used should be optimized for the corresponding usage. Iron-bearing flow-tubes strongly absorb the pump light. Color centers are caused (reversibly and irreversibly) for some laser materials due to short-wave radiation. This can be avoided by absorption of this radiation up to the corresponding critical wavelength by flow-tubes or additives in the coolant.

Water

Due to the strong absorption of water at longer wavelengths, the path for the pump light from excitation source to laser medium should be kept short. The spectral transmission of water is given in Table 14.1 and plotted in Fig. 14.1.

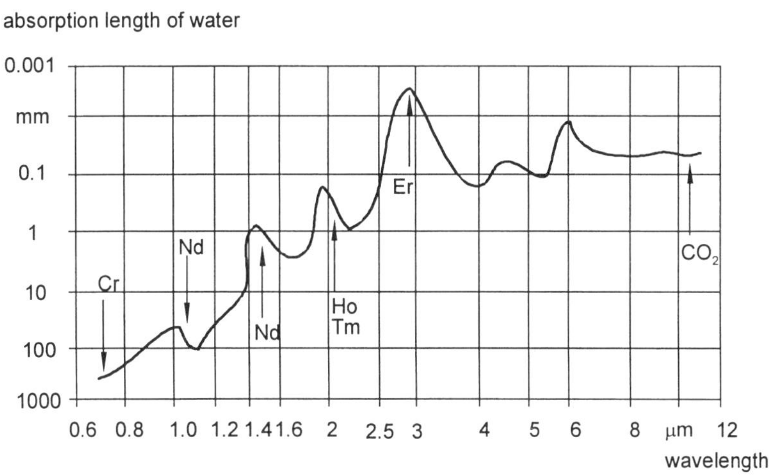

Fig. 14.1. Transmission of water, [14.1, 14.2]

Table 14.1 Transmission of water [14.1, 14.2]

λ / nm	400	500	600	700	800	900	950	1000	1050	1100
α / m^{-1}	0.08	0.03	0.2	0.58	2.1	7.0	40	35	13	20
1 cm thickness										
T / %	99.9	100	99.8	99.4	98.0	93.2	67.0	70.5	87.8	81.9
10 cm thickness										
T / %	99.2	99.7	98.1	94.4	81.3	49.7	1.8	3.0	27.3	13.5

λ = wavelength, α = absorption coefficient, T = transmission at the indicated thickness

Transparent Media

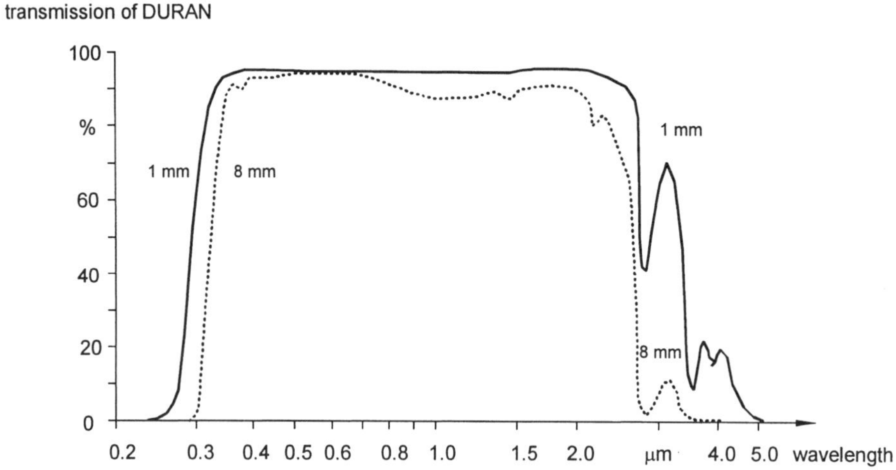

Fig. 14.2. Transmission of DURAN glass including reflection losses, [14.13])

Table 14.2 Transmission of diamond, [14.19]

0.1 mm thickness												
λ / nm	225	226	230	235	300	400	2500	3000	4000	4550	5500	>8000
T / %	45	56	61	65	68	70	71	70	68	62	69	71

λ = wavelength, T = transmission

Table 14.3 Properties of glasses

		DURAN[a)] Al_2O_3	Sapphire	ZERODUR	BK-7 SiO_2	Quartz
n		1.473 [c]	1.755 [d]	1.54	1.506 [d]	1.450
dn_r/dT	10^{-6}/°C		13	15.2 [b]	2.4	10.5 [b]
dn/dT	10^{-6}/°C			13.9 [b]	1.2	
c_p	W s g^{-1} °C^{-1}	0.88	0.75	0.82	0.858	0.75
k	W m^{-1}°C^{-1}	1.16	30	1.64	1.114	1.4
α	10^{-6}/°C	3.25	6.2	0.05	8.3	0.54
T_g	°C	530	2050		559	950
ρ	g/cm^3	2.23	3.99	2.53	2.51	2.2
ν		0.20		0.24		0.16
E	kN/mm^2		350	91	81	72
H_K	kN/mm^2		20			
u	10^3 m/s					6

a) DURAN = (81×SiO_2) + (13×B_2O_3)+ (4×Na_2O K_2O) + (2×Al_2O_3) [14.3]
b) at λ = 644 nm, c) at λ = 588 nm, d) at λ = 1060 nm,
References [14.3, 14.4, 14.5, 14.6, 14.7, 14.12, 14,14]

Table 14.4 Optical qualities of transparent plastic optics, [14.17]

	n_C(656 nm)	n_D(589 nm)	n_F(486 nm)	dn/dT 10^{-5}/°C	ν_d	T_{max} °C
PMMA	1.488	1.491	1.496	−8.5	61.4	92
PS	1.585	1.590	1.604	−12.0	31.1	82
NAS	1.558	1.53 – 1.56	1.575	−14.0	35	93
SAN	1.563	1.56 – 1.57	1.578		37.8	79 – 88
PC	1.581	1.586	1.598	−11 to −14.3	34.5	124
TPX	1.464	1.467	1.473		51.9	
ABS		1.538				
Nylon		1.535				82

PMMA = Plexiglas, polymethylmethacrylate PS = Polystyrene
NAS = Methyl SAN = Acrylonitrile
PC = Polycabonate TPX = Methylpentene ABS = Acrylnitril/Butadien/Stryrol

Table 14.5 Mechanical qualities of transparent plastic optics

	α 10^{-6}/°C	k W $m^{-1}K^{-1}$	c_p W s g^{-1} K^{-1}	ρ g/cm^3	λ cm^2/s	T_m °C	Ref.
PMMA	67.4	0.208					[14.17]
		0.18	1.5	1.17 – 1.2	1.0 – 1.03	90 – 105	[14.18]
PS	60 – 80	0.1 – 0.14					[14.17]
		0.16	1.3	1.04 – 1.05	1.17 – 1.18	88 – 100	[14.18]
NAS		0.188					[14.17]
SAN	65 – 67	0.18	1.3	1.08	1.28	200 – 265	[14.17] [14.18]
PC	66 – 70	0.21	1.2	1.2	1.46	140 – 150	[14.17] [14.18]
ABS	83	0.18	1.5	1.02 – 1.06	1.13 – 1.17	180 – 241	[14.17] [14.18]
Nylon	68						[14.17]

14.2 Materials for Manufacturing

Pump cavities and mountings for the optics cover a wide variety of different materials that should be matched in their qualities. Therefore, the different physical qualities of the common materials are listed in Table 14.4. In Fig. 14.3 the spectral reflection of metals is shown, some of which are used as reflector materials.

Table 14.6 Material parameters

		Invar NILO	Ceramic Al_2O_3	MACOR	Granite	Ref..
c_p	W s g^{-1} K^{-1}		500 J/k	0.9		[14.8,14.10]
k	W m^{-1} K^{-1}	10.0	30			[14.8,14.10] 14.11]
α	10^{-6} K^{-1}	1.5	8.5	9.7	8	[14.8,14.10] 14.11,14.16]
T_g	°C		1700	1000		[14.10,14.11]
ρ	g/cm^3	8.13	3.8	2.52	3.0	[14.8,14.10] 14.11,14.16]
ν		0.25	0.22			[14.9,14.10]
E	kN/mm^2	140	350	65 GPa	30 – 90	[14.9,14.10] 14.11,14.16]

MACOR (workable ceramic) is a registered trademark of Corning.

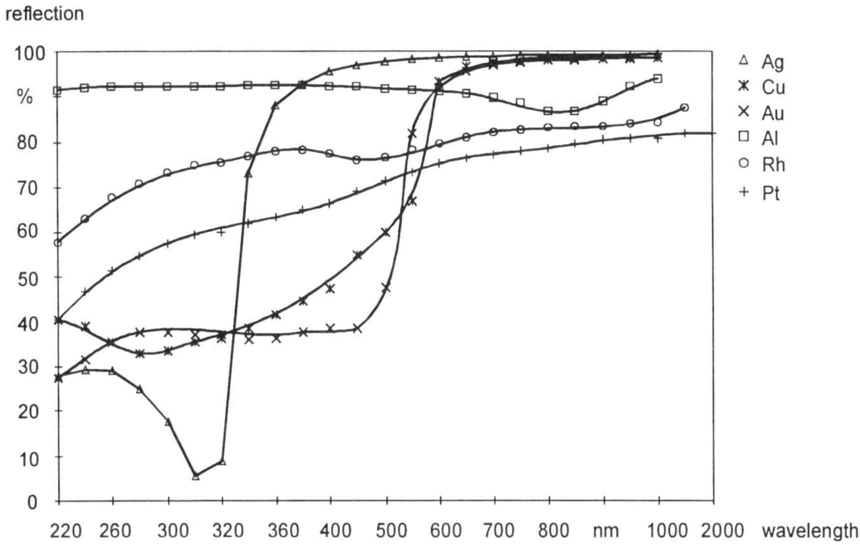

Fig. 14.3. Spectral reflection of various metals, [14.15]

15 Tables and Constants

15.1 Conversion Tables

Table 15.1 Length, [15.1]

Dimension		inch	feet	cm	m
1 inch	=	1	0.0833	2.540	0.0254
1 feet	=	12	1	30.48	0.3048
1 cm	=	0.394		1	0.01
1 m	=	39.37	3.281	100	1

1 mil = 1 milli inch. 1 Å = 0.1 nm

Table 15.2 Volume, [15.1]

Volume		gal	liter	cm^3
1 gal	=	1	3.785	3785
1 liter	=	0.264	1	1000

Table 15.3 Angle, [15.1]

Angle		Grad	min	sec	rad	mrad	μrad
1 Grad	=	1	60	3600	0.017	17.45	
1 min	=		1	60		0.29	
1 sec	=			1			4.85
1 rad	=	57° +	17' +	45 "	1		
1 mrad	=	-	3' +	26 "		1	
1 μrad	=	-	-	0.206"			1

Solid angle = area / radius^2. Dimension = steradian (sr). Full solid angle = 4π

Table 15.4 Force, [15.1]

Force	N	dyn	kp
1 N	1	10 5	0.102
1 dyn	10^{-5}	1	0.102×10^{-5}
1 kp	9.806	9.806×10^5	1

Table 15.5 Pressure, [15.1]

Druck	N/m^2	bar	kp/cm^2	atm	torr	mm(Wc)	psi
1 N/m^2 = Pa	1	10^{-5}	1.02×10^{-5}	9.87×10^{-6}	7.5×10^{-3}	0.102	
	145.1						
1 bar	10^5	1	1.02	0.987	$7.5\cdot10^2$	1.02×10^4	
1 kp/cm^2 = at	9.81×10^4	0.981	1	0.968	736	10^4	
1 atm	1.013×10^5	1.013	1.033	1	760	1.033×10^4	14.7
1 torr	1.333×10^2	1.333×10^{-3}	$1.36\cdot10^{-3}$	$1.316\cdot10^{-3}$	1	13.6	
1 mm(Wc)	9.81	9.81×10^{-5}	10^{-4}	9.68×10^{-3}	736×10^{-4}	1	
1 psi	6.89×10^3						1

(Wc) = water column, psi = pounds per square inch

Table 15.6 Energy, [15.1]

Energy	W s	cal	kW h	kp m	erg	eV
1 J = N m	1	0.2388	2.778×10^{-7}	0.102	10^7	6.242×10^{18}
1 cal	4.187	1	1.163×10^{-6}	0.4269	4.187×10^7	6.614×10^{19}
1 kW h	3.6×10^6	8.598×10^5	1	3.671×10^5	3.6×10^{13}	2.247×10^{25}
1 kp m	9.807	2.342	2.724×10^{-6}	1	9.807×10^7	6.122×10^{19}
1 erg	10^{-7}	2.388×10^{-8}	2.778×10^{-14}	1.020×10^{-8}	1	6.242×10^{11}
1 eV	1.602×10^{-19}	3.826×10^{-20}	4.450×10^{-26}	1.634×10^{-15}	1.602×10^{-12}	1

Table 15.7 Temperature, [15.1]

Temperature		Celsius	Réaumur	Fahrenheit
Celsius	$T_C =$	T_C	$T_R\times10/8$	$(T_F - 32)\times10/18$
Réaumur	$T_R =$	$T_C\times8/10$	T_R	$(T_F - 32)\times8/18$
Fahrenheit	$T_F =$	$T_C\times18/10 + 32$	$T_R\times18/8 + 32$	T_F

15.2 Special Conversions

Energy

1 Btu (British thermal unit) = 10558 W s

1 W s = 5.03404×10^{22} cm^{-1}

Weight

1 g = 0.002205 lbs

1 g/cm^3 = 62.43 lbs/ft^3

Pressure

145.5 MPa = 1 lb/in^2
1 Pa = 1 N/m^2 = 1.02×10^{-6} kg$_f$ /mm^2
1 kg$_f$ /mm^2 = 9.81 MPa

Heat conductivity

1 W/(m K) = 0.860 kcal/(m h °C) = 0.578 Btu/(h ft °F)
1 Btu/(inch h °R) = 20.75 W/(cm K) = 178.6 cal/(s cm K)

Electromagnetic units

1 Gauss = 10^{-4} for Vs/m^2
1 Tesla = 1 Vs/m^2 = 10^{-4} Vs/cm^2
1 Henry = 1 Vs/A
1 Farad = 1 As/V

Optical units

Nonlinear refractive index n_2
1 esu = 2.386×10^6 n_2 m^2/W

$$1\frac{\text{m}^2}{\text{W}} = \frac{42\times10^{-8}}{n_2}\text{esu}$$

Optical density D

$D = \log(1/T)$ $\quad T = 10^{-D}$ $\quad T$ = transmission
$D = \log(\text{e})\alpha d$ $\quad$ for $T = \text{e}^{\alpha d}$
α = absorption coefficient, d = material strength

For a test material thickness of d = 2.4 mm, one has $D_{2.4} = \alpha$ with log(e) = 0.43

1 Brewster = 1.02×10^6 cm^2/N

$\tilde{n}$ = wave number [cm^{-1}], $\tilde{n} = \frac{1}{\lambda} = \frac{\nu}{c} = \frac{E}{hc}$, $c = \lambda/\nu$ = light velocity

$\Delta\nu$ = frequency range, $\Delta\nu = \Delta\tilde{n}$

$\Delta\lambda$ = line width, $\Delta\lambda = \Delta\tilde{n}\lambda^2 = \Delta\nu\lambda^2 / c$, $\Delta\lambda / \lambda = \Delta\tilde{n} / \tilde{n} = \Delta\nu / \nu$

g = amplification, absorption, g[dB] = $-10\times\log(J_2/J_1)$
J_2/J_1 = intensity ratio
1 dB ≡ J_2/J_1 = 0.79, 3 dB ≡ J_2/J_1 = 0.5, 10 dB ≡ J_2/J_1 = 0.1

Reflection at perpendicular incident $R = \left(\frac{n-1}{n+1}\right)^2$

15.3 Physical Constants

$h = 6.626\times10^{-34}$	W s^2	Planck constant
$\hbar = 1.054\times10^{-34}$	W s^2	
$h = 4.135\times10^{-15}$	eV/Hz	$\hbar = h/2\pi$
$hc = 1.986\times10^{-23}$	W s cm	
$e_o = 1.602\times10^{-19}$	A s	electron charge
$m_o = 9.11\times10^{-28}$	g	electron mass
$c = 2.998\times10^{8}$	m/s	light velocity in the vacuum
$\varepsilon_0 = 8.854\times10^{-12}$	A s/(V m)	electric constant
$\mu_o = 1.257\times10^{-6}$	V s/(A m)	magnetic constant
$Z_o = 376.7$	Ω	vacuum impedance $Z_0 = \sqrt{\mu_0/\varepsilon_0}$
$g = 9.81$	m/s^2	gravity
$k_B = 1.381\times10^{-23}$	J/K	Boltzmann constant ($k_B = R/N_A$)
$k_B T = 4.14\times10^{-21}$	W s	for $T = 0$°C
$N_A = 6.022\times10^{23}$	1/mol	Avogadro constant (Loschmidt number)
$R = 8.3146$	W s/(°C mol)	gas constant (≈ 2 cal/(°C mol))
$T = -273.15$	°C	absolute temperature

15.4 Mathematical Constants

$\pi = 3.141\,593$	Pi
$e = 2.718\,282$	Euler number
$\sqrt{2} = 1.41421$	$\sqrt{3} = 1.73205$

References

Overview Literature

Charschan S.S. (ed.): Lasers in Industry (Western Electric Series 1972)
Gerrad A., Burch, J.M.: Introduction to Matrix Methods in Optics (Wiley, London 1975)
Kaminskii A.A.: Laser Crystals (Springer, Berlin 1990)
Kneubühl F.K., Sigrist M.W.: Laser (Teubner, Stuttgart 1988)
Koechner W.: Solid-State Laser Engineering (Springer, Berlin 1976)
Nonhof C.J.: Material Processing with Nd-Lasers (Electrochem. Publ. 1988)
Naumann H., Schröder G.: Bauelemente der Optik. Taschenbuch der technischen Optik 5. Aufl (Hanser Verlag, München 1987)
Penzkofer A.: Solid-State Lasers, Prog. Quantum Electron. **12** (1988) 291–427
Siegmann A.E.: An Introduction to Lasers and Masers (Mc Graw Hill, New York 1971)
Sutter E., Schreiber P., Ott G.: Handbuch Laser-Strahlenschutz (Springer, Berlin 1989)
Weber M.J. (ed.): Handbook of Laser Science and Technology. Chemical Rubber Comp. CRC Press Inc. Vol.1 (1982)
Yariv A.: Introduction to Optical Electronics (California Institute of Technology 1976)
Hodgson N., Weber H.: Optical Resonators (Springer, London 1997)
Saleh B.E.A., Teich M.C.: Fundamentals of Photonics (John Wiley & Sons, New York 1991)

Chapter 1

1.1 Koechner W.: Solid-State Laser Engineering (Springer, Berlin 1976)
1.2 Rigrod W.W.: Homogeneously Broadened CW Laser with Uniform Distributed Loss, IEEE J. Quantum Electron. **14** (1978) 377–381
1.3 Schindler G.M.: Optimum Output Efficiency of Homogeneously Broadened Lasers with Constant Loss, IEEE J. Quantum Electron. **16** (1980) 546–549
1.4 Findlay D., Clay R.A.: The Measurement of Internal Losses in 4-level Lasers, Phys. Lett. **20** (1966) 277–278
1.5 Mann K., Phillipps G.: Absorption- und Streulichtmessung bei 1,06 μm an ungepumpten Nd:YAG- und Cr, Nd:GGG-Laserstäben (Festkörperlaser-Institut Berlin, Interner Bericht, 1989)

1.6 Kuratev I.I., et al.: Mechanism of Short-Lived Absorption in YAG:Nd Crystals, Sov. J. Quantum Electron. **15** (1985) 861–862

1.7 Weber H.: Optimimierung von Nd-Lasern (Universität Kaiserslautern, Inst. für Physik, Interner Bericht Feb. 1987)

1.8 Hodgson N. Festkörperlaser-Institut Berlin. Personal communication 1988

1.9 Wildmann D., Junghans J., Jundt H.F.: Laserbeschriftung, Stand der Technik, Proc. Laser 87 Optoelectronic Munich (1987)

1.10 Baldwin G.D.: Output Power Calculations for a Continously Pumped Q-Switched YAG^{+3}:Nd Laser, IEEE J. Quantum Electron. **7** (1971) 220–224

1.11 Nonhof C.J., Keränen R.: Pulse to Pulse Instabilities in a Multimode Q-Switched ND:YAG-Laser, Proc. Laser 87 Optoelectronic Munich (1987) 332–338

1.12 Yariv A.: Energy and Power Considerations in Injection and Optically Pumped Lasers, Proc. IEEE Dec. (1963) 1723–1731

1.13 Hodgson N. Weber H.: Extraktionswirkungsgrade von Laseroszillatoren (Inst. für Physik, Universität Kaiserslautern April 1987)

1.14 Eicher J., Mann K., Weber H.: Untersuchung der Eigenschaften von Festkörperslablasersystemen, Zwischenbericht zum BMFT-Projekt 13 N 5306/1

1.15 Caird J.A., et al.: Measurements of Losses and Lasing Efficiency in GSGG:Cr, Nd and YAG:Nd Laser Rods, Appl. Opt. **25** (1986) 4294–4305

1.16 Sina B.K.: A New Method for the Estimation of Pumping Coefficient for a Ruby Laser, IEEE J. Quantum Electron. **15** (1979) 1083–1085

1.17 Laporta P., Magni V., Svelto O.: Comparative Study of the Optical Pumping Efficiency in Solid-State Lasers, IEEE J. Quantum Electron. **21** (1985) 1211–1218

1.18 Mindak M., Szydlak J.: Examples of Operating Characteristics and Power Balance in Pump Cavity of CW Nd:YAG Laser, Optica Applicata **13** (1983) 407–419

1.19 Kliewer M.L., Powell R.C.: Excited State Absorption of Pump Radiation as a Loss Mechanism in Solid-State Lasers, IEEE J. Quantum Electron. **25** (1989) 1850–1854

1.20 Jeys T.H.: Suppression of Laser Spiking by Intracavity Second Harmonic Generation, Appl. Opt. **30** (1991) 1011–1013

Chapter 2

2.1 Kogelnik H., Li T.: Laser Beams and Resonators, Proc. IEEE **54** (1966) 1312–1329

2.2 Kogelnik H.: Imaging of Optical Modes – Resonators with Internal Lenses, The Bell System Techn. J. March (1965) 455–494

2.3 Dembowski J., et al.: Resonators for High Power Lasers, Proc. 4th Symp. on Gas Flow and Chemical Lasers (1982) Stresa 541–548

2.4 Ripper G., Herziger G.: Werkstoffbearbeitung mit Laserstrahlung Teil 5: Fokussierung von Laserstrahlung am Beispiel des CO_2-Lasers, Feinwerktechnik Messtechnik **92** (1984) 297–302

2.5 Phillips R.L., Andrews, L.C.: Spot Size and Divergence for Laguerre Gaussian Beams of Any Order, Appl. Opt. **22** (1983) 643–644

2.6 Carter W.H.: Spot Size and Divergence for Hermite Gaussian Beams of Any Order, Appl. Opt. **19** (1980) 1027–1029

2.7 Iffländer R., Weber H.: Focusing of Multimode Laser Beams with Variable Beam Parameters, Optica Acta **33** (1986) 1083–1090

2.8 Halbach K.: Matrix Representation of Gaussian Optics, Am. J. Phys. **32** (1964) 90–108
2.9 Gerrad A., Burch J.M.: Introduction to Matrix Methods in Optics (Wiley, London 1975)
2.10 Weber H.: Optical Resonators – Basics and Applications, Vorlesungsmanuskript 3. Aufl. Univ. Kaiserslautern (1983)
2.11 Weber H.: Laserphysik I. Vorlesungsmanuskript 3. Aufl. WS 1978/79. Univ. Kaiserslautern
2.12 Weber H.: Ausbreitung und Fokussierung Gaußscher Strahlen, Interner Bericht. Univ. Kaiserslautern (1978)
2.13 Hodgson N., Weber H.: Optical Resonators (Springer, London 1997)

Chapter 3

3.1 Lörtscher J.P., Steffen J., Herziger G.: Dynamic Stable Resonators: A design procedure, Opt. Quantum Electron. **7** (1975) 505–514
3.2 Baues P.: Huygens' Principle in Inhomogeneous Isotropic Media and a General Integral Equation Applicable to Optical Resonators, Opto-Electron. 1 (1969) 37–44
3.3 Magni V.: Multielement Stable Resonators Containing a Variable Lens, J. Opt. Soc. Am. A **4** (1987) 1962-1969
3.4 Grau G.K.: Laserspiegel zur Auskopplung eines speziellen beugungsbegrenzten Parallelstrahls, AEÜ **20** (1966) 704–705
3.5 Iffländer R., Kortz H.P., Weber H.: Beam Divergence and Refractive Power of Directly Coated Solid-State Lasers, Opt. Commun. **29** (1979) 223–226
3.6 Kortz H.P., Iffländer R., Weber H.: Stability and Beam Divergence of Multimode Lasers with Internal Variable Lenses, Appl. Opt. **20** (1981) 4124–4134
3.7 Le Floch A., Lenormand J.M., Le Naour R., Taché J.P.: A Critical Geometry for Lasers with Internal Lenslike Effects, J. Phys. Lett. 43 (1982) L493–L498
3.8 Metcalf D. De Giovanni P., Zachorowski J., Leduc M.: Laser Resonators Containing Self-Focusing Elements, Appl. Opt. **26** (1987) 4508–4517
3.9 Driedger K.P., Iffländer R.M., Weber H.: Multirod Resonators for High-Power Solid-State Lasers with Improved Beam Quality, IEEE J. Quantum Electron. **24** (1988) 665–674
3.10 Weber H., Iffländer R., Seiler P.: High Power Nd-Lasers for Industrial Application, Proc. SPIE **650** (1986) 92–100
3.11 Driedger K.P., Lu B., Weber H.: Multimode Resonators, Insensitive Against Thermal Lensing, Opt. Acta **32** (1985) 847–854
3.12 Hauck R., Kortz H.P., Weber H.: Misalignment Sensitivity of Optical Resonators, Appl. Opt. **19** (1980) 598–601
3.13 Siegman A.E.: A Canonical Formulation for Analyzing Multielement Unstable Resonators, IEEE J. Quantum Electron. **12** (1976) 35–39
3.14 Anan'ev Y.A.: Unstable Resonators and their Applications (Review), Sov. J. Quantum. Electron. **1** (1972) 565–586
3.15 Hanna D.C., Laylock L.C.: An Unstable Resonator Nd-YAG Laser, Opt. Quantum. Electron. **11** (1979) 153–160
3.16 Lissak B., Ruschin S.: Transverse Pattern Modifications in a Stable Apertured Laser Resonator, Appl. Opt. **29** (1990) 767–771

3.17 Pipes l., Harvill L.R., Applied Mathematics for Engineers and Physicists (Mc Graw Hill, New York 1970)

Chapter 4

4.1 Krupke W.F.: Specific Heat Loading in Nd-Glas Lasers, LLNL-Report UCID-20531 DE36 002151 (1985)

4.2 Driedger K.P., et al.: Average Refractive Powers of an Alexandrit Laser Rod, Opt. Commun. **57** (1986) 403–406

4.3 Mangir M.S., Rockwell D.A.: Measurements of Heating and Energy Storage in Flashlamp-Pumped Nd:YAG and Nd-Doped Phosphate Laser Glasses, IEEE J. Quantum Electron. **22** (1986) 574–580

4.4 Blink J.A., et al.: Thermal Power Distribution in a Zig-Zag Slab Laser, LLNL. High-average-Power lasers 7. Laser program annual report (1986) 7-88 to 7-95

4.5 Mann K., Weber H.: Surface Heat Transfer Coefficient, Heat Efficiency, and Temperature of Pulsed Solid-State Lasers, J. Appl. Phys. **64** (1988) 1015–1021

4.6 Sumida D., Rockwell D.: Dependence of Cr Nd:GSGG Pumping Efficiency on Cr Concentration, HUGHES Research Lab., Malibu, CA (1986)

4.7 Sun Y-C., Weber H.: Bestimmung der Spannungsdoppelbrechung optisch gepumpter Laserstäbe und ihr Einfluß auf die Brechkraft, Interner Bericht Univ. Kaiserslautern (1981)

4.8 Koechner W.: Solid-State Laser Engineering (Springer, Berlin, Heidelberg 1976)

4.9 Tautz H.: Wärmeleitung und Temperaturausgleich (Chemie-Verlag, Weinheim 1971)

4.10 Hagen W.F.: Thermal Fracture of Laser Glasses and Crystals, LLL Internal Report LRD 87-170 / 6061T (1987)

4.11 Hodgson N., Weber H.: Measurement of Extraction Efficiency and Excitation Efficiency of Lasers, J. Mod. Opt. **35** (1988) 807–813

4.12 Brown D.C., Lee K.L.: Methods for Scaling High Average Power Laser Performance, Proc. SPIE **622** (1986) 30–41

4.13 Marion J.E.: Strengthening of Solid-State Laser Materials, Conf. Dig. CLEO (1985) THR 1.
Marion J.E.: Strengthening of Solid-State Laser Materials, Appl. Phys. Lett. **47** (1985) 694–696

4.14 Walling J.C.: Tunable Alexandrite Lasers: Development and Performance, IEEE J. Quantum Electron. **21** (1985) 1568

4.15 Foster J.D., Osterink L.M.: Thermal Effects in a ND:YAG Laser, J. Appl. Phys. **41** (1970) 3656–3663

4.16 Struve B., Fuhrberg P., Luhs W., Litfin G.: Thermal Lensing and Laser Operation of Flashlamp-Pumped Cr:GSAG, Opt, Commun. **65** (1988) 291–296

4.17 Horowitz, et al.: Thermal Lensing Analysis of Alexandrite Laser Rods by Moire Deflectometry, Appl. Opt. **23** (1984) 2229–2233

4.18 Jankiewicz Z. et al.: Analysis of the Thermal Focusing Effect in a CW Nd:YAG Laser, Optica Applicata **XV** (1985) 125–134

4.19 Murray J.E.: Pulsed Gain and Thermal Lensing of Nd:LiF4, IEEE J. Quantum Electron. **19** (1983) 488–490

4.20 Kim H.K., et al.: Thermal Problems of the Cr:Nd:GSGG Laser under Solar-Simulator Pumping, Conf Dig. CLEO (1988) TUM35

4.21 Reed E.: A Flashlamp-Pumped, Q-switched Cr:Nd:GSGG Laser, IEEE J. Quantum Electron. **21** (1985) 1625–1629
4.22 Kelly J.H., et al.: High Repetition Rate Cr:Nd:GSGG Active Mirror Amplifier, Opt. Lett. **12** (1987) 996–998
4.23 Martin W.S.: Multiple Internal Reflection Face-Pumped Laser, US Patent 3,633,126 (1972)
4.24 Hughes J.L.: High Power Continuous Wave Multi-Slab Laser Oscillator, Int. Patent WO 87/05160 (1987)
4.25 Hoffmann H.J.: Verbundmaterialien für die Lasertechnik und Optik, Offenlegungsschrift DE 36 17 362 A1 (1987)

Chapter 5

5.1 Takada Y., Saito H., Fujioka T.: New Type of Solid-State Laser for Several Kilowatts, Proc. SPIE **801** (1987) 62–66
5.2 Hughes J.L.: High Power Continous Wave Multi-Slab Laser Oscillator, Int. Patent WO 87/05160 (1987)
5.3 Huchital D.A., Steinberg G.N.: Pumping of Nd:YAG with Electrodeless Arc Lamps, IEEE J. Quantum Electron. **12** (1976) 1–9
5.4 Edler H.G., et al.: Festkörperlaser mittlerer Leistung zur Materialbearbeitung, Forschungsbericht zum BMFT-Projekt 13 N 52 08 6 (1985)
5.5 Hachfeld K.D.: The Engineering Art of Solid-State Laser Pump Cavity Design, Proc. SPIE **609** (1986) 55–77
5.6 Bowness C.: On the Efficiency of Single and Multiple Elliptical Laser Cavities, Appl. Opt. **4** (1965) 103–107
5.7 Kamiryo K., et al.: Optimum Design of Elliptical Pumping Chambers for Solid-Lasers, Jpn. J. Appl. Phys. **5** (1966) 1217–1226
5.8 Acharekar M.A., et al.: Tracing Rays Inside a YAG Laser Cavity, Laser Focus **11** (1981) 67-69
5.9 Eckardt R.C., et al.: Compact, Low-Cost Cryogenic Laser Head, Rev. Sci. Instrum. **55** (1984) 1945–1947
5.10 Maeda K., et al.: Concentration Dependance of Fluorescence Lifetime of Nd^{3+}-Doped $Gd_3Ga_5O_{12}$ Lasers, Jpn. J. Appl. Phys. **23** (1984) L759–L760
5.11 Neuroth N.: Eigenschaften und Anwendungen der Glaslaser, Laser und angewandte Strahlentechnik **4** (1970) 17–22
5.12 Yoshikawa S., Iwamoto K., Washio K.: Efficient Arc Lamps for Optical Pumping of Neodymium Lasers, Appl. Opt. **10** (1971) 1620–1623
5.13 Danielmeyer H.G.: Progress in Nd:YAG lasers, Lasers: A Series of Advances **4** (1976), Levine De muria (eds.) Marcel Dekker
5.14 Krupke F.W., Chase L.L.: Ground State Depleted (GSD) Solid-State Lasers: Principles, Characteristics and Scaling, Proc. SPIE **1040** (1989) 68–83
5.15 Martin W.S., Chernoch J.P., General Electric: Multiple Internal Reflection Face Pumped Laser, US Patent 3,633,126 (1972)
5.16 von Arb H.P., et al.: Laser avec systéme de refroidissement perfectionnè, Demande de brevet Europeen (1988) 0 296 512 A1
5.17 Bender D.A.: Thermal Design. High-average-power lasers. Laser Program annual report LLNL (1985) 9–41 (UCRL-50021-85)

5.18 Yoshida K., Yoshida H., Kato Y.: Characterization of High Average Power Nd:GGG Slab Lasers, IEEE J. Quantum Electron. **24** (1988) 1188–1192

5.19 Wada N., Abe M., Honda T.: $Nd^{3+}Gd_3Ga_5O_{12}$ Slab Laser, Conf. Dig. CLEO (1984) 78

5.20 Zapata L.E., et al.: Performance of a 500 Watt Nd:GGG Zigzag Slab Oscillator, Proc. SPIE **1223** (1990) 259–273

5.21 Arb von H.P., et al.: Properties of Pulsed Solid-State Slab lasers, Personal communication (1987)

5.22 Arb von H.P., et al.: High Average Power Slab Geometry Solid-State Lasers, Proc. SPIE **1021** (1988) 24–30

5.23 Kasai T., et al.: High Average Power Nd:YAG Slab Laser, Proc. SPIE **1040** (1989) 32–36

5.24 Dong S., et al.: Extraction Efficiency and Beam Quality of Prism-Mirror-Resonators for Slab-Lasers, Internal Report, Solid-State Laser Institute Berlin (1990)

5.25 Chernoch J.P.: Characteristics of a 1 kW Nd:YAG Faced-Pumped Laser, ICALEO 1990

5.26 Mann K., Weber H.: Surface Heat Transfer Coefficient, Heat Efficiency and Temperature of Pulsed Solid-State Lasers, J. Appl. Phys. **64** (1988) 1015–1021

5.27 Gibbson J.: Laser Water-Cooling Loops Deserve Attention, Laser Focus World **4** (1989) 123–129

5.28 Tuckerman D.B., Pease R.F.W.: High-Performance Heat Sinking for VLSI, IEEE Electron Devices Lett. **EDL-2** (1981) 126–129

Chapter 6

Overview

EG a. G Electro-Optics: Flashlamp Applications Manual (1982)

6.1 Littlechield J., Mossler D.: Knowledge of Arc-Lamp Aging and Lifetime Effects Can Help to Avoid Unpleasant Surprises, Laser Focus/Electro-Optics Nov (1988) 67 - 76

6.2 Marshak I.S.: Pulsed Light Sources (Consultant's Bureau. Plenium Publishing, New York 1984) 23

6.3 Schuda F.: Flashlamp Simmer Voltage, ILC engineering note 171 (1986)

6.4 Smith B.: An Overview of Flashlamps and CW Arc Lamps, Technical Bulletin 3. ILC Technology (1986)

6.5 VERRE ET QUARTZ: Standard High Pressure Linear Lamps, Company Brochure (1981)

6.6 Krypton Arc lamps for CW Neodymium Laser Pumping, ILC Bulletin 1215 (1985)

6.7 Walsh P.J.: Electrical Characterization of CW Xenon Arcs at Moderate Currents, J. Appl. Phys. **61** (1987) 4484–4491

6.8 Rashiah I.J., Tan B.C., Lee H.W.: Voltage-Current Relationship for Flashlamps: An Empirical Approach, Appl. Opt. **30** (1991) 485–488

6.9 Markiewicz J.P., Emmett J.L.: Design of flashlamp Driving Circuits, IEEE J. Quantum Electron. **11** (1966) 707

6.10 Hohlfeld R.G.: Self-Inductance Effects in Linear Flashtubes: An Extension to the Markiewicz and Emmett Theory, Appl. Opt. **22** (1983) 1986

6.11 Brown D.C.: High-Peak-Power Nd:Glass Laser Systems (Springer, Berlin 1981) 115
6.12 Richards J., et al.: Operation of Krypton-Filled Flashlamps at High Repetition Rates, Appl. Opt. **22** (1983) 1325
6.13 Erlandson A.C.: Flashlamps for High-Average-Power Lasers, Laser program annual report LLNL (1985) 9–77 (UCRL-50021-85)
6.14 Heraeus: Laseranregungslampen für Dauer- und Impulsbetreib, Company Brochure (1983)
6.15 VBI Technologies: Short and Long Arc Lamps, Technical Information
6.16 Witting H.L.: Acoustic Resonances in Cylindrical High-Pressure Arc Discharges, J. Appl. Phys. **49** (1978) 2680–2683
6.17 Lama W., Hammond T.: Arc-Acoustic Interaction in Rare Gas Flashlamps, Appl. Opt. **20** (1981) 765–769
6.18 Yoshikawa S., Iwamoto K., Washio K.: Efficient Arc Lamps for Optical Pumping of Neodymium Lasers, Appl. Opt. **10** (1971) 1620–1623
6.19 Greve P., et al.: Emmisionscharakteristik und Reabsorption von Krypton-Bogenlampen in Festkörperlasern, Feinwerktechnik und Meßtechnik **96** (1988) 245–247
6.20 Greve P., et al.: Emission Characteristic of High Power Kr/Xe-Flashlamps, Proc. SPIE **1021** (1988) 84
6.21 Kushner M.J.: Arc Expansion in Xenon Flashlamps, J. Appl. Phys. **57** (1985) 2486
6.22 Krönert U.: Final Report 13 EU 707 (1992) EUREKA-Project 226
6.23 Hollinger F.: Solid-State Laser Institute Berlin (1992) Personal Communication

Chapter 7

7.1 Berger J., et al.: Fiber-Bundle Coupled, Diode End-Pumped Nd:YAG Laser, Opt. Lett. **13** (1988) 306–308
7.2 Burnham R., et al.: High-Power-Diode-Pumped Solid-State Industrial Laser, Adv. Solid State Lasers OSA Conf. Dig. (1993) AMA3-1
7.3 Baer T.M.: High Efficiency Mode-Matched Solid-State Laser with Transverse Pumping and Cascaded Amplifier Stages, US Patent 4,894,839 (1990)
7.4 Comaskey B., et al.: High Average Power Diode Pumped Slab Laser, UCRL-JC-108447 Preprint (IEEE J. Quantum Electron.) (1991)
Comaskey B., et al.: Diode Pumped 275 Watt Average Power Nd:YAG Slab Laser, UCRL-JC-109185 Preprint (SPIE OE LADSE 92) (1992)
7.5 Baberg F., Luft J.: GaAlAs-Halbleiterlaser für hohe Leistungen, Siemens Components **26** (1988) 154–159
7.6 Greve P.: Optische Probleme diodengepumpter Festkörperlaser, Laser Magazin **6** (1988) 35–39
7.7 Sakamoto M., et al.: 120 W Output Power From Monolithic AlGaAs (800 nm) Laser Diode Array Mounted on Diamond Heatsink, Electron. Lett. **28** (1992) 197–199
7.8 Forrest G.T.: Diode-Laser Arrays Achieve Increased Power and Brightness: Laser Focus/ Electro Optics **8** (1988) 59–66
7.9 Siemens: Preliminary Data Sheet to a 5×12 Stripe array SFH 48 R1 (1989)
7.10 Harnagel G., et al.: Five Watt Continous-Wave AlGaAs Laser Diodes, Electron. Lett. **22** (1986) 605–606

7.11 Sakamoto M., et al.: 20 W Monolithic AlGaAs (810 nm) Laser Diode Arrays, Electron. Lett. **28** (1992) 178–180

7.12 Sakamoto M., et al.: 76 W CW Monolithic Laser Diode Arrays, Appl. Phys. Lett. **54** (1989) 2299–2300

7.13 Cross P.S., et al.: High Power Diode Laser Arrays: Lifetests and Applications, Laser Appl. **4** (1985) 89–91

7.14 Harnagel G.L., Cross P.S.: Laser Diode Pump Systems For Nd:YAG, Technical Information Spectra Diode Labs

7.15 High-Repetition-Rate, Diode-Pumped, Solid-State Slab Lasers, Energy Technology (June 1992) E&TR

7.16 Saleh B.E.A., Teich M.C.: Fundamentals of Photonics (John Wiley & Sons, New York 1991) 548

7.17 Laser Focus World April (1993) 200

7.18 Krupke W.F., Chase L.L.: Ground-state Depleted Solid-State Lasers: Principles, Characteristics and Scaling, Opt. Quantum Electron. **22** (1990) S1–S22

7.19 Krupke W.F.: New Challenges in Optical Materials Development (1993) University of California LLNL

7.20 Frey H., Kienel G.: Dünnschichttechnologie (VDI-Verlag, Düsseldorf 1987)

7.21 Petersen K.E.: Silicon as a Mechanical Material, Proc. IEEE **70** (1982) 420..

7.22 Sumitomo Data sheet EN-25R2 (1993.7) IV

7.23 Morris P.J., et al.: Operation of Resonantly Pumped Tm:Ho:YAG in Active Mirror Mode, Opt. Commun. **104** (1993) 97–101
Saruwatari M., Kimura T.: LED Pumped Lithium Neodymium Tetraphosphate Lasers, IEEE J. Quantum Electron. **12** (1976) 584–591

7.24 Brand T.: Festkörperlaseranordnung, Offenlegungsschrift DE 44 11 599 A1 (1994)
Brand T.: Anregungseffizienz und Heiztransferwirkungsgrad diodengepumpter Hochleistungs-Festkörperlaser, Dissertation TU-Berlin (1994)

7.25 Golla D., et al.: 15 W Single-Frequency Operation of a CW, Diode Laser-Pumped Nd:YAG Ring Laser, Opt. Commun. **98** (1993) 86–90
Golla D., et al.: 62-W CW TEM_{00} Nd:YAG Laser Side-Pumped by Fiber-Coupled Diode Lasers, Opt. Lett. **21** (1996) 210–212

7.26 Giesen A., et al.: New Concepts for High Power Pumped Solid-State Lasers, Personal communication

7.27 Zellmer H., et al.: High Power CW Neodymium-Doped Fiber Laser Operating at 9.2 W with High Beam Quality, Opt. Lett. **20** (1995) 578–580

7.28 Hoffmann H.J., et al.: Laser mit einem laseraktiven Festkörpermedium und mehreren Pumpquellen, Patent Anmeldung DE 37 26 279 C2 (1987)

7.29 Coyle D.B.: Design of a High-Gain Laser Diode-Array Pumped Nd:YAG Alternative Precessive Slab Amplifier, IEE J. Quantum Electron **27** (1991) 2327–2331

7.30 Clarkson W.A., Hanna D.C.: Two-Mirror Beam-Shaping Technique for High-Power Diode Bars, Opt. Lett **21** (1996) 375–377

7.31 Du et al.: Anordnung zur Führung und Formung von Strahlen eines geradlinigen Laserdiodenarrays, Patent DE 44 38 368 C2 (1996)

7.32 Talghader J., Smith J.S.: Optical Measurements Using Surface Emitting Cavitites, Conf. Dig. CLEO (1995) 188

Chapter 8

Overview

Mass B.: Power Conditioning for Flashlamps and Arc Lamps, Proc. SPIE **609** (1986) 42–54

8.1 Brown D.C., Nee T.N.: Design of Single Mesh Flashlamp Driving Circuits with Resistive Losses, IEEE Trans. Electron. Devices **ED-24** (1977)
8.2 Burbeck R.N., et al.: Laser Flash Tube Power Supply, UK Patent Appl. GB 2023 330 A (1979)

Chapter 9

9.1 Roussel, Phillipe, Wassenaar: Tête optique d'une installation pour l'observation et le traitement par rayonnement laser de l'oeil, Fascicule du Brevet 645 801 (1982)
9.2 Melles Griot: Optics Guide 2 (1981), Optics Guide 3 (1985) 36
Jenkins F.A., White, H.E.: Fundamentals of Optics (McGraw Hill Book Company, Tokyo 1975)
9.3 Goethals W.: Optical Fiber Multiplexer for Industrial Nd:YAG Lasers, SPIE Conference Hamburg (1988)
9.4 Dammann: Synthetic Digital-Phase Gratings – Design, Features, Applications, Philips Research Laboratories Hamburg
9.5 Killat,U., Rabe,G., Rave,W.: Binary Phase Gratings for Star Couplers with High Splitting Ratio, Fiber Integrated Opt. **4** (1982) 150–167
9.6 Nath G.: Hand-held Laser Welding of Metals Using Fiber Optics, Opt. Laser Technol. **10** (1974) 233–235
9.7 Notenboom G, Nonhof C, Schildbach K.: Beam Delivery Technology in Nd:YAG Laser Processing, Proc. LAMP, Osaka (May, 1987) 107–111
9.8 Goethals W.: Optischer Glasfaser-Multiplexer für industrielle Nd:YAG Laser, **10** Laser Magazin 4 (1988) 10–12
9.9 Jahn R.: Grundlagen der Faseroptik, Feinwerktechnik **12** (1970) 524–530
9.10 Gloge, D.: Bending Loss in Multimode Fibers with Graded and Ungraded Core Index, Appl. Opt. **11** (1972) 2506–2513
9.11 Boechat A.A.P, Su D., Hall D.R, Jones J.D.C.: Bend Loss in Large Core Multimode Optical Fiber Beam Delivery Systems, Proc. SPIE **1277** (1990) 176–187
9.12 Jahn R.: Selbstfokussierende Lichtleitfasern, Feinwerktechnik und Micronic **2** (1973) 56–64
9.13 Becker J., Bernhardt A.: Bestimmung der Zerstörschwellen optischer Komponenten aus der Zerstörwahrscheinlichkeit, Laser Optoelektron. **25** (1993) 49–54
Becker J., Bernhardt A.: Damage Threshold Determination According to ISO-DIS 11254 - First Results, Proc. SPIE **1781** (1992) 359–368
9.14 Penzkofer A.: Solid-State Lasers, Prog. Quantum Electron. **12** (1988)
9.15 Kuabara M, et al.: Nd:YAG Laser Marker Utilizing a Liquid-Crystal-Device Mask, Proc. SPIE **1223** (1990) 47–53
9.16 Huber R., Iffländer R.: Laser Setup, Offenlegungsschrift DE 41 02505 A1 (1991)

Chapter 10

10.1 Herziger G.: Werkstoffbearbeitung mit dem Laserstrahl, Feinwerktechnik Meßtechnik **91** (1983) 156

10.2 Steffen J.: Prozessoptimierung bei materialabtragenden Bearbeitungsproblemen mit Laserstrahlung, Feinwerktechnik Meßtechnik **7** (1979) 309–356

10.3 Steffen J.: Schweißen mit dem Laserstrahl, Feinwerktechnik Meßtechnik **88** (1980) 7–20

10.4 Doduco: Contacts and Contacts Materials, Data Book 2nd Ed. Doduco, Pforzheim (1977)

10.5 Handbook of Chemistry and Physics. 5th Edition (CRC Press Inc., Boca Raton Florida 1985)

10.6 Hulst A., Zuidema,R.: Joining of Copper and Aluminium Alloys by Laser, Digest of European Industrial Laser Forum (1993) BE-3326

10.7 Kasenbacher T.: Laserschweißbarkeit von Dentallegierungen, Diss. Klinik und Polyklinik für Zahn-, Mund- und Kieferkrankheiten der LMU- München (1992)

10.8 Landolt-Börnstein: Zahlenwerte und Funktionen. Eigenschaften der Materie in ihren Aggegatzuständen, 8. Teil Optische Konstante (Springer, Berlin 1962)

10.9 Nonhof C.J.: Material processing with Nd-lasers, Electrochem. Publ. (1988) 34, 128

10.10 Seka W., et al.: Photodiodearrays –A Convenient Tool for Laser Diagnostics, Rev. Sci. Instrum. **45** (1974) 1175–1176

10.11 Cohen M.I.: Laser Handbook II. (North-Holland Publishing, Amsterdam)

10.12 Treusch H.G.: Geometrie und Reproduzierbarkeit einer plasmaunterstützten Materialabtragung durch Laserstrahlung, Dissertation TH Darmstadt (1985)

10.13 Schäfer P.: Metalle gepulst flexibel und präzis schneiden, Laser Praxis Juni (1988) L55–L60

10.14 ILT FhG Fraunhofer-Institut für Lasertechnik: Final Report EU 226: High Power Solid-State Lasers, 13 EU 00659 Data and Basics for Material Processing with pulsed Solid-State Lasers (1990)

10.15 Bethge U., et al.: Schweisstechnischer Vergleich zwischen Hochleistungsfestkörper- und CO_2-laser, Laser München (1993) Vortrag

10.16 Dausinger F.: Strahlwerkzeug Laser: Energieeinkopplung und Prozeßeffektivität, (Teubner, Stuttgart 1995)

Chapter 11

11.1 Filer E.D., et al.: Theoretical Branching Ratios for the $^5I_7 \rightarrow {}^5I^7$ Levels of Ho^{3+} in the garnets $A_3B_2C_3O_{12}$ (A=Y, La, Lu, Gd; B = Al, Lu, SC, Ga; C = Al, Ga), OSA Proc. Adv. Solid State Lasers **6** (1990) 354–362

11.2 Hoffman, H.J., Hayden J.S.: Glasses as Active Materials for High Average Power Solid-State Lasers, Proc. SPIE **1021** (1988) 42–50

11.3 Milam D., Weber M.J., Glass, A.J.: Nonlinear Refractive Index of Fluoride Crystals, Appl. Phys. Lett. **31** (1977) 822–825

11.4 Penzkofer A.: Solid-State Lasers, Prog. Quantum Electron. **12** (1988) 366–369

11.5 Brown D.C.: High-Peak-Power Nd:Glass Laser Systems (Springer, Berlin 1981)

11.6 Boling N.L., et al.: Empirical Relationships for Predicting Nonlinear Refractive Index Changes in Optical Solids, IEEE J. Quantum Electron. **14** (1978) 601–608

11.7 Lai S.T.: Highly Efficient Emerald Laser, J. Opt. Soc. AM. B **4** (1987) 1286–1290
Lai S.T.: Review of Spectroscopic and Laser Properties of Emerald, Proc. SPIE **622** (1986) 146–150

11.8 Petermann K., Huber G.: Broad Band Fluorescence of Transition Metal Doped Garnets and Tungstates, J. Lumin. **31&23** (1984) 71–77

11.9 Chase L.L., et al.: Emission Cross Sections of the Mid-Infrared Laser Transitions of Ho^{3+}, Er^{3+} and Tm^{3+} in Fluoride and Oxid Crystals, Adv. Solid State Lasers Tech. Dig. (1991) 118–120

11.10 Chase L.L., et al.: Emission Cross Sections and Energy Extraction for the Mid-infrared Transitions of Er, Tm and Ho in Oxide and Fluoride Crystals, OSA Proc. Adv. Solid State Lasers **10** (1991) 161–165

11.11 Knights M.G., et al.: High-Efficiency Deep-Red Laser Pumped by Doubled Nd:YAG, IEEE J. Quantum Electron. **18** (1982) 163–166

11.12 Kaminskii A.A.: Laser Crystals (Springer, Berlin 1990) 327

11.13 Kaminskii A.A., et al.: $LiYF_4$:Pr^{3+} Laser at 639,5 nm with 30 J Flashlamp Pumping and 87 mJ Output Energy, Phys. Stat. Sol. **138** (1993) K45–K48
Eichler H.J, Liu B.: Gepulster $LiYF_4$: Pr^{3+}-Laser, TU-Berlin (1993)

11.14 Armagan G., et al.: Spectroscopic Characterization of Dynamical Process for Tm, HO:YAG Lasers, OSA Proc. Adv. Solid State Lasers **6** (1990) 144–149

11.15 Lacavora P., et al.: Room-Temperature Diode-Pumped Yb:YAG Laser, Opt. Lett. **16** (1991) 1089–1091

11.16 Lacavora P., et al.: Room-Temperature InGaAS Diode-Pumped Yb:YAG laser, Adv. Solid State Lasers Conf. Dig. (1991) 220–222

11.17 Payne S.A., et al.: Laser Properties of Yb-Doped Fluorapatite, Adv. Solid State Lasers Conf. Dig. (1992) Pd8-1– Pd8-4

11.18 Kaminskii A.A.: Achievements of Modern Crystal-Laser Physics, Ann Phys. **16** (1991) 639–706
Kaminskii A.A., et al.: New Data on Stimulated Emission of Crystals Containing Er^{3+} and Ho^{3+} Ions, Sov. Tech. Phys. Lett. **2** (1976) 308–310

11.19 Koechner W.: Solid-State Laser Engineering (Springer, Berlin1992)

11.20 Walling J.C., et al.: Tunable Alexandrite Lasers, IEEE. J. Quantum Electron. **16** (1980) 1302–1315

11.21 Stange H., et al.: Continuous Wave 1.6 μm Laser Action in Er Doped Garnets at Room Temperature, Appl. Phys. B **49** (1989) 269–273

11.22 Thogersen J., Bjerre N.: Multiphoton Absorption and Cooperative Upconversion Excitation in Er^{3+}-Doped Fibers, Opt. Lett. **18** (1993) 197–199

11.23 Arahira S., et al.: Successive Excited-State Absorption Through a Multistep Process in Highly Er^{3+}-Doped Fiber Pumped by a 1.48-μm Laser Diode, Opt. Lett **17** (1992) 1679–1681

11.24 Kurtz R., Fathe L., Birnbaum M.: New Laser Lines of Erbium in Yttrium Aluminium Garnet, OSA Proc. Adv. Solid State Lasers **6** (1990) 247–250

11.25 Handbook of Laser Science and Technology I (CRC Press, Boca Raton Florida 1982) 26–35

11.26 Hollinger FLI-Berlin (1993) Personal communication

11.27 Hodgson N., Golding D.J.: High Power 1.444 μm Nd:YAG Laser and its Medical Applications, Laser Optoelectronik **25** (1993) 38–47

11.28 Wong, S.K., Mathieu P., Pace P.: Eye-Safe Nd:YAG Laser, Appl. Phys. Lett **57** (1990) 650–652

11.29 Ohishi Y, et al.: Pr^{3+}-Doped Fluoride Fiber Amplifier Operating at 1.31 μm, Opt. Lett **16** (1991) 1747–1749

11.30 Neto J.A.M., et al.: Radiative and Nonradiative Properties of Prasodymium-Doped Glasses, Conf. Dig. CLEO (1993) 498

11.31 Petermann K., Universität Hamburg, Personal communication (1994)

11.32 Seelert W., Strauss, E.: Absolute Excited-State Absorption Cross Section and Fluorescence Quantum Efficiency of Cr^{3+}:Gadolinium Scandium Gallium Garnet, Opt. Lett. **12** (1987) 798–199

11.33 Petricevic A.S., Alfano R.R., Seas A.: Slope Efficiency Measurements of a Chromium-Doped Forsterite Laser, Opt. Lett. **16** (1991) 811–813

11.34 Barnes N.P., et al.: Operation of an Er:YLF Laser at 1.73 μm, IEEE J. Quantum Electron. **22** (1986) 337–343

11.35 Union Carbide. CZ Ruby Laser Rods, Data sheet (1986)

11.36 Walling J.C., et al.: Tunable Alexandrite Lasers, IEEE J. Quantum Electron. **16** (1980) 1302–1315

11.37 Miniscalco W.J., Quimby R.S.: General Procedure for the Analysis of Er^{3+} Cross Sections, Opt. Lett. **16** (1991) 258–260

11.38 Dätwyler M., Lüthy W., Weber H.P.: New Wavelengths of the $YALO_3$:Er Laser, IEEE J. Quantum Electron. **23** (1987) 158–159

11.39 Imai S., et al.: High Power Alexandrite Laser and its Applications, Conf. Dig. CLEO (1986) 106

11.40 Rapoport W.R., Samelson H.: Alexandrite Slab Laser, LASER 85 Las Vegas

11.41 Bowman S.R., et al.: Upconversion Studies of Flashlamp-Pumped Cr, Tm, Ho: YAG, OSA Proc. Adv. Solid State Lasers **10** (1991) 172–177

11.42 Imai S., et al.: A 20 W Cr^{3+}, Tm^{3+}, Ho^{3+}:YAG Laser, Opt. Laser Technol. **22** (1990) 351–353

11.43 Knights M.G., et al.: High power TEM_{00} 2-μm Laser, Conf. Dig. CLEO (1985) 94

11.44 Teichmann H., Duczynski E.W., Huber G.: Efficient Flashlamp Pumped Operation of a Cr, Tm, Ho: YAG Laser at 2.08 μm, Conf. Dig. CLEO (1988)
Teichmann H., Duczynski E.W., Huber G.: 17 J Ho-Laser at 2 Microns, Proc. SPIE **1021** (1988) 74–81

11.45 Meier J.V., et al.: Flashlamp-Pumped Cr^{3+}: GSAG Laser, IEEE J. Quantum Electron. **22** (1986) 2058–2064

11.46 Maeda K., et al.: Concentration Dependence of Fluorescence Lifetime of Nd^{3+}-Doped $Gd_3Ga_5O_{12}$ Lasers, Jpn. J. Appl. Phys. **23** (1984) L759–L760

11.47 Krupke W.F.: Spectroscopic, Optical and Thermo-Mechanical Properties of GSGG and its Laser Performance, UCRL 93853 Preprint (1985)

11.48 Hayakawa H., et al.: High Average Power Nd:$Gd_3Ga_5O_{12}$ Slab Laser, Jpn. J. Appl. Phys. **26** (1987) L1623–L1625

11.49 Jenssen H.P., et al.: Spectroscopic Properties and Laser Performance of Nd^{3+} in Lanthanum Beryllate, J. Appl. Phys. **47** (1976) 1496–1500

11.50 Tucker A.W., et al.: Stimulated-Emission Cross Section at 1064 and 1342 nm in Nd:YVO_4, J. Appl. Phys. **48** (1977) 4907–4911

11.51 DeShazer L.G., et al.: Laser Performance of Nd^{3+} and Ho^{3+} in YVO_4 and Nd^{3+} in Gadolinium Gallium Garnet (GGG), Dig. Tech. Pap. 8th Int. Quantum Electron. Conf. (1974) 683

11.52 Pollak T.M., et al.: CW Laser Operation of Nd:YLF, IEEE J. Quantum Electron. **18** (1982) 159–162

11.53 Thomas M.D., Chicklis E.P.: High power 1.3 μm Nd:YAG Laser, Conf. Dig. CLEO (1986) 214
11.54 Fuhrmann K., et al.: Effective Cross Section of the Nd:YAG 1.0641-μm Laser Transition, J. Appl. Phys. **62** (1987) 4041–4044
11.55 Shen H., et al.: High Power 1.314 μm Nd:YAG Laser, Opt. Laser Technology (Aug 1986) 193–197
11.56 Frauchiger J., Lüthy W.: Power Limits of a YAG:Er Laser, Opt. Laser Technology **19** (1987) 312–315
11.57 Mochizuki T., et al.: Development of High Power Solid-State Lasers at Hoya Corp.
11.58 Bass M., et al.: Room Temperature of the 50% Doped Er:YAG Laser at 2940 nm Conf. Dig. CLEO (1986) 3–7
11.59 Quarles G.J., et al.: High Efficiency 2.09 μm Flashlamp-Pumped Laser, Appl. Phys. Lett. **55** (1989) 1062–1064
Quarles G.J., et al.: Flash pumped, Room-temperature 2 μm Laser with 5% Slope Efficiency, Proc. SPIE **1223** (1990) 221–229
11.60 Fan T.Y., et al.: Spectroscopy and Diode Laser-Pumped Operation of Tm, Ho:YAG, IEEE J. Quantum Electron. **24** (1988) 924–933
Fan T.Y., Huber G., Byer R.L.: Continuous-Wave Operation at 2.1 μm of a Diode -Laser-Pumped, Tm Sensitized Ho:$Y_3Al_5O_{12}$ Laser at 300 K, Opt. Lett. 123 (1987) 678–680
11.61 Rotman S.R.: Nonradiative Energy Transfer in Nd:YAG – Evidence for the Correlated Placement of Ions, Appl. Phys. Lett. **54** (1989) 2053–2055
11.62 Jain R.K., et al.: Diode-Pumped 1.3 μm Nd:YVO_4 Laser, Conf. Dig. CLEO (1988) THB5
11.63 Shen H., et al.: New Advances in Nd:YAP-Lasers, Fujian Institute on the Structure of Matter. Fuzhou, Fujian China
11.64 Massey G.A.: Measurement of Device Parameters for Nd:$YAlO_3$ Lasers, IEEE J. Quantum Electron. **8** (1972) 669–674
11.65 Schütz I, Freitag I., Wallenstein R.: Minature Self-Frequency-Doubling CW Nd:YAB Laser Pumped by a Diode-Laser, Opt. Commun. **77** (1990) 221–225
11.66 Kaminskii, et al.: New Stimulated Emission Data of $Gd_3Ga_5O_{12}$:Nd^{3+}, Phys. Status Solidi **49** (1978) 305–311
11.67 Albrecht G.F., et al.: Measurements of Ti^{3+}:Al_2O_3 as a Lasing Material, Opt. Commun. **52** (1985) 401–404
11.68 Lacovara P., et al.: Flash-Lamp-Pumped Ti^{3+}:Al_2O_3 Laser Using Fluorescent Conversion, Opt. Lett. **10** (1985) 273–275
11.69 Sanchez A., et al.: Room-Temperature Continuous-Wave Operation of a Ti:Al_2O_3 Laser, Opt. Lett. **11** (1986) 363–364
11.70 Carts Y.A.: Titanium Sapphire's Star Rises, Laser Focus World **9** (1989) 73–88
Flashlamp Pumps Ti:Sapphire Laser: Laser Focus World **8** (1989) 21
11.71 Armagan G., et al.: Comparison of Spectroscopic Properties of Tm and Ho in YAG and YLF Crystals, OSA Proc. Adv. Solid State Lasers **10** (1991) 222–226
11.72 Conf. Dig. CLEO CthF-4 CthF-5 (1993) 409
11.73 Quarles G., Pinto J.F., Esterowitz L.: Broad Tunability of Flash-Pumped, Tm-Activated Garnet Lasers, OSA Proc. Adv. Solid State Lasers **10** (1991) 167–171
Quarles G.J., Marquardt C.L., Esterowitz L.: Efficient Room-Temperature Operation of a Flashlamp-Pumped Cr;Tm:YAG Laser at 2.014 μm, OSA Proc. **6** (1990) 150

11.74 DeLoach L.D., et al.: Evaluation of Absorption and Emission Properties of Yb^{3+} Doped Crystals for Laser Applications, University of California, LLNL Livermore (1993)

11.75 Giesen A., et al.: Scalable Concept for Diode-Pumped High-Power Solid-State Lasers, Appl. Phys. B (1994) accepted

11.76 Hovies F.E., et al.: Low-Level Relaxation of Nd:YAG, OSA Proc. **9** (1990) 227–229

11.77 Nakatsuka M., et al.: Nd:Quartz Laser Rod Produced by the Sol-Gel Method with no Additives, Conf. Dig. CLEO (1993) 590

11.78 Cao J.D., et al.: Five Watt Single Transverse Mode Neodymium-Doped Fiber Laser, Conf. Dig. CLEO (1993) 622

11.79 Hoffstädt A.: High Average Power Flash-Lamp-Pumped Ti:Sapphire Laser, Opt. Lett (1994) to be published

11.80 Marion J.E.: Fracture Mechanisms and Strengthening of Slab Lasers, Proc. SPIE **736** (1987) 2–12

11.81 Marion J.E., Pertica A.J.: Materials for High Average Power Lasers, Proc. SPIE **1040** (1989) 2–18

11.82 Esmeria J.M., et al.: Efficient Continous-Wave Lasing Operation of Nd:$KGd(WO_4)_2$ at 1.067 μm with Diode and Ti:Sapphire Laser Pumping, Opt. Lett **20** (1995) 1538–1540

11.83 Sumida D.S., Fan T.Y., Hutcheson R.: Spectroscopy and Diode-Pumped Lasing of Yb^{3+}-Doped $Lu_3Al_5O_{12}$ (Yb:LuAG), OSA Proc. Adv. Solid State Lasers **24** (1995) 348–350

11.84 Sumida D.S., Fan T.Y.: Radiation Trapping in Solid-State Laser Media and its Impact on Fluorescence Lifetime and Emission Cross Section Measurements, OSA Proc. Adv. Solid State Lasers **24** (1995) 542–544

Chapter 12

12.1 Walling J.C., et al.: Tunable Alexandrite Lasers: Development and Performance, IEEE J. Quantum Electron. **21** (1985) 1568–1581

12.2 Walling J.C.: Properties of Alexandrite lasers, Manuscript of the talk presented at the Nato meeting, Italy (1984)

12.3 Allied Synthetic Crystal Products, Charlotte, North Carolina: YAG Data sheet (1985)

12.4 Krupke W.F., et al.: Spectroscopic, Optical and Thermomechanical Properties of Neodymium- and Chromium-Doped Gadolinium Scandium Gallium Garnet, J. Opt. Soc. Am. B **3** (1986) 102–113

12.5 Chin T., et al.: Athermal Nd:BEL Lasers, Proc. SPIE **622** (1986) 53–60

12.6 Rapoport W.R, Chin T.: Laser and Laser Related Characteristics of Nd:BEL, Proc. SPIE **1040** (1989) 19–31

12.7 Allied Synthetic Crystal Products, Charlotte, North Carolina: Lanthanum Beryllate: A New Laser Host of Great Promise (1975)

12.8 Hoefer C.S., Kirby K.W., DeShazer L.G.: Thermo-Optic Properties of Garnet Laser Crystals, IRD Quarterly Report (1985). Hughes Research Labs, Malibu CA
Hoefer C.S., Kirby K.W., DeShazer L.G.: Thermo-Optic Properties of Gadolinium Garnet Laser Crystals, J. Opt. Soc. AM. B **5** (1988) 2327–2332

12.9 Kitaeva, et al.: The properties of Crystals with Garnet Structure, Phys. Status Solidi **A 92** (1985) 475–488
12.10 Wacker Chemitronic: Cr:Nd:GGG Laser Crystals, Data sheet (1989)
12.11 Lundt H., Wacker Chemitronic. Personal communication (1988)(1989)
12.12 Lee J.C., Jacobs S.D.: Refractive Index and Δn/ΔT of Cr:Nd:GSGG at 1064 nm, Appl. Opt. **26** (1987) 777–778
12.13 Payne J.P., Evans H.W.: Flashlamp-Pumped Lasing of Chromium-Doped GSG Garnet, Conf. Dig. CLEO (1986) 106
12.14 Litton Airtron Morris Plains NJ: GSGG Data sheet GSGG S-4/88 (1988)
12.15 Aubert, Crismatec (1991) Personal Communication
12.16 Kahn A., et al.: Preparation, Structure, Optical and Magnetic Properties of Lanthanide Aluminate Single Crystals ($LnMAl_{11}O_{19}$), J. Appl. Phys. **52** (1981) 6864–6867
12.17 Aminoff C.G., et al.: Characterization and Laser Properties of Lanthanum Magnesium Hexa-Aluminate Activated by Neodymium and Chromium, J. Lumin. **50** (1991) 21–29
12.18 Aminoff C.G., et al.: Optical Pumping of Helium with Arc Lamp Excited LNA Lasers, Rev. Phys. Appl. **24** (1989) 827–831
12.19 Union Carbide: Titanium-Doped Sapphire, Tunable Laser Crystals, Data sheet (1989)
12.20 Arsenjew P.A., et al.: Kristalle in der modernen Lasertechnik (Akademische Verlagsgesellschaft, Leipzig 1980)
12.21 Allied Synthetic Crystal Products Charlotte North Carolina: YAG Data sheet (1990)
12.22 Litton Airtron Morris Plains NJ: YAG Data sheet NDYAGS-4/88 (1988)
Litton Airtron Charlotte, NC: Calculated Cr, Tm, Ho: YAG compositions, Technical Brief 0813 (1992)
12.23 Union Carbide: YAG:Nd Data, Data sheet
Union Carbide: Tm:Ho:Cr:YAG laser rods, Roditi communication (1988)
12.24 Slack G.A., Oliver D.W.: Thermal Conductivity of Garnets and Phonon Scattering by Rare-Earth Ions, Phys. Rev. B **4** (1971) 592–609
12.25 Zeng Z., et al.: Measurement of the Refractive Index and Thermal Refractive Index Coefficients of Nd:YAP Crystal, Appl. Opt. **29** (1990) 1281–1286
12.26 Weber M.J., et al.: Czochralski Growth and Properties of $YAlO_3$ Laser Crystals, Appl. Phys. Lett. **15** (1969) 342–345
12.27 Keig G.A., DeShazer L.G.: Laserverhalten von Yttrium-Orthoaluminat bei Dotierung mit seltenen Erden, Laser Elektro Optik **3** (1972) 45–50
12.28 Heraeus: Nd:YAP Laser-Crystals, Data sheet (1981)
12.29 Young D.D., et al.: Holographic Interferometry Measurement of the Thermal Refractive Index and Thermal Expansion Coefficient of Nd:YAG and Nd:YAlO, IEEE J. Quantum Electron. (Aug 1972) 720–721
12.30 Litton Airtron Morris Plains NJ: YLF Data sheet YLFS-4/88 (1988)
12.31 Allik T.H., et al.: Crystallography, Spectroscopic Analysis, and Lasing Properties of Nd^{3+}:$Y_3Sc_2Al_3O_{12}$, Phys. Rev B **41** (1990) 21–29
12.32 Duczynski E.W., et al.: Efficient Flashlamp Pumped Cr,Nd:YSAG Laser, Proc. SPIE **1021** (1988) 69–73
12.33 Kuzmin O., (FIRN, Krasnodar, USSR): Personal communication (1992)
12.34 Litton, Airtron: Nd: Yttrium Vanadate, Technical Brief (1992)

12.35 Comaskey, et al.: Flash-Lamp-Pumped Laser Operation of Nd^{3+}:Y_2SiO_5 at 1.074 µm, Opt. Lett. **18** (1993) 2029–2031
12.36 Lin J.T.: Doubled Jeopardy: The Blue-Green Race's New Players, Laser Optronics **12** (1990) 40
12.37 Wang,S.C., Stone,R.E.: Characteristics of Neodymium Yttrium Aluminium Borate as a Diode-Pumped Laser Material, OSA Proc. Adv. Solid State Laser **6** (1990) 23–25
12.38 Chase L.L., et al.: Measurement of Thermal Lensing for the $LiCaAlF_6$:Cr^{3+} Laser Material, OSA Proc. Adv. Solid State Laser **6** (1990) 83–85
12.39 Ramet Ltd Moscow 109017: Product Data sheet (1993)
12.40 Optron Technology Ltd: Nd:KGW Data sheet (1993) 1000 Sofia Bulgaria, 67 Vassil Levsky Blvd
12.41 Devor D.P., DeShazer L.G., Pastor R.C.: YAG Quantum Efficiency and Related Radiative Properties, IEEE J. Quantum Electron. **25** (1989) 1863–1873
12.42 Lupei V., Lupei A., Georgescu S.: Effects of Energy Transfer on Quantum Efficiency of YAG:Nd, J. Appl. Phys. **66** (1989) 3792–3797
12.43 Alkor Techn. Inc. St. Petersburg 198035 Russia: Nd:KGW laser Rods, Data sheet (1993)
12.44 Castech. Crystals Data sheet (1995)
12.45 Bremlas: Data Sheet (1996) Lasertechnik Bremen

Chapter 13

13.1 Neuroth N.: Laser Glass: Status and Prospects, Opt. Eng. **26** (1987) 96–101
13.2 Schott Glass Techn. Inc.: APG-Laser Glass from Schott, Data sheet 2302/88 (1988)
13.3 Hayden J.S., et al.: Advances in Glasses for High Average Power Laser Systems, Proc. SPIE **1021** (1988) 36
13.4 Hoya: Data sheet (1986)
13.5 Hoya: Laser Glass, Data sheet E 9005-1A010D (1991)
Hoya: Data sheet (1985)
13.6 UCRL-50021-86: High-Average-Power Lasers 7. Laser Program. Annual Report (1986) 7-151
13.7 Cerqua K.A., et al.: Strengthened Glass for High Average Power Laser Applications, Proc. SPIE **736** (1987) 13–21
13.8 Jones W.B.: Slab Geometry Lasers, Handbook of Solid-State lasers. Cheo, P.K. ed. p 581
13.9 Jasbir S., et al.: Study of Thermal Effects in an Nd Doped Phosphate Glass Laser Rod, IEEE J. Quantum Electron. **22** (1986) 2259–^2265
13.10 Marion J.E., Weber M.J.: Phosphate Laser Glasses, Eur. J. Solid State Inorg. Chem. **28** (1991) 271–287
13.11 Linford G.J., et al.: Measurements and Modeling of Gain Coefficients for Neodymium Laser Glasses, IEEE J. Quantum Electron. **15** (1979) 510–523
13.12 Waxler R.M., Feldman A.: Piezooptic Coefficients of Four Neodymium-Doped Laser Glasses, Appl. Opt. **19** (1980) 2481–2482
13.13 Borik M.A., et al.: High-Power Pulse-Periodic Neodymium Glass Laser with a Plate-Shaped Active Element, Sov. J. Quantum Electron. **20** (1990) 332–336

13.14 Kigre: Glass Laser Rods, Data sheet Rods 500-10-90 (1990) Kigre Inc. Hilton Head Island
13.15 Kigre: QE-7S Erbium-Doped Phosphate Laser Glass, Data sheet LC500-10-89 (1989) Kigre Inc. Hilton Head Island
13.16 Schott: Laser glass from Schott, Schott Catalog 2301(1988)
13.17 Schott: Nd-Phosphat glasses for the Laser Technique, LG 703 and LG 706, Product Information Nr. 7515 (1979)
13.18 Klocek P.: Inorganic Optical Materials, Critical Rev. Opt. Sci. Technol. Proc. SPIE **CR64** (1996)

Chapter 14

14.1 Curcio J.A., Petty C.C.: The Near Infrared Absorption Spectrum of Liquid Water, J. Opt. Soc. Am. **41** (1951) 302
14.2 Friedmann E., et al.: Absorption Coefficient Instrument for Turbid Natural Waters, Appl. Opt. **19** (1980) 1688
14.3 Schott Ruhrglas: Information Sheet DURAN (1979)
14.4 Schott: TEMPAX Tafelglas, Data sheet Druckschrift Nr. 8841/1 d
14.5 Schott: ZERODUR Glaskeramik, Druckschrift Nr. 3131/1 d X/85
14.6 Schott: Optisches Glas, Data sheet Nr. 3111 d
14.7 Quartz & Silice: Silica Glass, Technical leaflet S 2
14.8 Wiggin Alloys: Produktbeschreibung.Veröffentlichung Nr. 3770G April (1983)
14.9 Wiggin Alloys: Data Sheet NILO. Publication Number 3721
14.10 Friedrichsfeld: Products Made from Oxide Ceramic, Data sheet 1078a 10 V 86 L
14.11 Morgan: Technical Ceramic, Morgan Matroc Lim. MML.86.01
14.12 Heraeus: Transparent and Opaque Fused Silica, Company brochure Q-A 1/112.1
14.13 Schott: Technische Gläser (Physikalische und chemische Eigenschaften) Company brochure
14.14 Badische Industrie-Edelstein-Gesellschaft. Company brochure
14.15 Driscoll W.G., et al. (eds): Handbook of Optics (1978)
Hass G.: Optical Properties of Metal, American Institute of Physics Handbook. McGraw-Hill.(3rd edn) 6–118
Jenkins F.A., White,,H.E.: Fundamentals of Optics (McGraw Hill Book Company, Tokyo 1975) 536
14.16 Zilling F.: Präzisions-Keramik für hohe Genauigkeiten, Feinwerktechnik Meßtechnik **7** (1989) Sonderteil ZM 96
14.17 Tribastone C., Teyssier C.: Design Plastic Optics for Manufacturing, Photonics Spectra (May 1991) 120
14.18 Haferkamp H., et al.: Thermographic Material Identification System for the Waste Sorting Process, SPIE Symposium, Orlando (1995)
14.19 Hopper C.: Diamond in Electronics and Optics: British Meeting, Laser Optoelektronik **26** (1994) 23

Chapter 15

15.1 Ardenne M.v.: Tabellen zur angewandten Physik, (VEB Deutscher Verlag der Wissenschaften, Berlin 1973)

Index

Springer Series in
OPTICAL SCIENCES

New editions of volumes prior to volume 60

1 **Solid-State Laser Engineering**
By W. Koechner, 5th revised and updated ed. 1999, 472 figs., 55 tabs., XII, 746 pages

14 **Laser Crystals**
Their Physics and Properties
By A. A. Kaminskii, 2nd ed. 1990, 89 figs., 56 tabs., XVI, 456 pages

15 **X-Ray Spectroscopy**
An Introduction
By B. K. Agarwal, 2nd ed. 1991, 239 figs., XV, 419 pages

36 **Transmission Electron Microscopy**
Physics of Image Formation and Microanalysis
By L. Reimer, 4th ed. 1997, 273 figs. XVI, 584 pages

45 **Scanning Electron Microscopy**
Physics of Image Formation and Microanalysis
By L. Reimer, 2nd completely revised and updated ed. 1998, 260 figs., XIV, 527 pages

Published titles since volume 60

60 **Holographic Interferometry in Experimental Mechanics**
By Yu. I. Ostrovsky, V. P. Shchepinov, V. V. Yakovlev, 1991, 167 figs., IX, 248 pages

61 **Millimetre and Submillimetre Wavelength Lasers**
A Handbook of cw Measurements
By N. G. Douglas, 1989, 15 figs., IX, 278 pages

62 **Photoacoustic and Photothermal Phenomena II**
Proceedings of the 6th International Topical Meeting, Baltimore, Maryland, July 31 - August 3, 1989
By J. C. Murphy, J. W. Maclachlan Spicer, L. C. Aamodt, B. S. H. Royce (Eds.), 1990, 389 figs., 23 tabs., XXI, 545 pages

63 **Electron Energy Loss Spectrometers**
The Technology of High Performance
By H. Ibach, 1991, 103 figs., VIII, 178 pages

64 **Handbook of Nonlinear Optical Crystals**
By V. G. Dmitriev, G. G. Gurzadyan, D. N. Nikogosyan, 3rd revised ed. 1999, 39 figs., XVIII, 413 pages

65 **High-Power Dye Lasers**
By F. J. Duarte (Ed.), 1991, 93 figs., XIII, 252 pages

66 **Silver-Halide Recording Materials**
for Holography and Their Processing
By H. I. Bjelkhagen, 2nd ed. 1995, 64 figs., XX, 440 pages

67 **X-Ray Microscopy III**
Proceedings of the Third International Conference, London, September 3-7, 1990
By A. G. Michette, G. R. Morrison, C. J. Buckley (Eds.), 1992, 359 figs., XVI, 491 pages

68 **Holographic Interferometry**
Principles and Methods
By P. K. Rastogi (Ed.), 1994, 178 figs., 3 in color, XIII, 328 pages

69 **Photoacoustic and Photothermal Phenomena III**
Proceedings of the 7th International Topical Meeting, Doorwerth, The Netherlands, August 26-30, 1991
By D. Bicanic (Ed.), 1992, 501 figs., XXVIII, 731 pages

Springer Series in
OPTICAL SCIENCES

70 **Electron Holography**
By A. Tonomura, 2nd, enlarged ed. 1999, 127 figs., XII, 162 pages

71 **Energy-Filtering Transmission Electron Microscopy**
By L. Reimer (Ed.), 1995, 199 figs., XIV, 424 pages

72 **Nonlinear Optical Effects and Materials**
By P. Günter (Ed.), 2000, 174 figs., 43 tabs., XIV, 540 pages

73 **Evanescent Waves**
From Newtonian Optics to Atomic Optics
By F. de Fornel, 2001, 277 figs., XVIII, 268 pages

74 **International Trends in Optics and Photonics**
ICO IV
By T. Asakura (Ed.), 1999, 190 figs., 14 tabs., XX, 426 pages

75 **Advanced Optical Imaging Theory**
By M. Gu, 2000, 93 figs., XII, 214 pages

76 **Holographic Data Storage**
By H.J. Coufal, D. Psaltis, G.T. Sincerbox (Eds.), 2000
228 figs., 64 in color, 12 tabs., XXVI, 486 pages

77 **Solid-State Lasers for Materials Processing**
Fundamental Relations and Technical Realizations
By R. Iffländer, 2001, 230 figs., 72 tabs., XVI, 350 pages

78 **Holography**
The First 50 Years
By J.-M. Fournier (Ed.), 2001, 266 figs., XII, 460 pages

79 **Mathematical Methods of Quantum Optics**
By R.R. Puri, 2001, 13 figs., XIV, 285 pages

80 **Optical Properties of Photonic Crystals**
By K. Sakoda, 2001, 85 figs., 23 tabs., X, 192 pages

81 **Photonic Analog-to-Digital Conversion**
By B. Shoop, 2001, 252 figs., 11 tabs., XIII, 328 pages

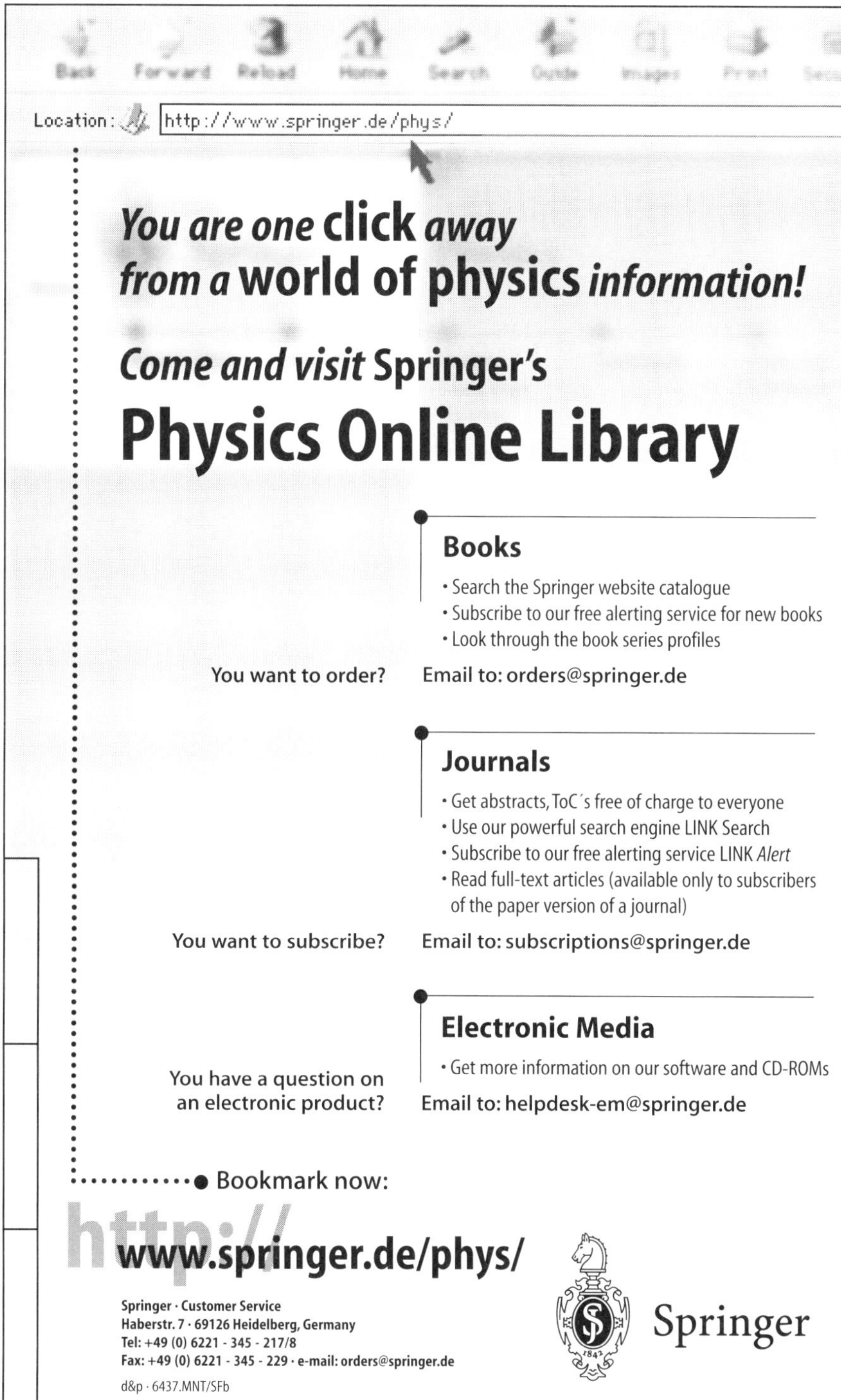

Druck: Strauss Offsetdruck, Mörlenbach
Verarbeitung: Schäffer, Grünstadt